Future Speech Interfaces with Sensors and Machine Intelligence

Future Speech Interfaces with Sensors and Machine Intelligence

Editors

Bruce Denby
Tamás Gábor Csapó
Michael Wand

MDPI • Basel • Beijing • Wuhan • Barcelona • Belgrade • Manchester • Tokyo • Cluj • Tianjin

Editors

Bruce Denby
Sorbonne University
Paris
France

Tamás Gábor Csapó
Budapest University of
Technology and Economics
Budapest
Hungary

Michael Wand
Università della Svizzera
Italiana, USI-SUPSI
Viganello
Switzerland

Editorial Office
MDPI
St. Alban-Anlage 66
4052 Basel, Switzerland

This is a reprint of articles from the Special Issue published online in the open access journal *Sensors* (ISSN 1424-8220) (available at: https://www.mdpi.com/journal/sensors/special_issues/FSI-SMI).

For citation purposes, cite each article independently as indicated on the article page online and as indicated below:

LastName, A.A.; LastName, B.B.; LastName, C.C. Article Title. *Journal Name* **Year**, *Volume Number*, Page Range.

ISBN 978-3-0365-6938-3 (Hbk)
ISBN 978-3-0365-6939-0 (PDF)

Contents

sensors

MDPI

Editorial

Future Speech Interfaces with Sensors and Machine Intelligence

Bruce Denby [1,*], Tamás Gábor Csapó [2] and Michael Wand [3,4]

[1] Institut Langevin, ESPCI Paris, PSL University, CNRS, Sorbonne Université, 75005 Paris, France
[2] Department of Telecommunications and Media Informatics, Budapest University of Technology and Economics, 1117 Budapest, Hungary
[3] Dalle Molle Institute for Artificial Intelligence USI-SUPSI, 6962 Viganello, Switzerland
[4] Institute for Digital Technologies for Personalized Healthcare, SUPSI, 6962 Viganello, Switzerland
* Correspondence: denby@ieee.org

Abstract: Speech is the most spontaneous and natural means of communication. Speech is also becoming the preferred modality for interacting with mobile or fixed electronic devices. However, speech interfaces have drawbacks, including a lack of user privacy; non-inclusivity for certain users; poor robustness in noisy conditions; and the difficulty of creating complex man–machine interfaces. To help address these problems, the Special Issue "Future Speech Interfaces with Sensors and Machine Intelligence" assembles eleven contributions covering multimodal and silent speech interfaces; lip reading applications; novel sensors for speech interfaces; and enhanced speech inclusivity tools for future speech interfaces. Short summaries of the articles are presented, followed by an overall evaluation. The success of this Special Issue has led to its being re-issued as "Future Speech Interfaces with Sensors and Machine Intelligence-II" with a deadline in March of 2023.

Keywords: multimodal speech; silent speech interfaces; lip reading; speech sensors

Citation: Denby, B.; Csapó, T.G.; Wand, M. Future Speech Interfaces with Sensors and Machine Intelligence. *Sensors* **2023**, *23*, 1971. https://doi.org/10.3390/s23041971

Received: 2 February 2023
Accepted: 7 February 2023
Published: 10 February 2023

1. Introduction

It was not long after the advent of digital computers in the 1950's that the idea of using computers to recognize speech began to be investigated. In the ensuing years, numerous techniques for treating the speech signal were developed by researchers worldwide, giving rise today to a wide variety of tools such as Automatic Speech Recognition (ASR) applications, powerful speech compression tools, Text-To-Speech (TTS) synthesis, as well as speaker identification and, more recently, diarization tools, to name only a few. Despite these enormous gains, though, we may rightfully speak of a new kind of revolution in speech processing today.

Indeed, while speech has always been something of a "specialist" field, requiring fluency in topics such as Mel Frequency Cepstral Coefficients (MFCC), Gaussian Mixture Model—Hidden Markov Models (GMM-HMM), and the like, the staggering growth of Machine Learning techniques and an increasing preference for Open Source solutions are today propelling speech processing into the mainstream. And concomitant with this "democratization" of speech processing is a desire to free Future Speech Interfaces from some inherent difficulties that have traditionally handicapped speech applications:

- Robustness: It is well known that speech understanding degrades rapidly in noisy conditions, both for human interaction and for machine communication.
- Privacy: Based on an audible acoustic signal, overheard speech can be merely a nuisance, a real source of interference, or even a troublesome security problem.
- Inclusivity: Some sectors of the population cannot use speech in traditional ways due to health issues. In addition, the complexity of producing high performance speech applications has often meant that less-frequently heard languages or dialects lack the kind of tools available, for example, for English or Mandarin.

- Fluidity: It has proven difficult to deploy automatic speech interfaces possessing the robustness, fluidity, and intricacy of genuine face-to-face human interactions.

The present Special Issue brings together eleven recent original research contributions addressing one or more of the above concerns. The basic approaches represented in the works fall into three broad categories:

- Audio-Visual Speech Recognition (AVSR): The key in AVSR is to combine speech modalities in order to improve robustness to noise, interference, and other environmental effects. A typical application might combine speech signals from a microphone with lip video taken by an external camera.
- Silent Speech Interfaces (SSI): Silent Speech Interfaces completely do away with an audio signal, either because the audio is unexploitable or because articulation was done silently, and perform ASR using other sensors as input—a camera, of course, but also electrical signals, ultrasound, etc. If visual input is used, the technique is called Visual Speech Recognition or VSR, including lipreading applications. SSI/VSR are useful for addressing privacy issues as well as in speech interfaces for persons unable to speak in a traditional manner.
- Novel interfaces, for example:
    - Low-resource language-specific algorithms to address specificities of particular languages or dialects, enhancing inclusivity in speech processing.
    - Avatar-like entities for more natural speech interaction.

We may also classify the contributions according to the types of tools adopted—or created:

- Sensors: The use of cameras has already been invoked above; however, particularly in SSIs, more exotic techniques such as surface electromyography (sEM), electromagnetic articulography (EMA), ultrasound or radar may be employed.
- Machine Learning (ML): Development of novel Machine Learning techniques to deal with the specific tasks arising in innovative speech interfaces, like stream-combining techniques for multimodal speech recognition, or adaptation techniques to combine data from different subjects.
- New tools: Some contributions benefit from the availability of neural synthesis techniques, recent AVSR databases, or Open Cloud speech processing modules; while others propose new image processing tools, for example in ultrasound image analysis

In what follows, we provide brief summaries of the eleven articles chosen for publication in the Special Issue Future Speech Interfaces with Sensors and Machine Intelligence. The articles are grouped by topic according to the categories described above. A conclusion as well as some prospects for the future follow the summaries. Indeed, due to its success and popularity, a second edition of the Special Issue, "Future Speech Interfaces with Sensors and Machine Intelligence II", was opened for submissions with a deadline in March 2023.

2. AVSR Articles

2.1. *Yu, Zeiler, and Kolossa*

The first article in the AVSR category is "Reliability-Based Large-Vocabulary Audio-Visual Speech Recognition" by Wentao Yu, Steffen Zeiler and Dorothea Kolossa, at the Institute of Communication Acoustics, at Ruhr University in Bochum, Germany. The authors propose a novel dynamic stream weighting technique for combining the audio and visual input streams for AVSR, in order to improve the robustness of AVSR in noisy conditions. Called the Dynamic Fusion Network (DNF), the approach employs aspects of existing decision and representation fusion strategies in a unified view using the posterior probabilities of the single-modality models as representations of the uni-modal streams. Implemented as a ML architecture leveraging off of standard audio and video reliability measure, the DNF is evaluated on the Oxford BBC LRS2 and LRS3 large vocabulary lipreading corpora. Remarkably, using DFN, Word Error Rates (WER) are improved compared to audio-only input by about 50% for a hybrid recognizer, and 43% for an End to End recognizer.

2.2. *Jeon and Kim*

A trio of papers by the team of Sanghun Jeon and Mun Sang Kim, at the Gwangju Institute of Science and Technology (GIST) in South Korea, features new contributions both in sensor development and the use of open cloud services. In the first article of the trio, "Noise-Robust Multimodal Audio-Visual Speech Recognition System for Speech-Based Interaction Applications", the authors target a virtual aquarium edutainment application, in which users instrumented with a lightweight audio-visual helmet interact in real time with the virtual aquarium information system. The approach leverages an existing pretrained Open Cloud Speech Recognition System (OCSR), for the audio channel, by coupling its outputs to features extracted in a bespoke visual speech recognition system. For the video channel, lip/face images sequences are analyzed with 3D Convolutional Neural Networks (3DCNN), augmented with a novel Spatial Attention Module, to produce feature vectors that are then concatenated with audio features before entering a Connectionist Temporal Classification (CTC) module. The visual recognition dataset was prepared by a team of volunteers, instrumented with the audio-visual helmet, who repeat a set of 54 commands. After training, the final evaluation step is carried out in an in situ virtual aquarium environment, using 4 different additive noise profiles. In a typical trial using the system, combined audio-visual features improved performance from 91% Word Accuracy Rate to 98%.

In a second contribution, the same team proposes "End-to-End Lip-Reading Open Cloud-Based Speech Architecture", an extension of the research described above. In this case, several OCSR, including Google, Microsoft, Amazon, and Naver, were evaluated, using as a training corpus 20 commands selected from the Google Voice Command Dataset v2, recited by the same team of volunteers as above, albeit with a standard microphone and remote camera rather than a special helmet. Furthermore, the noise scenario portfolio was extended to eight different environments, and a concatenation of three types of 3DCNN used in the feature extraction step. WAR values, measured over a range of audio Signal to Noise Ratios, varied according to the OCSR and noise profile used; however, on average, audio-visual recognition improved WAR by some 14% percentage points (on the scale of 0% to 100% WAR) compared to pure audio recognition. Based on performance, Microsoft Azure was chosen as the principal API for the detailed comparisons in the article.

In the final entry of this trio of articles, "End-to-End Sentence-Level Multi-View Lipreading Architecture with Spatial Attention Module Integrated Multiple CNNs and Cascaded Local Self-Attention-CTC", the focus is on rendering the visual input channel more robust through the inclusion of 4 different camera angles of the face and lips: frontal, 30°, 45°, and 60°. In addition, a modified version of the Spatial Attention Module cited in the first article of the trio, is employed, in order to enhance features in the specific case of words having similar pronunciations. The OuluVS2 dataset, which employs 40 speakers for training and 12 for testing, on digit, phrase, and TIMIT sentence corpora, was used to evaluate the proposed ML speech recognition architecture. Here, results using any of the single camera inputs improved upon baseline audio-only input by about 5%; whereas including the full complement of 4 cameras brought an overall gain of about 9%, indicating that the multi-view visual input approach is indeed a useful innovation.

3. SSI/VSR Articles

3.1. *Cao, Wisler, and Wang*

The first article in the field of Silent Speech interfaces is "Speaker Adaptation on Articulation and Acoustics for Articulation-to-Speech Synthesis", by Beiming Cao, Alan Wisler, and Jun Wang. This article concerns speech reconstruction from Electromagnetic Articulographic (EMA) data, where the position of articulators is directly measured using sensors attached to the articulators. The output of the system is the speech waveform (created from speech features using the Waveglow vocoder), and the contribution of this study is speaker adaptation in both EMA input and acoustic target space. This is a key requirement for creating large SSI systems, since it is practically impossible to collect large amounts of data from a single speaker. Using Procrustes matching in the EMA space and Voice Conversion in the acoustic space achieves significant improvements over the speaker-independent baseline, measured using the objective Mel Cepstral Distance (MCD) criterion.

3.2. *Csapó et al.*

The collection of SSI articles continues with "Optimizing the Ultrasound Tongue Image Representation for Residual Network-Based Articulatory-to-Acoustic Mapping", by Tamás Gábor Csapó, Gábor Gosztolya, László Tóth, Amin Honarmandi Shandiz, and Alexandra Markó. Here ultrasound tongue images (UTI) of the vocal tract are used as input for a speech reconstruction system; which offers a way to capture vocal tract information very different from EMA considered in the study above, having its own set of advantages and challenges. In particular, the data needs to be interpreted using image processing techniques, where state-of-the-art systems work best on raw input data. While classical UTI systems perform data preprocessing which is adapted for manual inspection (e.g., by a medical doctor), new systems also provide access to raw data. In this study, raw and preprocessed input are directly compared using a standard underlying UTI-to-speech system based on a multilayer ResNet architecture. While no significant differences between the two input types could be ascertained, it is shown that it is possible to reconstruct speech of optimal quality using rather small input images, thus allowing to use smaller neural networks which are faster to train on large amounts of input data.

3.3. *Ferreira et al.*

The third SSI article is "Exploring Silent Speech Interfaces Based on Frequency-Modulated Continuous-Wave Radar", by David Ferreira, Samuel Silva, Francisco Curado, and António Teixeira. Here the input consists of features obtained from a radar sensor. Unlike for the systems presented in the previous articles, where speech is directly reconstructed as audio waveform, the goal in this study is to obtain textual output, i.e., to perform speech recognition from radar sensor data. 13 different words are distinguished with an accuracy of 88% in a speaker-dependent setup and 82% in a speaker-independent setup; in particular the latter result is noteworthy since speaker discrepancies are a known cause of problems in many SSI systems. A further advantage of the radar sensor is the contactless recording, which it shares with video-based methods, but not with systems based on electrical biosignals.

3.4. *Jeon, Elsharkaway, and Kim*

The fourth, and last "classical" SSI article, is "Lipreading Architecture Based on Multiple Convolutional Neural Networks for Sentence-Level Visual Speech Recognition", by Sanghun Jeon, Ahmed Elsharkawy, and Mun Sang Kim. In contrast with Audiovisual speech recognition, as exposed above, the system presented here uses only visual input, namely the video part of the well-known GRID audiovisual speech corpus. The GRID dataset is a relatively small dataset, but it presents a very relevant challenge: namely, a large number of words are very short (e.g., the letters of the alphabet) and thus difficult to recognize. In this study, the authors develop a specific architecture based on convolutional neural networks to mitigate this problem, obtaining accurate prediction even for short visual-acoustic units. here.

3.5. *Wrench and Balch-Tomes*

Finally, the study "Beyond the Edge: Markerless Pose Estimation of Speech Articulators from Ultrasound and Camera Images Using DeepLabCut" by Alan Wrench and Jonathan Balch-Tomes pursues an objective different from the papers presented above, namely, the goal is to estimate the position of speech articulators from ultrasound images without the explicit use of any form of markers or objects attached to the subject's face. This is in stark contrast to methods like EMA (which we have above in the first SSI publication), where sensors are directly attached to a person's articulators. Innovative image processing techniques are used for the task, just a small amount of hand-labeled images are required for training the system.

4. Novel Interface Articles

4.1. Oneață et al.

The paper "FlexLip: A Controllable Text-to-Lip System" by Dan Oneață, Beáta Lőrincz, Adriana Stan, and Horia Cucu deals with creating lip landmarks from textual input. These landmarks can then be used to generate natural lip contours for speech, for example for generation of animated movies and videos. The contribution of this paper is a flexible modular architecture which disentangles the text-to-speech component from the final generation of lip contours. This makes the system amenable to fast adaptation to new speakers, which does not only involve adapting the audio generation component, but also requires fine tuning the lip shapes to the new speaker. Based on several objective measures, the system performs on par with monolithic baseline systems trained on much larger corpora.

4.2. Baniata, Ampomah, and Park

The final paper in the issue, "A Transformer-Based Neural Machine Translation Model for Arabic Dialects That Utilizes Subword Units" by Laith H. Baniata, Isaac. K. E. Ampomah, and Seyoung Park, deals with the task of Machine Translation (MT). In the specific case of the Arab language and the multitude of its dialects, it has been observed that MT systems perform badly since many words appear very infrequently in available text corpora. This paper tackles the problem by introducing a transformer-based model which encodes such scarce words by putting together linguistically relevant sub-units (word pieces). The system is successfully evaluated on multiple translation tasks from Arabic vernacular dialects to standard Arabic.

5. Discussion and Conclusions

The special Issue "Future Speech Interfaces with Sensors and Machine Intelligence" has thus brought together a wide and varied palette of contributions to speech interface technology, from AVSR, for enhancing audio speech with other sensors and new techniques, to VSR and SSI, which seek to provide speech processing even in the absence of a viable acoustic signal, through brand new types of interfaces for multimodal speech processing and for low-resource languages. The published articles have in most cases made important improvements compared to the state of the art; while others have advanced the state of the art to new frontiers.

Using computers for audio processing to facilitate man's interaction with machines has been around for many years, and tools such as high quality audio recording, speech compression, automatic speech recognition, and speech synthesis including text-to-speech have become industry standards and have reached a level of sophistication now taken for granted. Advanced speech interfaces such as multimodal, silent, lip-reading and the like, began as science fiction dreams in the style of the lip-reading HAL-9000 computer in Stanley Kubrick's 1968 classic, 2001 Space Odyssey. Research in VSR began to emerge in the 1980's [1], developing through the 2000s [2], with SSI being formally introduced in 2010 [3], and a resurgence in lip-reading occurring in the 2000-teens [4]. The special issue provides a snapshot of the current state-of-the art in future speech interfaces that use sensors and machine intelligence. The progress in the past several years has been astounding, as amply illustrated in the collection of articles provided.

As a "specialty" field, nonetheless, novel speech interfaces like those presented here have not always received the same amount of attention in the research community as the more "core" technologies. This is in part due to the added difficulty of handling non-acoustic speech signals, as discussed in the context of SSI in [5]. As such, novel speech interface technologies quite naturally lag behind somewhat as far as some of the latest developments enjoyed in core speech technologies. In particular, we may reference the stunning recent developments in the audio-visual speech representations of Deepfakes [6] as well as the hugely powerful new language models used in ChatGPT [7,8].

At the same time, some more recent arrivals to the field of future speech interfaces are now making use of generative AI techniques [9,10]. Prompted by ChatGPT and other recent advances, other researchers have also stressed that future developments in AI—and by extension, speech—will need to be based on a concerted effort joining together academia, industry, and governments [11]. Research reports along these lines, simply as an example, would be welcome contributions to the second edition of our special issue: "Future Speech Interfaces with Sensors and Machine Intelligence II", which is currently open for submissions.

Author Contributions: All authors contributed equally to the article. All authors have read and agreed to the published version of the manuscript.

Institutional Review Board Statement: Not applicable.

Informed Consent Statement: Not applicable.

Data Availability Statement: No data were generated or analyzed in creating this article.

Acknowledgments: The authors wish to thank the editorial staff of Sensors for their help in making this Special Issue a success.

Conflicts of Interest: The authors declare no conflict of interest.

References

1. Petajan, E.D. Automatic lipreading to enhance speech recognition. In Proceedings of the IEEE Communications Society Global Telecommunications Conference, Atlanta, GA, USA, 26–29 November 1984.
2. Potamianos, G.; Neti, C.; Gravier, G.; Garg, A.; Senior, A.W. Recent advances in the automatic recognition of audiovisual speech. *Proc. IEEE* **2003**, *91*, 1306–1326. [CrossRef]
3. Denby, B.; Schultz, T.; Honda, K.; Hueber, T.; Gilbert, J.; Brumberg, J. Silent speech interfaces. *Speech Commun.* **2010**, *52*, 270–287. [CrossRef]
4. Chung, J.; Zisserman, A. Lip Reading in the Wild. In *Computer Vision—ACCV 2016. Lecture Notes in Computer Science*; Lai, S.H., Lepetit, V., Nishino, K., Sato, Y., Eds.; Springer: Cham, Switzerland, 2017; Volume 10112. [CrossRef]
5. Ji, Y.; Liu, L.; Wang, H.; Liu, Z.; Niu, Z.; Denby, B. Updating the Silent Speech Challenge benchmark with deep learning. *Speech Commun.* **2018**, *98*, 42–50. [CrossRef]
6. Kietzmann, J.; Lee, L.; McCarthy, I.; Kietzmann, T. Deepfakes: Trick or treat? *Bus. Horiz.* **2020**, *63*, 135–146. [CrossRef]
7. Brown, T.; Mann, B.; Ryder, N.; Subbiah, M.; Kaplan, J.; Dhariwal, P.; Neelakantan, A.; Shyam, P.; Sastry, G.; Askell, A.; et al. Language Models are Few-Shot Learners. In *Advances in Neural Information Processing Systems 33 (NeurIPS 2020)*; Larochelle, H., Ranzato, M., Hadsell, R., Balcan, M.F., Lin, H., Eds.; Neural Information Processing Systems Foundation, Inc. (NeurIPS): San Diego, CA, USA, 2020; ISBN 9781713829546.
8. Shen, Y.; Heacock, L.; Elias, J.; Hentel, K.; Reig, B.; Shih, G.; Moy, L. ChatGPT and Other Large Language Models Are Double-edged Swords. *Radiology* **2023**, 230163. [CrossRef] [PubMed]
9. Mira, R.; Vougioukas, K.; Ma, P.; Petridis, S.; Schuller, B.; Pantic, M. End-to-End Video-To-Speech Synthesis using Generative Adversarial Networks. *IEEE Trans. Cybern.* **2020**, 1–13. [CrossRef] [PubMed]
10. Ma, P.; Petridis, S.; Pantic, M. Visual speech recognition for multiple languages in the wild. *Nat. Mach. Intell.* **2022**, *4*, 930–939. [CrossRef]
11. Whittlestone, J.; Clark, J. Why and How Governments Should Monitor AI Development. *arXiv* **2021**, arXiv:2108.12427v2.

sensors

MDPI

Article

Reliability-Based Large-Vocabulary Audio-Visual Speech Recognition

Wentao Yu *, Steffen Zeiler and Dorothea Kolossa

Institute of Communication Acoustics, Ruhr University Bochum, 44801 Bochum, Germany; steffen.zeiler@rub.de (S.Z.); dorothea.kolossa@rub.de (D.K.)
* Correspondence: wentao.yu@rub.de

Abstract: Audio-visual speech recognition (AVSR) can significantly improve performance over audio-only recognition for small or medium vocabularies. However, current AVSR, whether hybrid or end-to-end (E2E), still does not appear to make optimal use of this secondary information stream as the performance is still clearly diminished in noisy conditions for large-vocabulary systems. We, therefore, propose a new fusion architecture—the decision fusion net (DFN). A broad range of time-variant reliability measures are used as an auxiliary input to improve performance. The DFN is used in both hybrid and E2E models. Our experiments on two large-vocabulary datasets, the Lip Reading Sentences 2 and 3 (LRS2 and LRS3) corpora, show highly significant improvements in performance over previous AVSR systems for large-vocabulary datasets. The hybrid model with the proposed DFN integration component even outperforms *oracle* dynamic stream-weighting, which is considered to be the theoretical upper bound for conventional dynamic stream-weighting approaches. Compared to the hybrid audio-only model, the proposed DFN achieves a relative word-error-rate reduction of 51% on average, while the E2E-DFN model, with its more competitive audio-only baseline system, achieves a relative word error rate reduction of 43%, both showing the efficacy of our proposed fusion architecture.

Keywords: audio-visual speech recognition; hybrid models; end-to-end recognition; reliability measures; decision fusion net

Citation: Yu, W.; Zeiler, S.; Kolossa, D. Reliability-Based Large-Vocabulary Audio-Visual Speech Recognition. *Sensors* **2022**, *22*, 5501. https://doi.org/10.3390/s22155501

Academic Editors: Zahir M. Hussain

Received: 6 May 2022
Accepted: 17 July 2022
Published: 23 July 2022

1. Introduction

When people converse in noisy environments, they often subconsciously focus on the speaker' s lips to obtain supplementary information. It was also shown in [1] that the integration of visual information is of great benefit to human listening and comprehension. Even in clean speech, simply seeing the speakers articulatory movements influences perception, which is impressively demonstrated by the McGurk effect [2]. Machine audio-visual speech recognition (AVSR) is partly inspired by the genuine ability of humans to integrate audio-visual information, and its history reaches back into the late 1990s [3]. Multiple studies have provided evidence for dramatic improvements regarding small-vocabulary AVSR tasks when compared to their audio-only speech recognition counterparts with otherwise equivalent set-ups [4–7].

Nevertheless, AVSR remains difficult for large-vocabulary tasks, e.g., in large-vocabulary lip-reading tasks, with many pairs of phonemes corresponding to identical visemes. This fact makes many words almost indistinguishable to a vision-only system, as for example "do" and "to". This intrinsic difficulty makes it difficult to improve the lip-reading performance and furthermore could worsen the AVSR performance on large- or open-vocabulary tasks. On the other hand, current AVSR stream-fusion strategies, whether for hybrid or end-to-end (E2E) models, still do not seem to integrate the additional information stream optimally, and thus word error rates (WERs) have long remained unsatisfactory in noisy conditions [3,8,9].

Decision fusion is regarded an effective fusion strategy for AVSR. Individual decisions of multiple classifiers' are integrated into a single joint decision. Decision fusion covers many different forms, such as dynamic stream-weighting [10] or state-based decision fusion (SBDF), e.g., in [11–14]. In [15], the output logits of the single-modality networks were fed into a fully connected layer. Instead of fusing decisions, representation fusion is an alternative fusion approach for AVSR, e.g., via multi-modal attentions [16] or via gating [17,18]—for example in [18], which proposed the gated multi-modal unit to dynamically fuse different feature streams. Another example for representation fusion is in [19–21], which used deep feed-forward networks to first create and secondly fuse audio and video representations.

Inspired by the decision and representation fusion strategies, in this work, based on [22,23], a unified view of both fusion strategies is presented, using the posterior probabilities $p(\mathbf{s}|\mathbf{o}_t^{\mathrm{i}})$ of $i = 1 \ldots M$ single-modality models as representations of the uni-modal streams. This new viewpoint opens up a variety of exciting possibilities, centered around these single-modality representations. On the one hand, new multi-modal models can be built from multiple pre-trained uni-modal ASR models. On the other hand, optimal stream integration networks can be learned. These can utilize the reliability information inherent in the posterior probabilities and may also incorporate longer temporal context into their fused stream outputs.

In this paper, we compare the performance of the proposed fusion network in both hybrid and E2E models. Two large-vocabulary datasets, the Lip Reading Sentences 2 and 3 (LRS2 and LRS3) corpora [9,24] are used in our experiments. To analyze the performance in different noise conditions, realistic noise and reverberation are added to all the acoustic data. Our baseline models are introduced in Section 2. Section 3 describes the proposed model structure in both hybrid and E2E models. Our models rely on a range of reliability measures that are used as auxiliary inputs to inform the fusion network. These measures are detailed in Section 4. Section 5 provides the experimental details and our results for both hybrid and E2E models are demonstrated in Section 6. The lessons learned are discussed in Section 7, which also provides perspectives for future work.

2. Fusion Models Furthermore, Baselines

Many fusion strategies are available in AVSR research. This section provides a brief introduction to the various fusion strategies that are used as baseline models for this work. In all baselines, M single-modality models are combined. $\mathbf{o}_t^i$ are the features of stream i, where $i = 1, \cdots, M$. Further details are given in Section 5.2.

2.1. Hybrid Baselines

Hybrid speech recognition models have been studied for many years [25]. Although hybrid models have the disadvantage of higher complexity, they show excellent results in many studies–for example in [26]—and are still the model of choice for low-resource settings. They also provide a convenient interface for many fusion strategies, the most widely used of which are described in the following.

2.1.1. Early Integration

Early integration simply fuses the information of all input streams at the level of the input features via

$$\mathbf{o}_t = [(\mathbf{o}_t^1)^T, \cdots, (\mathbf{o}_t^{\mathrm{M}})^T]^T. \tag{1}$$

Here, superscript T denotes the transpose.

2.1.2. Dynamic Stream Weighting

For the fusion of different information streams, stream weighting is a successful and theoretically sound approach. It addresses the problem that the various streams may be reliable and informative in distinct ways. Consequently, many researchers employ the strategy of weighting different modalities [6,14,27]. Many operate static weights; for example, Ref. [28] trained audio and video speech recognizers separately, and the different model state posteriors were combined with constant stream weights λ^i according to

$$\log \widetilde{p}(s|\mathbf{o}_t) = \sum_{i}^{M} \lambda^i \cdot \log p(s|\mathbf{o}_t^i). \tag{2}$$

Here, $\log p(s|\mathbf{o}_t^i)$ is the log-posterior of state s in stream i at time t, and $\log \widetilde{p}(s|\mathbf{o}_t)$ is its estimated combined log-posterior.

However, determining optimal weights is a difficult endeavor that has significant consequences for the overall system quality [29]. In different environmental conditions, the performance of the different streams varies greatly. Specifically, the visual information may be more useful in good lighting conditions, yet audio information is most beneficial in frames with high SNRs. Therefore, the weights ought to be optimized dynamically for the best performance and to reliably prevent any instances of *catastrophic fusion*.

As a baseline approach, we therefore re-implemented *dynamic* stream weighting [30], which is realized through a weighted combination of the DNN state posteriors of all modalities:

$$\log \widetilde{p}(s|\mathbf{o}_t) = \sum_{i}^{M} \lambda_t^i \cdot \log p(s|\mathbf{o}_t^i). \tag{3}$$

The dynamic stream weights λ_t^i are predicted by a feedforward network from the estimated reliability indicators, as discussed in detail in Section 4.

Many studies have shown that reliability information is of great benefit to multi-modal integration [5,6,31,32]. Reliability indicators enhance system performance by informing the integration model about the degree of reliability in the separate information streams across time. This approach to integrated stream information can effectively and significantly improve the recognition accuracy in lower signal-to-noise ratios (SNRs).

In contrast to many other strategies, such as [10,33,34], reliability-based stream integration does not suffer from wide disparities in audio and video model performance. This is greatly beneficial to our case as we wish to design a system that least avoids any performance degradation due to the inclusion of multiple streams and that ideally profits from the visual modality under all, even under clean, acoustic conditions.

2.1.3. Oracle Weighting

As an interesting reference point, so-called *oracle* stream weights [30] were also implemented. These oracle weights are computed by minimizing the cross-entropy with the ground-truth forced alignment information, which is obtained from the clean acoustic data set. Since this method requires the ground-truth text transcription of the test set, this is not strictly a baseline but, rather, it defines a theoretical upper performance bound for dynamic stream-weighting approaches. The computed oracle stream weights λ_t^i are used to calculate the estimated log-posterior through Equation (3).

2.2. End-to-End Baselines

End-to-end speech recognition is drawing a great deal of attention and has quickly gained widespread popularity for AVSR tasks [35–37]. End-to-end models typically predict text sequences directly from signals. In this work, we select the sequence-to-sequence (S2S) transformer model (TM) [38] with connectionist temporal classification (CTC) [39] as a baseline, denoted by TM-CTC [9].

This joint model has achieved high performance in many different tasks [9,40]. In the TM-CTC model, the CTC component learns to align features and transcriptions explicitly, which is helpful for model convergence [41]. The E2E AVSR model in [9] trains the transformer and CTC separately. The transformer combines the audio and video context vectors to realize the information stream integration, and, in the CTC part, the transformer audio and video encoder outputs are simply concatenated.

In this work, we re-implemented the same structure, with the difference that the model was trained with the joint CTC/transformer strategy, serving as our E2E AVSR baseline model [41]. This joint training strategy leads to better overall performance for the AVSR task than the separate training in [9]. For the joint TM-CTC optimization, the training stage uses an objective function that linearly combines the CTC and S2S objectives

$$L = \alpha \cdot \log p_{ctc}(\mathbf{s}|\mathbf{o}) + (1-\alpha)\log p_{s2s}(\mathbf{s}|\mathbf{o}), \tag{4}$$

with $\mathbf{s}$ as the states and α as the constant hyper-parameter. During decoding, an RNN language model $p_{LM}(\mathbf{s})$ is also used; thus, the decoder optimizes the objective:

$$\log p^*(\mathbf{s}|\mathbf{o}) = \alpha \log p_{ctc}(\mathbf{s}|\mathbf{o}) + (1-\alpha) \log p_{s2s}(\mathbf{s}|\mathbf{o}) + \theta \log p_{LM}(\mathbf{s}), \tag{5}$$

where θ controls the contribution of the language model.

3. System Overview

Our proposed decision fusion net (DFN) can be employed both in hybrid and E2E models. Both model architectures are introduced briefly in the following.

3.1. Hybrid System

In hybrid speech recognition systems, the ASR task is split into two constituent phases: an estimation of state posteriors from the extracted acoustic features and a decoding stage that utilizes these posteriors in finding an optimal path by a graph search through a decoding graph. This graph can be obtained and decoded efficiently on the basis of weighted finite state transducers (WFSTs) [42]. Thus, the hybrid structure provides a natural interface for stream fusion at the level of the estimated pseudo-posteriors of all modalities $p(\mathbf{s}|\mathbf{o}_t^i)$.

For our hybrid AVSR model, all modalities are therefore dynamically combined through the proposed DFN (Figure 1). The state posteriors of each modality represent the instantaneous feature input of the DFN. Different reliability indicators are also used as auxiliary inputs, which help in estimating the multi-modal log-posteriors $\log \widetilde{p}(\mathbf{s}|\mathbf{o}_t)$ for the decoder. In the hybrid system, we investigate $M = 3$ single-modality models, one acoustic and two visual. The estimated posterior $\log \widetilde{p}(\mathbf{s}|\mathbf{o}_t)$ is computed via

$$\log \widetilde{p}(\mathbf{s}|\mathbf{o}_t) = \mathrm{DFN}([p(\mathbf{s}|\mathbf{o}_t^{\mathrm{A}})^T, p(\mathbf{s}|\mathbf{o}_t^{\mathrm{VA}})^T, p(\mathbf{s}|\mathbf{o}_t^{\mathrm{VS}})^T, \mathbf{R}_t^T]^T), \tag{6}$$

where $p(\mathbf{s}|\mathbf{o}_t^{\mathrm{A}})$, $p(\mathbf{s}|\mathbf{o}_t^{\mathrm{VA}})$ and $p(\mathbf{s}|\mathbf{o}_t^{\mathrm{VS}})$ are the state posteriors of the audio model and of the appearance-based and a shape-based video model, respectively. $\mathbf{R}_t$ is the vector of all reliability measures at time t as detailed in Section 4.

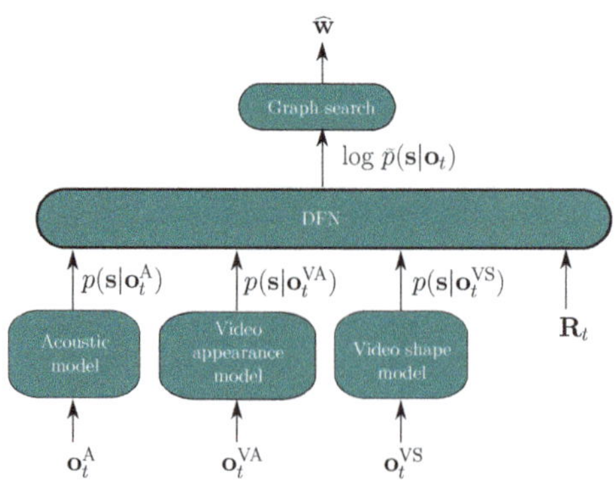

Figure 1. Audio-visual fusion based on the DFN, applied to one stream of audio and two streams of video features.

The hybrid AVSR fusion model is trained with the cross-entropy loss

$$\mathcal{L}_{\mathrm{CE}} = -\frac{1}{T}\sum_{t=1}^{T}\sum_{s=1}^{S} p^*(s|\mathbf{o}_t) \cdot \log \widetilde{p}(s|\mathbf{o}_t). \tag{7}$$

Here, $p^*(s|\mathbf{o}_t)$ is the goal state probability of state s, calculated by forced alignment of the clean acoustic training data. The estimated vector of log-posteriors $\log \widetilde{p}(\mathbf{s}|\mathbf{o}_t)$ is obtained from Equation (6). Finally, the decoder utilizes these estimated log-posteriors to find the optimum word sequence by graph searching through the decoding graph [43].

3.2. E2E System

Our E2E AVSR model is based on the TM-CTC model, which combines a transformer model (TM) and a connectionist temporal classification (CTC) model through Equation (4) during the training stage and through Equation (5) in the decoding stage. In all E2E experiments, $M = 2$ modalities are considered, one acoustic and one visual ($\mathbf{o}^{\mathrm{A}}$ and $\mathbf{o}^{\mathrm{VI}}$ in Figure 2). The following sections describe the encoder and decoder architecture, which both needed modifications for our proposed stream integration approach.

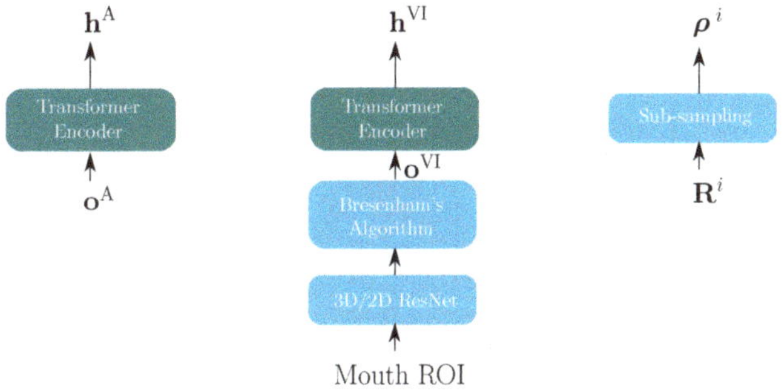

Figure 2. Audio encoder (**left**), video encoder (**middle**) and reliability measure encoder (**right**) for both modalities $i \in$ A, VI. The blue blocks are used to align video features with audio features; the turquoise block shows the transformer encoder.

3.2.1. Encoder Architecture

The structure of the conventional transformer encoder is depicted in Figure 3. The features are first fed into a sub-sampling block comprised of two 2D convolution layers with a kernel size of 3 and stride of 2, which are used to decrease the computational effort. The input has dimension [batch, 1, N_f, d_f], where N_F is the number of frames and d_f is the input feature dimension. With two 2D convolution layers and a feed-forward layer, the sub-sampling layer reduces the sequence length from N_F to $N_F/4$ and changes the feature size d_f to a common dimension $d_{att} = 256$. A stack of 12 encoder blocks, consisting of a multi-head self-attention and a fully connected feed-forward layer, yields the desired encoder output $\mathbf{h}^i$ for each modality.

Figure 2 depicts all encoders in the E2E system—an audio encoder, a video encoder and a reliability encoder. As described in [41], for a joint TM-CTC model, the output sequence of the transformer encoder is used in both the transformer and the CTC decoder. The video features are extracted according to [9] via a pre-trained spatio-temporal visual front-end [44] (the 3D/2D ResNet in Figure 2). The extracted video features are then passed through the transformer encoder. Due to the different frame rates of the audio and video features, a Digital Differential Analyzer (comparable to Bresenham's algorithm [45]) is used to optimally replicate the video features to achieve the same sequence length.

In the multi-head self-attention block in Figure 3, the queries **Q**, keys **K** and values **V** are identical. The attention transform matrix [38] of every attention head with index j is computed via

$$\mathbf{T}_j = \text{softmax}\left(\frac{\left(\mathbf{W}_j^Q\mathbf{Q}^T\right)^T\left(\mathbf{W}_j^K\mathbf{K}^T\right)}{\sqrt{d_k}}\right). \tag{8}$$

The attention is computed as

$$\boldsymbol{\alpha}_j = \text{attention}_j(\mathbf{Q},\mathbf{K},\mathbf{V}) = \mathbf{T}_j\left(\mathbf{W}_j^V\mathbf{V}^T\right)^T, \tag{9}$$

where $\mathbf{W}_j^*$ are the learned parameters, $d_k = \frac{d_{att}}{h}$ and h is the number of attention heads. In the attention mechanism, the attention transform matrix $\mathbf{T}_j$ indicates the relevance of the current keys for the current queries. $\mathbf{T}_j$ is of size$N_Q \times N_K$, where N_Q and N_K are the lengths of **Q** and **K**, respectively. A fully connected layer is used in the self-attention block to project the concatenated outputs of all heads $\boldsymbol{\alpha}_j$. Finally, the output of the self-attention block is input to a feed-forward layer, which yields the encoder output $\mathbf{h}^i$.

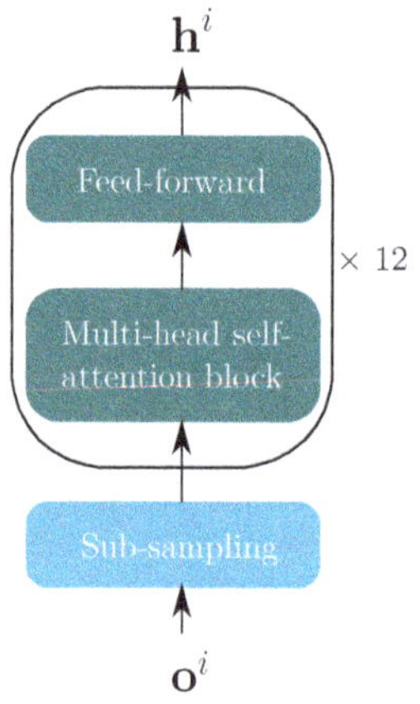

Figure 3. Transformer encoder for both modalities $i \in$ A, VI. The blue block shows the sub-sampling, whereas the turquoise blocks comprise the the transformer encoder.

3.2.2. Decoder Architecture

Figure 4 shows the TM-CTC decoder components for each stream. As in the baseline model [9], the CTC decoder consists of a stack of six multi-head self-attention blocks and the output layer. The transformer decoder is comprised of a stack of six decoder blocks, each containing a multi-head attention block. For each decoder, the keys (**K**) and values (**V**) are the encoder outputs $\mathbf{h}^i$—both of size $(N_F/4) \times 256$. The queries (**Q**) come from the previous decoder block and are transformed by a multi-head self-attention block. **Q** is a $N_T \times 256$ matrix, where N_T represents the length, or the number of tokens, of the transcription. In the decoder, the attention transform matrix $\mathbf{T}_j$ is of size $N_T \times N_F/4$, which transforms the sequence length from $N_F/4$ to N_T. Hence, the length of the transformer posteriors is N_T.

Our goal is to integrate the stream-wise posteriors given all the stream reliability measures. Fortunately the integration step for the CTC model is straightforward, because the stream-wise posteriors $p_{ctc}(\mathbf{s}|\mathbf{o}^i)$ are already temporally aligned with the reliability metrics $\boldsymbol{\rho}^{\,i}$—both of length $N_F/4$.

In contrast, the integration for the transformer remains difficult. The reliability metrics $\boldsymbol{\rho}^{\,i}$ in Figure 2, are of length $N_F/4$; however, we expect them to temporally match the token-by-token posteriors $p_{s2s}(\mathbf{s}|\mathbf{o}^i)$. Therefore, a transformation from the linear time domain of length $N_F/4$ to length N_T is necessary at this point. As shown in Figure 4, there are six multi-head attention blocks in the transformer decoder, and each block has its own attention transform matrix $\mathbf{T}_j^i$. Here, the transform matrix in the final block of modality i is reused to transform the length of $\boldsymbol{\rho}^{\,i}$ from $N_F/4$ to N_T. The transformed reliability attention of head j ($\tilde{\boldsymbol{\rho}}_j^i$) is computed by

$$\tilde{\boldsymbol{\rho}}_j^i = \mathbf{T}_j^i \cdot \left(\mathbf{W}_j^{i\rho}(\boldsymbol{\rho}^{\,i})^T\right)^T. \tag{10}$$

The final reliability embedding vector $\tilde{\boldsymbol{\rho}}^{\,i}$ is obtained by projecting a concatenation of all heads of the transformed reliability attentions via a fully connected layer.

Figure 5 shows the topology of the multi-modal fusion for the E2E model. The posterior probabilities from all modalities are the inputs, and the corresponding reliabilities $\boldsymbol{\rho}^{\,i}$, or their embeddings $\widetilde{\boldsymbol{\rho}}^{\,i}$ are used to estimating the multi-modal log-posteriors $\log \widetilde{p}(\mathbf{s}|\mathbf{o})$, for both the CTC and the S2S model. Finally, the estimated log-posteriors from both transformer and CTC model are combined through Equation (4) in the training stage and via Equation (5) in the decoding stage.

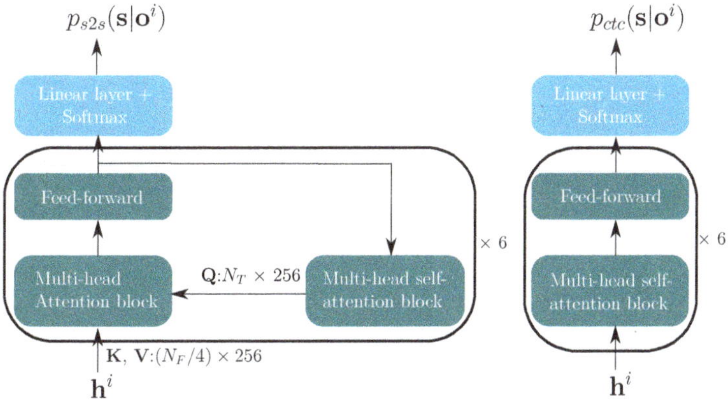

Figure 4. Transformer decoder (**left**) and CTC decoder (**right**) for both modalities $i \in$ A, VI.

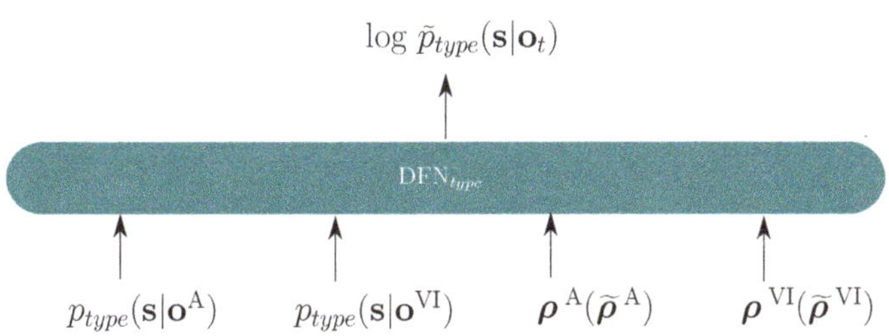

Figure 5. DFN fusion topology for E2E model, $type \in s2s, ctc$.

4. Reliability Measures

As stated before, in this work, we aim to fuse stream-wise posteriors into joint posteriors according to the respective stream reliabilities. Therefore, a variety of reliability measures are extracted to inform the integration model of the time varying reliability of the separate streams. Although the reliabilities for the hybrid and E2E models are similar, there are some subtle differences. These will be discussed in more detail in the following part.

4.1. Reliabilities for the Hybrid Model

For the dynamic stream weighting in our proposed DFN hybrid model, both model-based and signal-based reliability measures (e.g., see Table 1) are extracted; most of them were previously introduced in [30].

Table 1. Overview of reliability measures.

Model-Based	Signal-Based	
	Audio-Based	Video-Based
Entropy Dispersion Posterior difference Temporal divergence Entropy and dispersion ratio	MFCC ΔMFCC SNR f_0 Δf_0 voicing probability	Confidence IDCT Image distortion

To obtain the model uncertainty information, a number of model-based measures are extracted, i.e., entropy, dispersion, posterior difference, temporal divergence, entropy- and dispersion-ratio. The model-based measures consider the audio and video models separately. All these measures are derived from the log-posterior probabilities of their respective single-modality models.

Signal-based measures are used to estimate the signal quality in each stream. They can be subdivided into audio- and video-based measures. The audio reliability measures are the first five MFCC coefficients with their temporal derivatives ΔMFCC, again as in [30]. The signal-to-noise ratio (SNR) is an important indicator related to the intelligibility of the audio signal. However, due to the acoustic data augmentation with realistic noise, conventional SNR estimation is not able to provide adequate results.

For this reason, the deep learning approach DeepXi [46] is used here to estimate the frame-wise SNR. Furthermore, as pitch appears to influence the reliability of acoustic features, specifically of MFCC [47,48], the estimated pitch f_0 and its temporal derivative, Δf_0, are also used as reliability indicators. The probability of voicing [48] is also a valuable reliability indicator, which is computed from the Normalized Cross-Correlation Function (NCCF) values for each frame.

For the video stream, OpenFace [49] is used for face detection and facial landmark extraction. Here, the confidence of the face detector in each frame is considered as a video signal quality indicator. The Inverse Discrete Cosine Transform (IDCT), as well as the image distortion estimates, are also included and computed as in [30].

4.2. Reliabilities for the E2E Model

The E2E model focuses on signal-based reliability measures, e.g., the confidence of the face detector. Additionally, some Facial Action Units (AUs) [49,50] about the chin, jaw and lip movements (AU12, AU15, AU17, AU23, AU25 and AU26) were also selected to help to improve the performance of the visual model. Different from the hybrid model, the E2E model does not use the image distortion estimates as part of the reliability measures, as our experimental results indicated these estimates to be detrimental to performance in initial experiments. More detailed analyses and discussions can be found in Section 6.1. The audio-based reliability measures comprise the first five MFCC coefficients, estimated SNR, the pitch f_0 and its first temporal derivative as well as the probability of voicing.

5. Experimental Setup

This section introduces the databases and the feature extraction for both streams and it details our experimental setup.

5.1. Dataset

The Oxford-BBC Lip Reading Sentences (LRS) 2 and 3 corpora [9,24] were selected for our experiments, see Table 2 for their statistics.

Table 2. Characteristics of the utilized datasets.

Subset	Utterances	Vocabulary	Duration [hh:mm]
LRS2 pre-train	96,318	41,427	196:25
LRS2 train	45,839	17,660	28:33
LRS2 validation	1082	1984	00:40
LRS2 test	1243	1698	00:35
LRS3 pre-train	118,516	51 k	409:10

The hybrid model experiments used the LRS2 corpus. All acoustic, visual and AV models were trained with the combined LRS2 pre-train and training set. To compare the performance of our proposed E2E model with the baseline model [9], the LRS3 corpus pre-train set was also used in the E2E experiments. In AVSR tasks, the acoustic model is always in a dominant position. To analyze the performance in different noise environments and counter the audio-visual model imbalance, we applied data augmentation. The acoustic noise data comes from the MUSAN noise corpus [51]. For the hybrid model dataset, the acoustic data was augmented with the ambient noise, which contains noises, such as wind, footsteps, paper rustling and rain as well as indistinct crowd noises. SNRs were randomly selected from −9 to 9 dB in steps of 3 dB, where the SNRs are computed by:

$$\text{SNR}_{dB} = 10\log_{10}\frac{P_{signal}}{P_{noise}} \tag{11}$$

with P_{signal} and P_{noise} as the signal and noise energy, respectively.

Since the LRS2 dataset does not contain highly reverberant data, the acoustic data was artificially reverberated by convolutions with measured impulse responses. These impulse responses also came from the MUSAN corpus. The E2E model training set augmentation was the same as that in hybrid model, with ambient noise and SNRs were between −9 and 9 dB. The video sequences were augmented with random cropping and horizontal flips with a 50% probability. To check the robustness of our model, new acoustic noise conditions that are unseen in the training data were added to the test set. Both ambient and music noise were used, from −12 to 12 dB. Similarly, Gaussian blur and salt-and-pepper noise were also applied to the visual data for the test set. The acoustic data augmentation was realized through a Kaldi Voxceleb example recipe.

5.2. Features

Both our hybrid and the E2E models used log-mel features together with the estimated pitch f_0 and its derivative, Δf_0, and the voicing probability as the audio features. The frame size was 25 ms with a 10 ms frameshift. The Kaldi hybrid model extracts audio features with 40 triangular mel filters, while in the ESPnet E2E model, the number of mel-frequency bins is 80.

For both systems, OpenFace [49] was used for face detection and facial landmark extraction. The speaker's face was detected at 25 frames per second. The digital differential analyzer, which uses the Bresenham algorithm, was used to align the audio and video streams. In the hybrid model, two kinds of video features were extracted: The video appearance model (VA) used 43-dimensional IDCT coefficients of the gray-scale region of interest (ROI) as features, where the mouth ROI was extracted from the facial mouth landmarks with a rectangular box.

The video shape model (VS), in contrast, is based on the 34-dimensional non-rigid shape parameters described in [49]. For the E2E model, the mouth ROI was fed directly into a pre-trained video model [44], which first performed 3D convolutions on the image sequence and then utilized a 2D ResNet to extract the final facial feature representation.

5.3. Hybrid Model Implementation Details

In the hybrid model, the Kaldi toolkit [52] was used for speech recognition. The LRS2 pre-train and training set were used together for model training. The hybrid model starts with HMM-GMM training, which follows the standard Kaldi AMI recipe, i.e., monophone training followed by triphone training. Afterwards, a linear discriminate analysis (LDA) stacks the context of features to obtain discriminative short-term features. Finally, the speaker adaptive training (SAT) is used to compensate the speaker characteristics. Each step produces a better forced alignment based on the current model for later network training. The subsequent HMM-DNN training used the nnet2 p-norm network [53] recipe, which is efficiently parallelizable.

The estimated log-posteriors $\log p(\mathbf{s}|\mathbf{o}_t^i)$ for each stream were obtained from each trained single modality. As shown in Figure 6, the posteriors of all modalities were the inputs for our proposed decision fusion net (DFN). The corresponding reliability measures were used to estimating the multi-modal log-posteriors $\log \widetilde{p}(\mathbf{s}|\mathbf{o}_t)$, which was finally used in graph searching through a decoding graph to obtain the best word sequence. In the hybrid model, all modalities were trained separately. To ensure that all modalities search through the same decoding graph, the phonetic decision tree was shared between all single modalities. For this reason, the number of states for each modality was identical—specifically 3856.

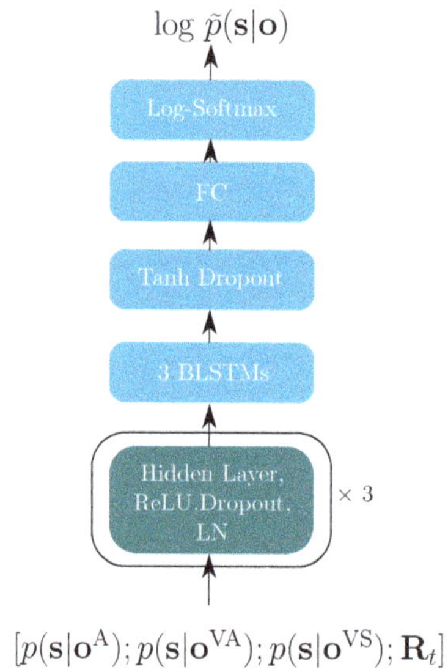

Figure 6. Decision fusion net structure for the hybrid model. The turquoise block indicates the successively repeated layers.

For the hybrid model, there were 41 reliability indicators, therefore, the input of the DFN was $(3 \times 3856 + 41) = 11{,}609$ dimension. The three hidden layers in Figure 6 contain 8192, 4096 and 1024 units, respectively, each followed by a ReLU activation function, layer normalization (LN) and with a dropout rate of 0.15. After hidden layers are three BLSTM layers with 1024 memory cells for each direction, with the tanh activation function. A fully connected (FC) final layer projects the data to the output dimension of 3856. A log-softmax function finally yields the log-posteriors.

To avoid overfitting, we applied early stopping and check every 7900 iterations. When the validation loss did not decrease for 23,700 iterations, the training was stopped. Finally, the trained model was evaluated on the test set. To evaluate the effect of bi-directional inference, two experiments with the proposed DFN strategy were conducted. The first one used the BLSTM-DFN—exactly as shown in Figure 6. The second employed an LSTM-DFN, replacing the BLSTM layers with LSTM layers.

The initial learning rate was 0.0005, and this was decreased by 20% if the validation loss did not reduce in the early stopping check. The batch size was 10. The DFN model fine-tuning was based on the PyTorch library [54] with the ADAM optimizer. The training was performed with a GeForce RTX 2080 Ti GPU. Each single-modality model and the early integration training took around 7 days. A complete training of the BLSTM-DFN or LSTM-DFN stream integration model ran for approximately 15 days.

E2E Model Implementation Details

To compare the performance between our proposed E2E AVSR model and the baseline model, all E2E models, which were trained by ESPnet, were pre-trained on the same data, the LRS2 and LRS3 pre-train set. However, training with such an enormous dataset is time-consuming. To save computational effort, in the pre-training stage, the parameters of the ResNet video feature extractor were frozen, which is the same as in the baseline model [9]. Then, in the training stage, all parameters, including those of the ResNet, were fine-tuned on the LRS2 training set. To improve the performance, our proposed TM-CTC AVSR model was initialized with the audio- and video-only model, which were trained separately.

All ESPnet E2E models share the same language model, which always predicts one character at a time and receives the previous character as its input. It was implemented as a unidirectional four-layer recurrent network, with each layer having 2048 units. This work was based on a pre-trained language model, which was trained on the LibriSpeech corpus [55].

As shown in Figure 7, in the E2E model, the single-modality posteriors are the inputs and, together with the corresponding reliability information, they are used to estimate the multi-modal log-posteriors, $\log \widetilde{p}(\mathbf{s}|\mathbf{o})$, for both the CTC and the S2S model. Both DFN_{ctc} and DFN_{s2s} in Figure 7 start with three hidden layers, which have 8192, 4096 and 512 units, each using the ReLU activation function and layer normalization (LN).

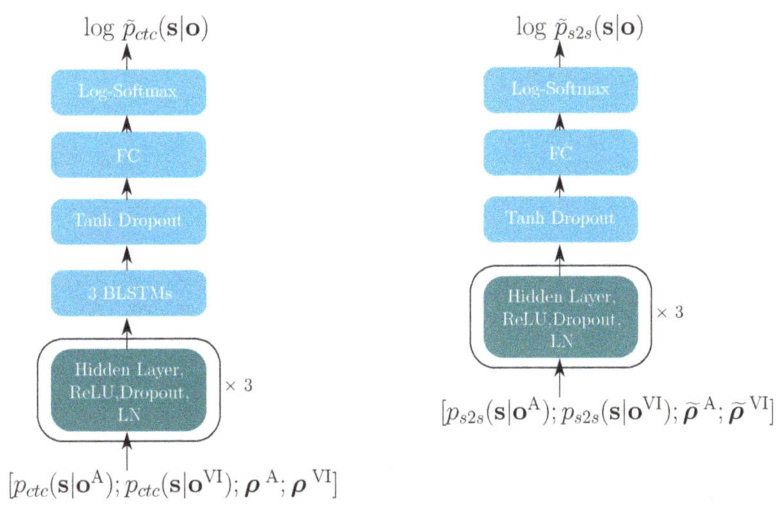

Figure 7. DFN_{ctc} (**left**) and DFN_{s2s} (**right**). The turquoise blocks indicate the successively repeated layers.

The dropout rate was 0.15. DFN_{ctc} contained three BLSTM layers with 512 memory cells for each direction, using the tanh as their activation function. BLSTM layers for the DFN_{s2s} were also tested; however, this resulted in overfitting. Similarly to the hybrid model, again, the final layer was realized as a fully connected (FC) layer followed by a log-softmax function, which gives us the estimated log-posteriors. In Equations (4) and (5), the language model contribution parameter θ is 0.5; α is 0.3. $h = 4$ heads were used in the attention blocks. The transformer-learning factor controls the learning rate. In the pre-training stage, the factor was 5.0, while in the fine-tuning stage, it was 0.05.

The ESPnet E2E models were trained by NVIDIA's Volta-based DGX-1 multi-GPU system with seven Tesla V100 GPUs, each with 32 GB memory. All single-modality models were trained for 100 epochs. The AVSR baseline model and our proposed model were pre-trained for 65 epochs and fine-tuned for 10 epochs.

6. Results

In this section, we compare the performance of our experimental results based on the hybrid and E2E models.

6.1. Hybrid Model

The performance of all hybrid baseline models and our fusion strategies are first shown in this part. In the following, some intuitive exemplary decoding results of our experiments are given in Table 3. Comparing all results, the proposed BLSTM-DFN had better performance compared with the other baseline strategies.

Table 3. Decoding results for three exemplary sentences S1, S2 and S3. *RT* represents the reference transcription; *AO* is audio only model; *EI* is early integration; *CE* and *MSE* represent dynamic stream weighting with CE and MSE as loss functions; *OW* is the oracle stream-weighting; and *LSTM-DFN* and *BLSTM-DFN* are variants of our proposed integration model.

	Type	Result
S1	RT	However, what a surprise when you come in
	AO	However, what a surprising coming
	EI	However, what a surprising coming
	CE	However, what a surprising coming
	MSE	However, what a surprising coming
	OW	However, what a surprising coming
	LSTM-DFN	However, what a surprising coming
	BLSTM-DFN	However, what a surprise when you come in
S2	RT	I'm not massively happy
	AO	I'm not mass of the to
	EI	Some more massive happy
	CE	I'm not massive into
	MSE	I'm not massive into
	OW	I'm not mass of the happiest
	LSTM-DFN	I'm not massive it happened
	BLSTM-DFN	I'm not massively happy
S3	RT	Better street lighting can help
	AO	Benefit lighting hope
	EI	However, the street lighting and hope
	CE	Benefit lighting hope
	MSE	Benefit lighting hope
	OW	In the street lighting hope
	LSTM-DFN	However, the street lighting and hope
	BLSTM-DFN	Better street lighting can help

The estimated log-posterior probabilities for the target state sequence, $\log \widetilde{p}(s_t^*|\mathbf{o}_t)$, are plotted in Figure 8 to show the discriminative power of different models. Larger log-posterior probabilities indicate that the estimated state is closer to the target state. As expected, the BLSTM-DFN produced larger log-posteriors on the reference states, compared to the other fusion strategies. This corresponds with the better performance of the BLSTM-DFN that was observed on this example.

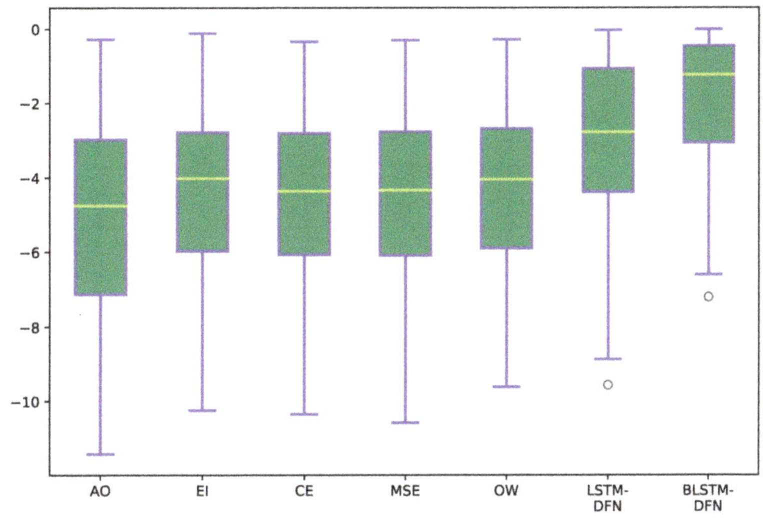

Figure 8. Estimated log-posteriors of sentence S2 for the target state s_t^*, with additive noise at −9 dB. All abbreviations are the same as in Table 3. The whiskers show the maximum and minimum values; the upper and lower bounds of the green blocks represent the respective 25th and 75th percentile; the yellow line in the center of the green block indicates the median.

Figure 9 gives an overall comparison of the performance of the audio-only model and AVSR models in different noise conditions. Our proposed fusion strategy improved the Word Error Rate (WER) in every SNR environment and even for the clean acoustic data. In worse SNR conditions, the proposed DFN reduced the WER over 10%. The DFN with BLSTM layers outperformed the—realistically unachievable—oracle weighting (OW) in many cases, while the latter is based on the ground-truth transcription information of the test set and could be considered as the upper limit for the dynamic stream-weighting method (as described in Equation (3)).

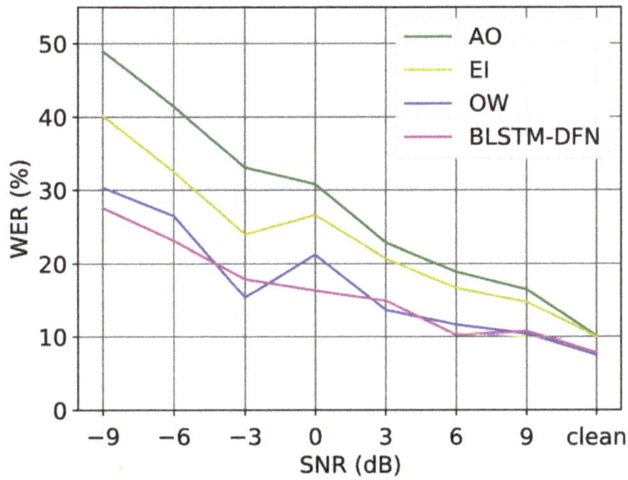

Figure 9. WER (%) on the test set of the LRS2 corpus in different noise conditions.

Table 4 gives the detailed results of all our experiments under additive noise. The average WERs of the visual models exceeds 80%, which means that lipreading is still difficult for the large-vocabulary task. One potential reason is that the video input is highly correlated in each frame, making the GMM model challenging to train. We also aimed to improve the performance of the visual models by using the pre-trained spatio-temporal visual front-end from [44] to extract high-level visual features but without seeing improvements.

Table 4. Word error rate (%) on the LRS2 test set under additive noise.

Model \ dB	−9	−6	−3	0	3	6	9	Clean	Avg.
AO	48.96	41.44	33.07	30.81	22.85	18.89	16.49	10.12	27.83
VA	85.83	87.00	85.26	88.10	87.03	88.44	88.25	88.10	87.25
VS	88.11	90.27	87.29	88.88	85.88	85.33	88.58	87.10	87.68
EI	40.14	32.47	23.96	26.59	20.67	16.68	14.76	10.02	23.16
MSE	46.48	37.79	27.45	27.47	19.52	16.58	15.09	9.42	24.98
CE	45.79	37.14	26.32	28.03	19.40	16.68	14.76	9.42	24.65
OW	30.33	26.47	**15.41**	21.25	**13.66**	11.66	**10.45**	**7.54**	17.10
LSTM-DFN	33.30	27.22	21.26	21.25	19.17	13.97	15.84	10.32	20.29
BLSTM-DFN	**27.55**	**23.11**	17.89	**16.35**	14.93	**10.25**	10.78	7.84	**16.09**

Early integration (EI) showed a relative WER reduction of 16.78%; however, the improvement was not as significant as the proposed DFN approach. Comparing the BLSTM-DFN and the LSTM-DFN, the former showed the better performance for non-real-time decoding. Both the LSTM- and BLSTM-DFN used recurrent layers with 1024 cells. A BLSTM-DFN using 512 memory cells per layer was also tested to balance the number of the model parameters. The average WER of this was 16.14%, which is still better than that of the LSTM-DFN with 1024 cells.

We tested the improvements that we were seeing for statistical significance, comparing in each case, with the audio-only model by using the NIST Scoring Toolkit SCTK (https://github.com/usnistgov/SCTK, accessed on 28 October 2021). All results are summarized in Table 5. As can be seen, the BLSTM-DFN yielded highly significant improvements over the audio-only model (AO). In contrast, the early integration model, EI, only considerably improved the performance at lower SNR conditions (at SNRs < 3 dB).

Table 5. Asterisks indicate a statistically significant difference compared with the audio-only model (AO). *** denotes $p \leqslant 0.001$, ** shows $0.001 < p \leqslant 0.01$, * corresponds to $0.01 < p \leqslant 0.05$, and ns indicates results where $p > 0.05$.

Model \ dB	−9	−6	−3	0	3	6	9	Clean	Avg.
EI	***	***	***	*	ns	ns	ns	ns	***
MSE	*	***	***	ns	*	**	**	ns	***
CE	ns	***	***	ns	*	**	**	ns	***
OW	***	***	***	***	***	***	***	***	***
LSTM-DFN	***	***	***	***	*	***	ns	ns	***
BLSTM-DFN	***	***	***	***	***	***	***	*	***

Far-field AVSR (by artificially reverberating the audio data through convolutions with measured impulse responses) was also evaluated. According to Table 6, the BLSTM-DFN still outperformed the other fusion strategies; however, in this case, it did not reach the performance of oracle weighting (which uses oracle knowledge for optimal weighting, see Section 2.1.3). One reason for this may be an insufficient amount of reverberant acoustic training signals—while the (non-realistic, upper-bound) OW setup requires few parameters to be estimated, the DFN actually learns an optimal, non-linear fusion strategy, for which more data may be required.

As can also be seen, all audiovisual models significantly improved the performance compared with the AO model. Here, again, the improvement of early integration was inferior to the other proposed models, rendering DFN as the most effective of all practical approaches. It can also be noted that the unidirectional LSTM-DFN was successful for this dataset, which would thus allow for real-time implementations as well. Overall, the introduced DFN was generally superior to instantaneous dynamic stream weighting.

Table 6. Far-field AVSR WER (%) and statistically significance compared with the AO model on the LRS2 dataset. *** denotes $p \leqslant 0.001$, ** shows $0.001 < p \leqslant 0.01$

AO	EI	MSE	CE	OW	LSTM-DFN	BLSTM-DFN
23.61	19.15 (**)	19.54 (***)	19.44 (***)	**12.70** (***)	15.67 (***)	15.28 (***)

It is also interesting to analyze which kinds of reliability measures are the most informative and effective. Therefore, after comparing the performance between our proposed model and the baseline models, we also conducted experiments, in which we utilized different reliability measure sets in our proposed BLSTM-DFN model. Both model-based and signal-based reliabilities were taken into consideration. Table 7 lists the experimental results based on different reliability indicator groups.

Our experimental results indicate that image distortion estimates were actually detrimental to performance ($\mathbf{R}^{\mathrm{V}}$ and *All* in Table 7). Consequentially, we repeated the BLSTM-DFN model training without these estimates ($\mathbf{R}^{\tilde{\mathrm{V}}}$ and $\tilde{All}$ in Table 7). Both audio- and video-based reliability indicators were able to improve the model performance. The audio-based measures outperformed the video-based measures on average. However, combining both audio- and video-based measures led to the best performance ($\tilde{All}$), achieving a relative word-error-rate reduction of 50.59% compared to the audio-only model.

Table 7. BLSTM-DFN word error rates (%) on the LRS2 test set under additive noise. *All*: apply all reliability indicators as shown in Table 1; $\mathbf{R}^{\mathrm{A}}$: all audio-based reliability indicators; $\mathbf{R}^{\mathrm{V}}$: all video-based reliability indicators; $\mathbf{R}^{\tilde{\mathrm{V}}}$: using the video-based reliability indicators, excluding the image distortion estimates; $\tilde{All}$: using all reliability indicators except for image distortion estimates; *None*: proposed model without reliabilities. *Avg*: Average performance, together with the significance of improvements (compared with *None*). ^{ns}: not significant and ***: $p \leqslant 0.001$.

R \ dB	−9	−6	−3	0	3	6	9	Clean	*Avg.*
All	27.55	23.11	17.89	16.35	14.93	10.25	10.78	7.84	16.09 ^{ns}
$\mathbf{R}^{\mathrm{A}}$	23.39	**17.96**	14.51	15.68	**12.97**	8.44	10.67	**6.94**	13.82 ***
$\mathbf{R}^{\mathrm{V}}$	98.12	98.50	98.76	98.22	99.43	98.79	99.46	98.81	98.76
$\mathbf{R}^{\tilde{\mathrm{V}}}$	25.97	21.23	17.66	17.58	14.24	10.85	**9.70**	7.54	15.60 ^{ns}
None	24.48	21.70	17.55	18.35	16.07	9.35	12.07	8.43	16.00
$\tilde{All}$	**22.20**	18.52	**14.40**	**15.46**	13.66	**8.04**	9.91	7.84	**13.75** ***

We also tested the improvements that were obtained when adding reliability information for their statistical significance. While the visual reliabilities slightly boosted the performance relative to the model without reliability information (*None*), these improvements were not statistically significant. This stands in contrast with the effect of acoustic reliability indicators, which provided highly significant improvements by themselves as well as in combination.

6.2. E2E Model

To compare the performance of the hybrid model and the E2E model directly, and an additional audio-only model was trained on the LRS2 corpus. The E2E audio-only model yielded a WER of 3.7%, while the hybrid audio-only model showed a WER of 11.28%. Table 8 shows the experimental results in all noise conditions. As expected, the audio-only model outperformed the video-only model. Comparing the performance between the baseline by [9] and our proposed AVSR model, our introduced DFN resulted in a better performance in all noise environments. Even in clean acoustic conditions, the proposed model clearly reduced the WER.

On average, the new system gained a relative word error rate reduction of 43% compared to the audio-only setup and 31% compared to the audio-visual end-to-end baseline. Table 9 also shows the results of the NIST statistical significance tests between different model setups.Our work compares the AV baseline and the DFN with the audio-only model and shows the difference between the AV baseline and the proposed DFN, all in different noise augmentation types.

The AV baseline only significantly improved the performance compared with the AO model in lower noise conditions (SNR < 0 dB). In contrast, our proposed DFN model substantially outperformed both the AO recognizer and the AV baseline, not only in most noise environments but also in clean acoustic conditions. It was also effective at information integration with blurred or noisy video data, again significantly improving over audio-only recognition as well as over the AV baseline model.

Table 8. Performance of the audio-visual and uni-modal speech recognition (WER [%]). AO: audio only. VO: video only. AV: AV baseline [9]. DFN: proposed DFN fusion. m: music noise. a: ambient noise. vc: clean visual data. gb: visual Gaussian blur. sp: visual salt-and-pepper noise.

Model \ dB	−12	−9	−6	−3	0	3	6	9	12	Clean	*Avg.*
AO (m)	18.9	13.7	11.2	8.4	6.3	6.8	4.5	4.1	4.3	4.2	8.2
AO (a)	25.7	23.4	18.5	11.6	8.2	9.0	5.9	3.8	4.4	4.2	11.5
VO (vc)	58.7	61.0	61.7	69.6	69.6	63.5	64.6	63.6	66.6	61.9	64.1
VO (gb)	66.6	69.2	71.0	68.5	68.5	71.1	62.7	69.4	67.6	66.9	68.2
VO (sp)	68.5	72.5	73.7	70.1	70.1	70.6	68.3	69.1	73.1	67.9	70.4
AV (m.vc)	14.6	11.8	6.4	7.9	7.9	6.3	5.2	4.4	3.4	4.0	7.2
DFN (m.vc)	**11.1**	**8.7**	**5.5**	**4.8**	**4.8**	**4.5**	**3.6**	**3.3**	**2.2**	**2.4**	**5.1**
AV (a.vc)	19.1	19.0	14.3	7.3	6.3	6.0	5.7	4.5	4.9	4.0	9.1
DFN (a.vc)	**14.3**	**11.9**	**8.1**	**4.8**	**4.0**	**5.4**	**3.7**	**2.8**	**3.6**	**2.4**	**6.1**
AV (a.gb)	20.6	18.9	15.0	7.7	6.8	7.5	5.9	3.9	4.8	4.0	9.5
DFN (a.gb)	**14.9**	**12.8**	**9.4**	**5.2**	**4.2**	**5.5**	**3.8**	**3.0**	**4.1**	**2.6**	**6.6**
AV (a.sp)	19.5	19.9	15.3	7.7	7.2	6.3	5.6	4.4	4.6	4.3	9.5
DFN (a.sp)	**15.4**	**12.8**	**9.9**	**5.2**	**4.7**	**5.5**	**3.4**	**2.6**	**4.0**	**2.5**	**6.6**

For the E2E model, we also tested the effect of the different groups of reliability measures. Again, both model-based and signal-based reliabilities were taken into consideration. Table 10 shows that the models with the audio- or video-based reliability indicators ($\mathbf{R}^{A}$ and $\mathbf{R}^{V}$) outperformed those without reliability measures (None). The audio-based reliabilities were, again, more effective than the video-based measures, particularly in high-SNR conditions.

Furthermore, as in the hybrid model, combing the audio- and video-based reliability indicators delivered the best performance (*All* in Table 10). The last column in Table 10 shows the results of a statistical significance test of those improvements. The audio-based reliability measures are clearly more effective than the visual ones. Similarly to the hybrid model in Table 7, using all reliability measures jointly led to the best overall performance, with highly significant improvements in comparison to the case without reliability information.

Table 9. Statistical significance tests, comparing the results of different model setups *** denotes $p \leqslant 0.001$, ** shows $0.001 < p \leqslant 0.01$, * corresponds to $0.01 < p \leqslant 0.05$, and ns indicates results where $p > 0.05$; the other abbreviations are described in Table 8.

Model \ dB	−12	−9	−6	−3	0	3	6	9	12	Clean	*Avg.*
AO-AV (m.vc)	*	ns	***	ns	ns	ns	ns	ns	ns	ns	***
AO-DFN (m.vc)	***	***	***	**	ns	**	ns	ns	*	***	***
AV-DFN (m.vc)	**	**	ns	**	***	*	ns	*	ns	**	***
AO-AV (a.vc)	***	**	**	**	ns	**	ns	ns	ns	ns	***
AO-DFN (a.vc)	***	***	***	***	***	***	**	ns	ns	***	***
AV-DFN (a.vc)	**	***	***	*	*	ns	**	ns	ns	**	***
AO-DFN (a.gb)	***	***	***	***	***	***	*	ns	ns	**	***
AV-DFN (a.gb)	***	***	***	*	**	*	**	ns	ns	*	***
AO-DFN (a.sp)	***	***	***	***	**	**	***	ns	ns	**	***
AV-DFN (a.sp)	*	***	***	*	*	ns	**	*	ns	**	***

Table 10. Performance of the proposed E2E DFN fusion (WER [%]), based on the different E2E reliability indicator configurations. Among these, $\mathbf{R}^A$ applies only audio-based reliability indicators and $\mathbf{R}^V$ applies only video-based reliability indicators. *None*: proposed model without reliability information; *All*: use all reliability indicators. Other abbreviations as defined in Table 8. *Avg*: Average performance, together with the significance of improvements (compared with *None*). ^{ns}: not significant, ***: $p \leqslant 0.001$, **: $0.001 < p \leqslant 0.01$ and *: $0.01 < p \leqslant 0.05$.

Model \ dB	−12	−9	−6	−3	0	3	6	9	12	Clean	*Avg.*
$\mathbf{R}^A$ (m.vc)	11.2	9.4	6.5	**4.3**	5.4	5.5	**3.6**	**3.1**	2.3	**2.4**	5.4 *
(a.vc)	14.9	14.5	10.0	6.6	4.2	5.8	4.3	**2.8**	**2.8**	**2.4**	6.8 ^{ns}
(a.gb)	16.4	14.3	10.7	6.3	4.8	6.0	4.6	**3.0**	**2.6**	**2.5**	7.1 **
(a.sp)	17.1	15.7	11.3	6.6	**4.4**	6.1	4.5	2.8	**2.9**	**2.5**	7.4 ^{ns}
$\mathbf{R}^V$ (m.vc)	**10.1**	**8.5**	6.2	5.3	5.3	5.6	3.7	**3.1**	2.6	2.7	5.3 *
(a.vc)	**14.3**	14.9	11.0	6.4	5.6	6.6	5.2	3.3	3.6	2.7	7.4 ^{ns}
(a.gb)	16.4	15.2	11.3	6.9	4.9	6.4	4.7	3.6	3.4	2.6	7.5 ^{ns}
(a.sp)	16.1	15.0	11.4	6.6	5.3	6.1	5.1	3.1	3.4	**2.5**	7.5 ^{ns}
None (m.vc)	11.8	8.8	6.7	7.5	6.0	5.6	**3.6**	3.6	3.0	3.7	6.0
(a.vc)	14.9	15.2	11.3	6.0	5.2	5.9	5.6	3.8	3.3	3.7	7.5
(a.gb)	17.2	15.1	12.6	6.8	5.7	6.3	6.6	4.4	3.6	3.6	8.2
(a.sp)	16.7	16.6	12.4	6.1	6.0	5.9	5.7	3.4	3.4	3.5	8.0
All (m.vc)	11.1	8.7	**5.5**	4.8	**4.8**	**4.5**	**3.6**	3.3	**2.2**	**2.4**	**5.1** **
(a.vc)	**14.3**	**11.9**	**8.1**	**4.8**	**4.0**	**5.4**	**3.7**	**2.8**	3.6	**2.4**	**6.1** ***
(a.gb)	**14.9**	**12.8**	**9.4**	**5.2**	**4.2**	**5.5**	**3.8**	**3.0**	4.1	2.6	**6.6** ***
(a.sp)	**15.4**	**12.8**	**9.9**	**5.2**	4.7	**5.5**	**3.4**	**2.6**	4.0	**2.5**	**6.6** ***

7. Conclusions

Large-vocabulary end-to-end speech recognition still faces a number of difficulties. However, as our experiments have shown, fusing the audio and video stream can bring a significant benefit to this task. For realizing those benefits, stream integration is a key possibility. Here, to optimally combine the audio and video information, a new decision fusion net (DFN) was proposed. This architecture utilized the posterior probabilities of the acoustic and visual model as stream representations for integration. Corresponding reliability measures of both streams were used to guide the DFN in estimating optimal multi-modal posteriors.

This fusion strategy was applied on both the conventional hybrid model, using the Kaldi toolkit, and on the joint CTC/transformer E2E model, based on the ESPnet toolkit. Comparing both experimental setups, the proposed DFN with reliability measures showed notable improvements in all noise conditions. In the hybrid AVSR setup, our system resulted in a relative word-error-rate reduction of 51% over audio-only recognition, also outperforming all baseline models.

Our proposed model was even superior to oracle stream weighting, which is considered a theoretical upper bound for instantaneous stream weighting approaches. In the joint CTC/transformer E2E architecture, the proposed model again surpassed the audio-only system, as well as the AV baseline models, achieving a relative word-error-rate reduction of 43% compared to the audio-only setup and 31% compared to the audio-visual end-to-end baseline.

Future work on stream integration still needs to answer many open questions. While our architecture is highly effective when sufficient training data is available for all conditions, we believe that information integration will truly come into its strengths when encountering new conditions that are unseen in training. In such scenarios, we also believe that uncertainty information and well-calibrated models will be essential. If all of these are appropriately designed, however, we are optimistic that information integration can pave the way towards robust models that are capable of operating successfully in unseen environments and capitalizing on their potential for multi-modal disambiguation and self-guided adaptation.

Author Contributions: Conceptualization, W.Y., S.Z. and D.K.; methodology, W.Y., S.Z. and D.K.; investigation, W.Y.; writing-original draft preparation, W.Y., S.Z. and D.K.; writing—review and editing, W.Y., S.Z. and D.K.; funding acquisition, S.Z. and D.K. All authors have read and agreed to the published version of the manuscript.

Funding: This research was funded by the German Research Foundation DFG under grant number KO3434/4-2.

Institutional Review Board Statement: Not applicable.

Informed Consent Statement: Not applicable.

Data Availability Statement: The databases used in this article are LRS2 and LRS3. For details, please refer to [9,24], respectively.

Conflicts of Interest: The authors declare no conflict of interest.

References

1. Crosse, M.J.; DiLiberto, G.M.; Lalor, E.C. Eye can hear clearly now: Inverse effectiveness in natural audiovisual speech processing relies on long-term crossmodal temporal integration. *J. Neurosci.* **2016**, *36*, 9888–9895. [CrossRef] [PubMed]
2. McGurk, H.; MacDonald, J. Hearing lips and seeing voices. *Nature* **1976**, *264*, 746–748. [CrossRef] [PubMed]
3. Potamianos, G.; Neti, C.; Luettin, J.; Matthews, I. *Audio-Visual Automatic Speech Recognition: An Overview. Issues in Visual and Audio-Visual Speech Processing*; MIT Press: Cambridge, MA, USA, 2004; Volume 22, p. 23.
4. Wand, M.; Schmidhuber, J. Improving speaker-independent lipreading with domain-adversarial training. *arXiv* **2017**, arXiv:1708.01565.
5. Meutzner, H.; Ma, N.; Nickel, R.; Schymura, C.; Kolossa, D. Improving audio-visual speech recognition using deep neural networks with dynamic stream reliability estimates. In Proceedings of the 2017 IEEE International Conference on Acoustics, Speech and Signal Processing (ICASSP), New Orleans, LA, USA, 5–9 March 2017; pp. 5320–5324.
6. Gurban, M.; Thiran, J.P.; Drugman, T.; Dutoit, T. Dynamic modality weighting for multi-stream hmms inaudio-visual speech recognition. In Proceedings of the Tenth International Conference on Multimodal Interfaces, Chania, Crete, Greece, 20–22 October 2008; pp. 237–240.
7. Kolossa, D.; Chong, J.; Zeiler, S.; Keutzer, K. Efficient manycore chmm speech recognition for audiovisual and multistream data. In Proceedings of the Eleventh Annual Conference of the International Speech Communication Association, Makuhari, Chiba, Japan, 26–30 September 2010.
8. Thangthai, K.; Harvey, R.W. Building large-vocabulary speaker-independent lipreading systems. In Proceedings of the 19th Annual Conference of the International Speech Communication Association, Hyderabad, India, 2–6 September 2018; pp. 2648–2652.

9. Afouras, T.; Chung, J.S.; Senior, A.; Vinyals, O.; Zisserman, A. Deep audio-visual speech recognition. *IEEE Trans. Pattern Anal. Mach. Intell.* **2018**, 1. [CrossRef]
10. Stewart, D.; Seymour, R.; Pass, A.; Ming, J. Robust audio-visual speech recognition under noisy audio-video conditions. *IEEE Trans. Cybern.* **2013**, *44*, 175–184. [CrossRef] [PubMed]
11. Abdelaziz, A.H.; Zeiler, S.; Kolossa, D. Learning dynamic stream weights for coupled-hmm-based audio-visual speech recognition. *IEEE/ACM Trans. Audio Speech Lang. Process.* **2015**, *23*, 863–876. [CrossRef]
12. Potamianos, G.; Neti, C.; Gravier, G.; Garg, A.; Senior, A.W. Recent advances in the automatic recognition of audiovisual speech. *Proc. IEEE* **2003**, *91*, 1306–1326. [CrossRef]
13. Luettin, J.; Potamianos, G.; Neti, C. Asynchronous stream modeling for large vocabulary audio-visual speech recognition. In Proceedings of the 2001 IEEE International Conference on Acoustics, Speech, and Signal Processing, Salt Lake City, UT, USA, 7–11 May 2001; Volume 1, pp. 169–172.
14. Nefian, A.V.; Liang, L.; Pi, X.; Liu, X.; Murphy, K. Dynamic bayesian networks for audio-visual speech recognition. *EURASIP J. Adv. Signal Process.* **2002**, *2002*, 1–15. [CrossRef]
15. Wand, M.; Schmidhuber, J. Fusion architectures for word-based audiovisual speech recognition. In Proceedings of the 21st Annual Conference of the International Speech Communication Association, Shanghai, China, 25–29 October 2020; pp. 3491–3495.
16. Zhou, P.; Yang, W.; Chen, W.; Wang, Y.; Jia, J. Modality attention for end-to-end audio-visual speech recognition. In Proceedings of the IEEE International Conference on Acoustics, Speech and Signal Processing (ICASSP), Brighton, UK, 12–17 May 2019; pp. 6565–6569.
17. Yu, J.; Zhang, S.X.; Wu, J.; Ghorbani, S.; Wu, B.; Kang, S.; Liu, S.; Liu, X.; Meng, H.; Yu, D. Audio-visual recognition of overlapped speech for the LRS2 dataset. In Proceedings of the IEEE International Conference on Acoustics, Speech and Signal Processing (ICASSP), Barcelona, Spain, 4–8 May 2020; pp. 6984–6988.
18. Arevalo, J.; Solorio, T.; Montes-y Gomez, M.; González, F.A. Gated multimodal networks. *Neural Comput. Appl.* **2020**, *32*, 10209–10228. [CrossRef]
19. Zhang, S.; Lei, M.; Ma, B.; Xie, L. Robust audio-visual speech recognition using bimodal DFSMN with multi-condition training and dropout regularization. In Proceedings of the IEEE International Conference on Acoustics, Speech and Signal Processing (ICASSP), Brighton, UK, 12–17 May 2019; pp. 6570–6574.
20. Wand, M.; Schmidhuber, J.; Vu, N.T. Investigations on end-to-end audiovisual fusion. In Proceedings of the IEEE International Conference on Acoustics, Speech and Signal Processing (ICASSP), Calgary, AB, Canada, 15–20 April 2018; pp. 3041–3045.
21. Riva, M.; Wand, M.; Schmidhuber, J. Motion dynamics improve speaker-independent lipreading. In Proceedings of the IEEE International Conference on Acoustics, Speech and Signal Processing (ICASSP), Barcelona, Spain, 4–8 May 2020; pp. 4407–4411.
22. Yu, W.; Zeiler, S.; Kolossa, D. Fusing information streams in end-to-end audio-visual speech recognition. In Proceedings of the IEEE International Conference on Acoustics, Speech and Signal Processing (ICASSP), Brno, Czech Republic, 30 August–3 September 2021; pp. 3430–3434.
23. Yu, W.; Zeiler, S.; Kolossa, D. Large-vocabulary audio-visual speech recognition in noisy environments. In Proceedings of the IEEE 23rd International Workshop on Multimedia Signal Processing (MMSP), Tampere, Finland, 6–8 October 2021; pp. 1–6.
24. Afouras, T.; Chung, J.S.; Zisserman, A. LRS2-TED: A large-scale dataset for visual speech recognition. *arXiv* **2018**, arXiv:1809.00496.
25. Bourlard, H.A.; Morgan, N. *Connectionist Speech Recognition: A Hybrid Approach*; Springer: Berlin/Heidelberg, Germany, 2012; Volume 247.
26. Lüscher, C.; Beck, E.; Irie, K.; Kitza, M.; Michel, W.; Zeyer, A.; Schlüter, R.; Ney, H. RWTH ASR systems for LibriSpeech: Hybrid vs. attention–w/o data augmentation. *arXiv* **2019**, arXiv:1905.03072.
27. Heckmann, M.; Berthommier, F.; Kroschel, K. Noise adaptive stream weighting in audio-visual speech recognition. *EURASIP J. Adv. Signal Process.* **2002**, *2002*, 1–14. [CrossRef]
28. Yang, M.T.; Wang, S.C.; Lin, Y.Y. A multimodal fusion system for people detection and tracking. *Int. J. Imaging Syst. Technol.* **2005**, *15*, 131–142. [CrossRef]
29. Kankanhalli, M.S.; Wang, J.; Jain, R. Experiential sampling in multimedia systems. *IEEE Trans. Multimed.* **2006**, *8*, 937–946. [CrossRef]
30. Yu, W.; Zeiler, S.; Kolossa, D. Multimodal integration for large-vocabulary audio-visual speech recognition. In Proceedings of the 28th European Signal Processing Conference (EUSIPCO), Amsterdam, The Netherlands, 18–21 January 2021; pp. 341–345.
31. Hermansky, H. Multistream recognition of speech: Dealing with unknown unknowns. *Proc. IEEE* **2013**, *101*, 1076–1088. [CrossRef]
32. Vorwerk, A.; Zeiler, S.; Kolossa, D.; Astudillo, R.F.; Lerch, D. Use of missing and unreliable data for audiovisual speech recognition. In *Robust Speech Recognition of Uncertain or Missing Data*; Springer: Berlin/Heidelberg, Germany, 2011; pp. 345–375.
33. Seymour, R.; Ming, J.; Stewart, D. A new posterior based audio-visual integration method for robust speech recognition. In Proceedings of the Ninth European Conference on Speech Communication and Technology, Lisbon, Portugal, 4–8 September 2005.
34. Receveur, S.; Weiß, R.; Fingscheidt, T. Turbo automatic speech recognition. *IEEE/ACM Trans. Audio Speech Lang. Process.* **2016**, *24*, 846–862. [CrossRef]
35. Chan, W.; Jaitly, N.; Le, Q.; Vinyals, O. Listen, attend and spell: A neural network for large vocabulary conversational speech recognition. In Proceedings of the IEEE International Conference on Acoustics, Speech and Signal Processing (ICASSP), Shanghai, China, 20–25 March 2016; pp. 4960–4964.

36. Son Chung, J.; Senior, A.; Vinyals, O.; Zisserman, A. Lip reading sentences in the wild. In Proceedings of the IEEE Conference on Computer Vision and Pattern Recognition, Honolulu, HI, USA, 21–26 July 2017; pp. 6447–6456.
37. Higuchi, Y.; Watanabe, S.; Chen, N.; Ogawa, T.; Kobayashi, T. Mask CTC: Non-autoregressive end-to-end ASR with CTC and mask predict. *arXiv* **2020**, arXiv:2005.08700.
38. Vaswani, A.; Shazeer, N.; Parmar, N.; Uszkoreit, J.; Jones, L.; Gomez, A.N.; Kaiser, Ł.; Polosukhin, I. Attention is all you need. In *Advances in Neural Information Processing Systems*; Neural Information Processing Systems Foundation: Long Beach, CA, USA, 2017; pp. 5998–6008.
39. Kawakami, K. Supervised Sequence Labelling with Recurrent Neural Networks. Ph.D. Thesis, Technical University of Munich, Munich, Germany, 2008.
40. Baevski, A.; Zhou, Y.; Mohamed, A.; Auli, M. Wav2vec 2.0: A framework for self-supervised learning of speech representations. *Adv. Neural Inf. Process. Syst.* **2020**, *33*, 12449–12460.
41. Nakatani, T. Improving transformer-based end-to-end speech recognition with connectionist temporal classification and language model integration. In Proceedings of the Proc. Interspeech, Graz, Austria, 15–19 September 2019.
42. Mohri, M.; Pereira, F.; Riley, M. Speech recognition with weighted finite-state transducers. In *Springer Handbook of Speech Processing*; Springer: Berlin/Heidelberg, Germany, 2008; pp. 559–584.
43. Povey, D.; Hannemann, M.; Boulianne, G.; Burget, L.; Ghoshal, A.; Janda, M.; Karafiát, M.; Kombrink, S.; Motlíček, P.; Qian, Y.; et al. Generating exact lattices in the WFST framework. In Proceedings of the IEEE International Conference on Acoustics, Speech and Signal Processing (ICASSP), Kyoto, Japan, 25–30 March 2012; pp. 4213–4216.
44. Stafylakis, T.; Tzimiropoulos, G. Combining residual networks with LSTMs for lipreading. *arXiv* **2017**, arXiv:1703.04105.
45. Sproull, R.F. Using program transformations to derive line-drawing algorithms. *ACM Trans. Graph.* **1982**, *1*, 259–273. [CrossRef]
46. Nicolson, A.; Paliwal, K.K. Deep learning for minimum mean-square error approaches to speech enhancement. *Speech Commun.* **2019**, *111*, 44–55. [CrossRef]
47. Dharanipragada, S.; Yapanel, U.H.; Rao, B.D. Robust feature extraction for continuous speech recognition using the MVDR spectrum estimation method. *IEEE Trans. Audio Speech Lang. Process.* **2006**, *15*, 224–234. [CrossRef]
48. Ghai, S.; Sinha, R. A study on the effect of pitch on LPCC and PLPC features for children's ASR in comparison to MFCC. In Proceedings of the Twelfth Annual Conference of the International Speech Communication Association, Florence, Italy, 27–31 August 2011.
49. Baltrušaitis, T.; Robinson, P.; Morency, L.P. Openface: An open source facial behavior analysis toolkit. In Proceedings of the IEEE Winter Conference on Applications of Computer Vision (WACV), Lake Placid, NY, USA, 7–10 March 2016; pp. 1–10.
50. Sterpu, G.; Saam, C.; Harte, N. How to teach DNNs to pay attention to the visual modality in speech recognition. *IEEE/ACM Trans. Audio Speech Lang. Process.* **2020**, *28*, 1052–1064. [CrossRef]
51. Snyder, D.; Chen, G.; Povey, D. Musan: A music, speech, and noise corpus. *arXiv* **2015**, arXiv:1510.08484.
52. Povey, D.; Ghoshal, A.; Boulianne, G.; Burget, L.; Glembek, O.; Goel, N.; Hannemann, M.; Motlicek, P.; Qian, Y.; Schwarz, P.; et al. The kaldi speech recognition toolkit. In Proceedings of the IEEE 2011 Workshop on Automatic Speech Recognition and Understanding, Waikoloa, HI, USA, 11–15 December 2011.
53. Zhang, X.; Trmal, J.; Povey, D.; Khudanpur, S. Improving deep neural network acoustic models using generalized maxout networks. In Proceedings of the IEEE International Conference on Acoustics, Speech and Signal Processing (ICASSP), Florence, Italy, 4–9 May 2014; pp. 215–219.
54. Paszke, A.; Gross, S.; Massa, F.; Lerer, A.; Bradbury, J.; Chanan, G.; Killeen, T.; Lin, Z.; Gimelshein, N.; Antiga, L.; et al. Pytorch: An imperative style, high-performance deep learning library. *Adv. Neural Inf. Process. Syst.* **2019**, *32*, 8026–8037.
55. Panayotov, V.; Chen, G.; Povey, D.; Khudanpur, S. Librispeech: An asr corpus based on public domain audio books. In Proceedings of the IEEE International Conference on Acoustics, Speech and Signal Processing (ICASSP), South Brisbane, QLD, Australia, 19–24 April 2015; pp. 5206–5210.

sensors

MDPI

Article

Noise-Robust Multimodal Audio-Visual Speech Recognition System for Speech-Based Interaction Applications

Sanghun Jeon and Mun Sang Kim *

Center for Healthcare Robotics, Gwangju Institute of Science and Technology (GIST), School of Integrated Technology, Gwangju 61005, Korea

* Correspondence: munsang@gist.ac.kr; Tel.: +82-10-9126-4628

Abstract: Speech is a commonly used interaction-recognition technique in edutainment-based systems and is a key technology for smooth educational learning and user–system interaction. However, its application to real environments is limited owing to the various noise disruptions in real environments. In this study, an audio and visual information-based multimode interaction system is proposed that enables virtual aquarium systems that use speech to interact to be robust to ambient noise. For audio-based speech recognition, a list of words recognized by a speech API is expressed as word vectors using a pretrained model. Meanwhile, vision-based speech recognition uses a composite end-to-end deep neural network. Subsequently, the vectors derived from the API and vision are classified after concatenation. The signal-to-noise ratio of the proposed system was determined based on data from four types of noise environments. Furthermore, it was tested for accuracy and efficiency against existing single-mode strategies for extracting visual features and audio speech recognition. Its average recognition rate was 91.42% when only speech was used, and improved by 6.7% to 98.12% when audio and visual information were combined. This method can be helpful in various real-world settings where speech recognition is regularly utilized, such as cafés, museums, music halls, and kiosks.

Keywords: deep learning; audiovisual speech recognition; lipreading; multimodal interaction; edutainment; virtual aquarium

Citation: Jeon, S.; Kim, M.S. Noise-Robust Multimodal Audio-Visual Speech Recognition System for Speech-Based Interaction Applications. *Sensors* **2022**, *22*, 7738. https://doi.org/10.3390/s22207738

Academic Editors: Bruce Denby, Tamás Gábor Csapó and Michael Wand

Received: 8 September 2022
Accepted: 10 October 2022
Published: 12 October 2022

1. Introduction

Recently, with the rapid development of deep learning, the educational, experiential, and auxiliary situations in which various deep learning technologies are applied have increased [1–4]. In particular, in terms of edutainment, various recognition technologies based on deep learning are being developed to recognize interactions such as speech, gestures, eye and head tracking, and even physical objects. The term edutainment, a portmanteau of education and entertainment, is an approach designed to be simultaneously educational and fun, and is widely used in robot platforms, museums, science centers, and aquariums [5–7]. Edutainment techniques that simultaneously provide both education and fun reportedly have an effect on learning outcomes, and numerous systems have consequently been developed [3,8–10]. They are also useful for interacting with the elderly [1].

Speech is one of the most commonly used interaction-recognition techniques in edutainment-based systems, and is a key technology for smooth educational learning and interaction between the system and the user [10–14]. Speech involves the perception of both auditory and visual information and is the most commonly used human engagement and communication mode. Janowski et al. [15] compared gestures and speech experimentally for four separate interaction tasks (navigation, selection, dialogue, and manipulation) in a virtual environment. They reported that in object manipulation, both speech and gestures were preferred. However, despite the increasing preference for interaction using speech, the application of interaction using speech to real environments is quite limited owing to the various noise disruptions that can occur in real environments. Therefore, core technology that is robust to difficult acoustic scenarios and various noises in real environments is essential. Consequently, in this study, an audio and visual information-based multimode interaction system is proposed that enables virtual aquarium systems that use speech to interact to be robust to ambient noise. Figure 1 gives a conceptual overview of the proposed system being utilized for interaction via speech in a virtual aquarium.

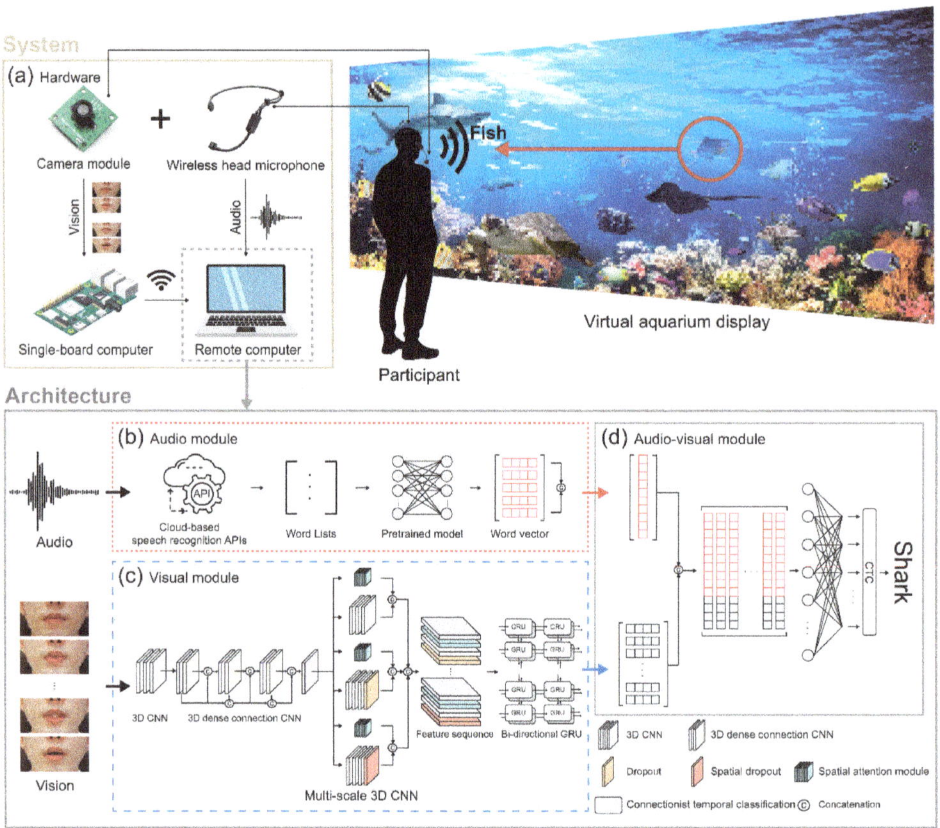

Figure 1. Composition of the proposed system and an example usage scenario. (**a**) Hardware components of the system and a virtual aquarium interaction usage scenario; (**b**) audio module for generating word vectors; (**c**) visual extraction module for extracting feature vectors; (**d**) audiovisual module for classification.

Multisensory integration has an important effect on human communication, and each sensory organ transmits specific sensory information [16]. The human nervous system is organized into multiple nonoverlapping sensory organs and uniquely processes the received input, which enables it to sense very reliably. Depending on the received input information, information from different senses can be connected to each other, thereby synergistically enhancing the ability to recognize and evaluate [17,18].

People can improve their ability to understand speech in noisy environments or where it is otherwise difficult to understand speech by detecting the movements of the tongue and teeth and the area around the speaker's lips. In other words, when the speech signal is not clear, visual speech information has a significant positive effect on speech comprehension [18–23]. As such, humans tend to process visual information first when visual and auditory information are received simultaneously. This phenomenon is called "synesthesia" or the "McGurk effect" in cognitive psychology [24]. Therefore, we judge many things visually to the extent that the sound we hear depends on what we see. For example, if people see a person's face expressing "ba" and hear the sound "ga," many of them will infer the third sound "da" that combines the two. The fusion approach can contribute to robust recognition of speech interactions in real life and overcome the problem of auditory and visual ambiguity of words with similar pronunciation in noisy environments.

This work offers a noise-robust system for use in virtual aquariums by integrating a deep neural network-based end-to-end visual speech recognition architecture with an open cloud speech recognition (OCSR) API system (Figure 1). The performance of this system is superior to that of single-mode systems that use either audio or visual speech recognition technologies. For audio-based speech recognition, a pretrained model is used to express word vectors for a list of words that have been identified by the speech API. Meanwhile, vision-based speech recognition uses a new deep neural network-based lipreading architecture consisting of end-to-end neural subnetworks. We consecutively combined three 3D convolutional neural networks (CNNs) for feature sequence extraction, shortening the training time by reducing the number of parameters, and referring to the existing 2D DenseNet [25] to suppress overfitting. The 3D densely connected CNN is composed of the components of a multichannel 3D CNN to extract the multichannel features of different levels. A bidirectional gated recurrent unit (GRU) followed by a linear layer is used to overcome the scarce visual information caused by the time-series input data and to obtain specific image features. A vector matrix is formed by connecting the word and feature vector output to the audio- and visual-based model. By inserting a SoftMax layer at each time step and then applying the connectionist temporal classification (CTC) loss function [26] to all time steps, the concatenated vector matrix is trained to acquire predictive words.

We also compared the accuracy and efficiency of the proposed system with existing single-mode techniques for extracting visual features and multiple OCSR APIs from the collected dataset. The results of extensive evaluations verified that the proposed system achieved faster convergence speed, higher computational efficiency, and superior performance. Thus, our system, which integrates the existing OCSR API system with an end-to-end lipreading architecture using visual information, is widely utilizable for diverse speech-based interaction settings such as cafés, music halls, and virtual aquariums.

The main contributions of this paper are as follows:

- We propose a novel audiovisual speech recognition system using multimode interaction for virtual aquarium applications that is robust to ambient noise;
- We compare the accuracy and efficiency of the proposed system with those of existing single-mode techniques for extracting visual features and multiple OCSR APIs from a collected dataset;
- We show that the proposed system provides faster convergence speed, higher computational efficiency, and superior performance compared with the existing OCSR API system.

The remainder of this paper is organized as follows. Section 2 of this paper gives an overview of research related to this study. Section 3 provides the details of each element of the proposed system. Section 4 provides information on the datasets collected, data augmentation techniques, and experimental setup. Section 5 presents the training procedure, convergence rate, optimization, and performance evaluation results of the proposed system. Sections 6 and 7 discuss the experimental results according to the research objectives, suggest directions for future research, and provide our conclusions.

2. Related Work

Google has released an open cloud-based speech API [27] along with various functions and application scenarios for finance, automobiles, and hospitals. It has advanced capabilities such as distinct speaker identification, automated speech language detection, data logging, and speech-to-text conversion with multiple channels, as well as support for over 125 languages. IBM's Watson AI service [28] supports 11 languages and provides services such as dialogue dialing of the speaker and word extraction and filtering, along with the ability to customize specific volume conditions by applying the system according to the usage environment. However, to use the pretrained language model, there is also a difficulty, in that it needs to be tuned to the customer management domain, and this requires adding a new corpus. In addition, Microsoft offers the speech-to-text capabilities of the Speech Service, part of Azure Cognitive Services [29], as a cloud service package. According to the official website, the Speech Service includes real-time speech synthesis, asynchronous synthesis of long audio, prebuilt neural network speech, and viseme technology. Xiong et al. [30] reported that it achieved human-level accuracy for the first time in a switchboard test in 2017. Amazon Alexa [31] is an artificial intelligence (AI) smart personal assistant with a speech-activated platform that enables voice interaction and Q&A. Alexa has various other capabilities, including the ability to play music, set alarms, and obtain weather information. It can also be used to operate smart home technology devices.

Owing to rapid improvements in the field, deep learning has lately delivered good performance in various applications in diverse research domains, including VSR systems. Deep learning-based algorithms outperform traditional prediction methods. For example, the systems proposed by Ji et al. [32] and Petridis and Pantic [33] distinguish different visemes by integrating the traditional approach with a CNN, and add time information after obtaining the CNN output using the hidden Markov model framework. Wand et al. [34] and Cooke et al. [35] integrated histograms of oriented gradients (HoG) with long short-term memory (LSTM) to evaluate a GRID benchmark dataset consisting of short phrases. In addition, the LSTM classifier was trained on the OuluVS and AVLetters datasets [35] using the discrete cosine transform, and word prediction evaluation was performed. Noda et al. [36] applied the sequence-to-sequence (seq2seq) model to lipreading. The model has a deep speech recognition architecture capable of recognizing and predicting the output of full input sequences. In addition, performance evaluation was performed on a benchmark dataset composed of real words by integrating all the audiovisual information.

Assael et al. [37] introduced LipNet, an end-to-end deep learning model, and it was trained and its performance evaluated on the GRID corpus, a sentence-level dataset. The GRID corpus was divided into overlapped and unseen-speaker database structures, and it achieved word error rates of 4.8% and 11.4%, respectively. However, the evaluation was performed with the same database, and the experienced human lip reader had a low success rate of 47.7%. Fenghour et al. [38] presented a deep learning network model for viseme-to-word translation that used an attention-based GRU and enhanced the performance of predicting spoken sentences by reaching a word accuracy of 79.6%. Li et al. [39] proposed an efficient two-stream model for learning dynamic information. The model extracts static characteristics from a single frame and dynamic information between multiple frame sequences using two distinct channel capacity CNN streams. Utilizing a more effective convolutional structure for each component in the front-end model yielded an 8% improvement. Xu et al. [40] implemented and evaluated a digit sequence prediction and an architecture similar to the CTC cascaded model on audiovisual datasets. Deep learning-based approaches are more resilient to huge data and visual ambiguity than conventional information-extraction techniques, and they can extract more precise, detailed, and accurate information from audiovisual data.

The evaluation and comparison of the recognition performance of previously available OCSR APIs have been the focus of several studies [41–43]. Contrastingly, the goal of this study is to enhance the interactive performance in a virtual aquarium by incorporating visual information into the already-existing OCSR API system. Compared with the traditional OCSR API system, the method performs better recognition and reduces the error rates in noisy environments. Consequently, in this work, we present an interaction model that, in contrast to the current interaction approach that only utilizes auditory information, employs audiovisual information based on deep learning and uses a system with a low error rate even in noisy environments.

3. Architecture of the Proposed System

Our proposed deep learning audiovisual speech recognition interaction system, which is integrated with the open cloud-based speech API, is shown in Figure 1. As shown in Figure 1b, human speech is input and the recognized words are converted into word vectors using a pretrained word embedding model. Simultaneously, as shown in Figure 1c, a face video matching the human speech is received and extracted as a sequence feature vector. Subsequently, the audiovisual module (Figure 1d) predicts a word or sentence by integrating the word and sequence feature vectors output from the audio and visual modules (Figure 1b,c).

3.1. *Audio Module*

In the audio speech recognition module, the speech API receives the user's speech via a microphone and sends it to an open cloud-based recognition system. The open cloud-based speech recognition API is a publicly available API for developing applied speech recognition systems. It has a speech recognition engine created by collecting large amounts of speech data via a cloud computing service and learning on large amounts of data with high-performance computing. Cloud firms provide these speech recognition engines so that anyone can use them without difficulty via the voice recognition Open API, saving significant amounts of development time, effort, and expense.

Figure 2 is a block diagram of the proposed audio speech recognition module; it utilizes two general algorithms. The human speech is input via a local device such as a microphone, and the recorded speech is delivered to an open cloud server provided by a commercial engine, Microsoft Azure, for further processing [44]. Through speech recognition by the OCSR API, which is closed-source, it is possible to quickly implement a user-optimized speech recognition system and has the usability advantage of being immediately applicable to various fields. Therefore, developers using application speech recognition systems should choose the appropriate OCSR API based on the capabilities of the system. In addition, the performance of the OCSR API is constantly updated by many companies that provide API services, and it depends on the study date and the type of training data. Therefore, we used the Microsoft Azure API, which has been proven to be superior in the results of previous studies [45]. In addition, the existing speech API can be replaced if another speech API with better performance is released and is not affected by performance changes over time. The word lists output through the OSCR API are output as individual word vectors using Google's pretrained Word2Vec embedding model. These word vector representations may have hundreds of dimensions in the Word2Vec model, also known as word embeddings [46,47]. For academic use, Google offers a Word2Vec model that has been pretrained on 100 billion words from the Google News corpus, producing 3 million 300-dimensional word embeddings. Therefore, a list of words is output, transformed into a 300-dimensional vector, and concatenated into a single vector.

3.2. *Visual Module*

Figure 3 is a detailed schematic of the proposed visual speech recognition module. It consists of three modules: feature extraction, sequence processing, and transcription. The feature extraction module consists of three CNNs: 3D CNN to extract fine motion around the lip, 3D dense connection CNN to reduce model parameters and prevent overfitting, and multiscale 3D CNN to extract rich features with different level features. The sequence-processing module uses Bi-GRU to comprehend a wide sequential feature context, and the transcription module combines the local self-attention mechanism in a cascaded approach to compensate for the shortcomings of the CTC loss function, focusing only on local information in the nearby frame.

3.2.1. Feature Extraction Module

Figure 4a shows that the 2D CNN collects encoded information about single-image data and transforms the information into a 2D feature map to calculate spatial-dimensional features. However, a 2D CNN cannot extract motion information in consecutive frames extracted from video.

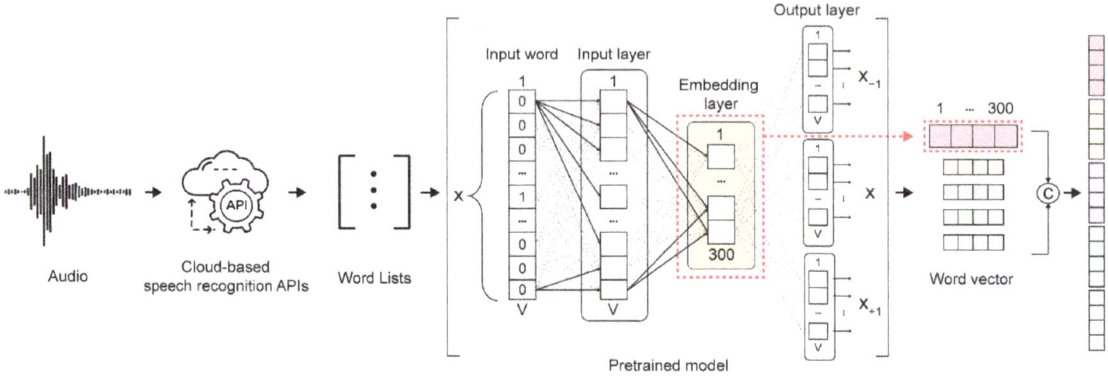

Figure 2. Block diagram of the proposed audio speech recognition module.

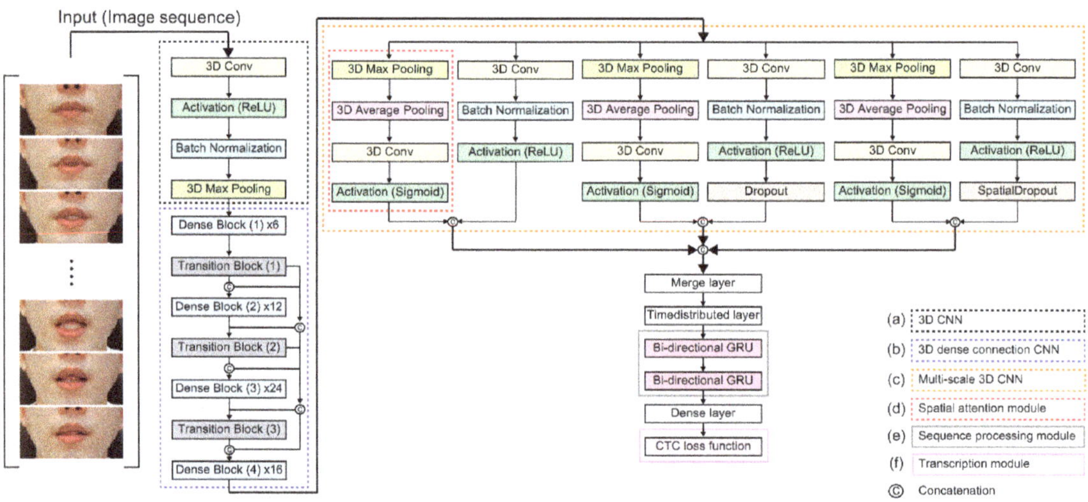

Figure 3. Detailed schematic of the proposed visual speech recognition module.

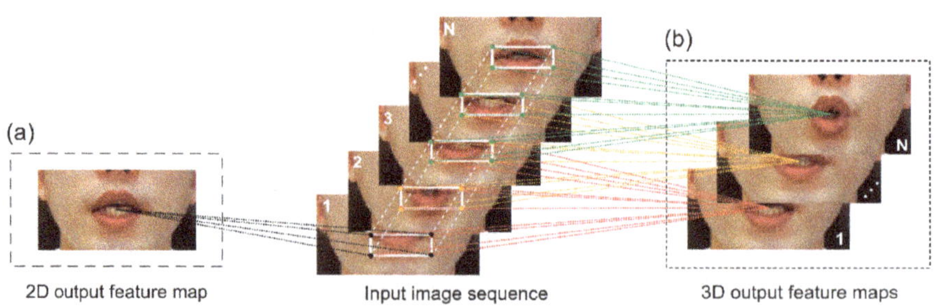

Figure 4. (a) Two-dimensional convolution operation. (b) Three-dimensional convolution operation.

As shown in Figure 4b, we need a 3D CNN that can simultaneously calculate spatial and temporal dimensional features to detect various peripheral information such as tongue and tooth movement information around the lips. The 3D CNN is a widely used technique to detect spatial and temporal information in time-series sequence data and has been proven to be effective in extracting spatial and temporal information in numerous studies [32,37]. In this study, CNN layers comprising 64 3D kernels of size 3 × 7 × 7 were constructed to extract and encode visual feature information into input sequential lip data, and combined with the batch normalization layer and ReLU. Subsequently, the spatial scale of the 3D feature map was reduced by connecting the max-pooling 3D layer (Figure 3a). The details of the proposed model hyperparameters are presented in Table A1.

Following the 3D CNN, a 3D dense-connection CNN is used to reduce the parameters of the model to save processing resources and effectively prevent overfitting. With this method, relationships between several linked layers are generated, facilitating network depth, vanishing gradients, and full functional utilization (Figure 5). We extend the 3D volumetric feature extraction task by referring to the existing 2D densely connected CNN structure composed of the l-th layer of the nonlinear transformation H_l. The output of the l-th layer of the existing 2D structure can be represented by x_l (Equation (1)), where $x_0, x_1, \ldots, x_{1-1}$ are generated in the previous layer and $[\ldots]$ denotes the concatenation operation [24].

$$x_l = H_l([x_0, x_1, \ldots, x_{1-1}]) \quad (1)$$

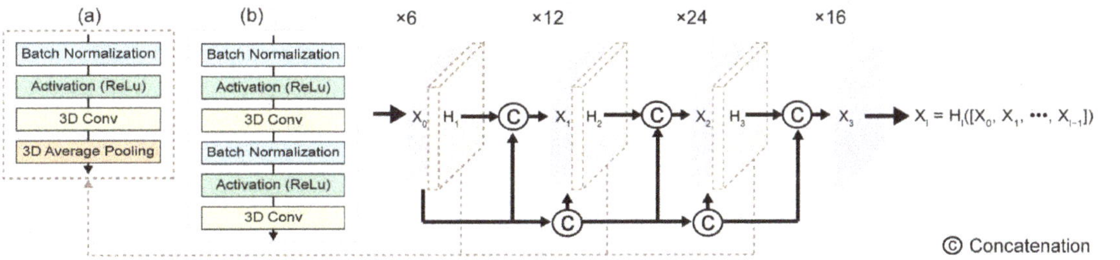

Figure 5. Detailed 3D dense-connection CNN architecture. (**a**) Three-dimensional transition layer; (**b**) three-dimensional dense block.

This approach utilizes two modules: a 3D transition layer module and a 3D dense block module. The feature maps processed in the 3D CNN are reduced by the bottleneck layers (Figure 5b) and then multichannel feature volumes are integrated. As the previous feature information still exists, subsequent layers are applied only to a few feature volumes, and the hyperparameter ta controlling the degree of compression is also included in the transition layer (Figure 5a) to increase compressibility. Thus, reduced growth rates can be achieved by using bottlenecks and transition layer tiers in succession. The dense block structure is doubly connected in the following order: batch normalization (BN) layer, activation function (ReLU), and 3D convolutional layer (Figure 5b). The transition layer connected after the dense block structure has the same structure as the dense block structure, and an average-pooling 3D layer of 2 × 2 × 2 is additionally connected (Figure 5a). The 3D convolutional layer used for the dense block structure is 3 × 1 × 1, and is composed of 3 × 3 × 3 3D convolutional layers of the transition layer.

By implementing them in various sizes and depths, we coupled multiscale 3D CNNs to extract various layers of spatial and visual information. In the multiscale 3D CNN, multiple convolutional layers of different level sizes can generate different level features based on different depths and filters, and this strategy can be used to extract richer feature information with a layered approach (Figure 3c). The proposed multiscale 3D CNN architecture is shown in Figure 6; it is divided into four structures. The three modules are composed of different kernel sizes based on the structure of Figure 6b. The first module, Figure 6b, connects to a 3D convolution layer with a 3D kernel size of 32 in order of batch normalization and activation functions (ReLU). The second module (Figure 6c) and third module (Figure 6d) add standard and spatial dropouts with 3D convolution layers of different 3D kernel sizes of 64 and 96, and then add activation functionality (ReLU).

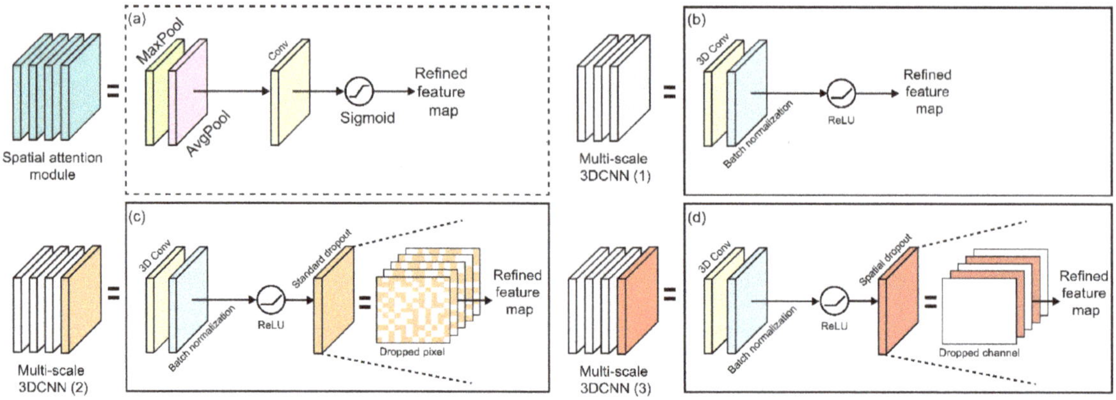

Figure 6. Multilayer 3D CNN architectures: (**a**) spatial attention module; (**b**) first architecture; (**c**) second architecture with standard dropout approach; (**d**) third architecture with spatial dropout approach.

The function of the standard dropout is to prevent strongly correlated activations in the image feature map by randomly dropping pixels, thereby preventing overfitting and overtraining, which affect the CNN performance improvement. Therefore, it plays an important role in small benchmark datasets compared to large datasets such as image classification datasets [48]. The spatial dropout of the third module outperforms and performs in strongly spatially correlated image classification by dropping the corresponding channel rather than pixels [49,50]. It is particularly effective in extracting the fine motion features of the lips, teeth, and tongue with strong spatial correlation. Additionally, to focus on the location of the information section and complement the attention channel, all three modules are combined with spatial attention modules of the same structure (Figure 6a). The spatial attention module focuses on utilizing interspace interactions to better select the most identifiable and useful portions of the input image [51]. It initially runs max-pooling and average-pooling operations along the channel axis and then connects them to create an efficient feature descriptor that calculates spatial attention. Therefore, the output of each multiscale 3D CNN and spatial attention module is merged and concatenated.

3.2.2. Sequence Processing Module

Because the feature extraction module extracts only fixed short viseme-level features, it is difficult to distinguish the longer context information of random-length time-series input data. We use a GRU that learns to propagate and control the flow of time-series data information using update and reset gates [52]. The GRU is derived from the LSTM unit that determines which information should be conveyed and which should be ignored, allowing for the use of update and reset gates to solve the gradient loss problem. A bidirectional GRU is configured with the feature sequence of the feature extraction module as input to provide forward and backward information so that both networks can obtain rich information.

3.2.3. Transcription Module

We use the CTC method, which does not require end-to-end alignment of deep neural networks and parameterizes the distribution of a label token sequence using a loss function. The marginal distributions created at each time step of the temporal module are conditionally independent of CTC. This is because it restricts the use of autoregressive connections to handle the inter-time-step dependencies of the label sequence. When the probabilities of the language model are ambiguous, the CTC models are decoded using a beam search approach to restore label temporal dependency.

4. Experimental Evaluation

4.1. Dataset, Data Preprocessing, and Augmentation

To evaluate the proposed model, we constructed a new dataset for the interaction of the virtual aquarium by referring to the Word Choice part of the most used speech recognition commands in IoT or real life in Google Speech Command Dataset V2 [53]. Currently, because most benchmark datasets are audio-based or consist of datasets used in real-world applications, datasets for the interaction of virtual aquariums are insufficient. Therefore, we constructed the dataset ourselves to evaluate the proposed model (Table 1).

Table 1. Collected speech commands datasets: (a) control commands; (b) marine life; (c) numbers; (d) emotions.

		(a) Control Commands (20)		
front	back	side	over	under
inside	outside	top	center	bottom
right	left	on	off	up
down	go	start	pause	stop
		(b) Marine Life (15)		
turtle	fish	shark	crab	dolphin
jellyfish	octopus	whale	starfish	shrimp
coral	squid	otter	lobster	
		(c) Numbers (11)		
one	two	three	four	five
six	seven	eight	nine	ten
zero				
		(d) Emotions (8)		
anger	fear	anticipation	surprise	joy
sadness	trust	disgust		

To perform aquarium interaction in a virtual environment, data were collected by dividing them into two categories—an operation command to manipulate the system and a command to control objects in the virtual environment. As the unit for operating the system is not a complete sentence but a word or a short phrase, the collected words are generally useful as commands for automobile and robot applications [53].

For data collection, 40 participants (20 males and 20 females; average age: 29.14 years) who were familiar with speech recognition equipment were recruited and ultimately compensated with gift certificates worth USD 20. To ensure balance in the data, 10 males and 10 females were native English speakers, and the other 10 males and 10 females were bilingual advanced English-speaking participants. The participants wore a head-mounted device that combined a webcam with a camera and microphone (Figure 7). Each participant stared at the display (Figure 8b) located in front and repeated the 54 keyword lists 100 times at 2 s intervals in sequence.

The collected video underwent frame extraction and was output as shown in Figure 8c. We collected a total of 216,000 video clips for audio and video information. We used a Logitech webcam (C920 HD PRO WEBCAM) with the following specifications for data collection: video resolution of 1920 × 1080 (FHD), frame rate of 30 fps, and a stereo microphone. The face-facing webcam fixture was made with a 3D printer. The video data were recorded for 2 s with a resolution of 640 × 480 pixels at 30 fps, and the audio was stereo with a sample rate of 44,100 Hz.

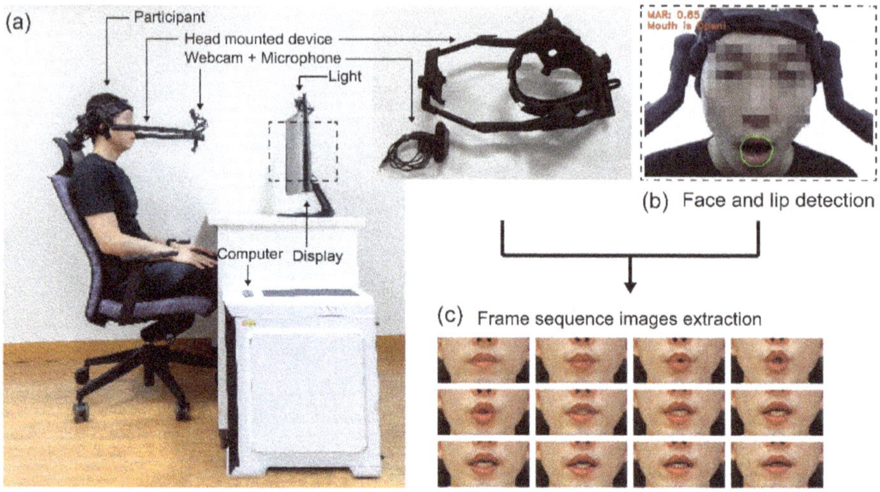

Figure 7. Data-recording environment and image extraction procedure: (**a**) Data-recording environment; (**b**) during data collection, the participant's face information is visible to the camera, and the green line demarcates the lip detection area; (**c**) frame sequence images extracted from the acquired video.

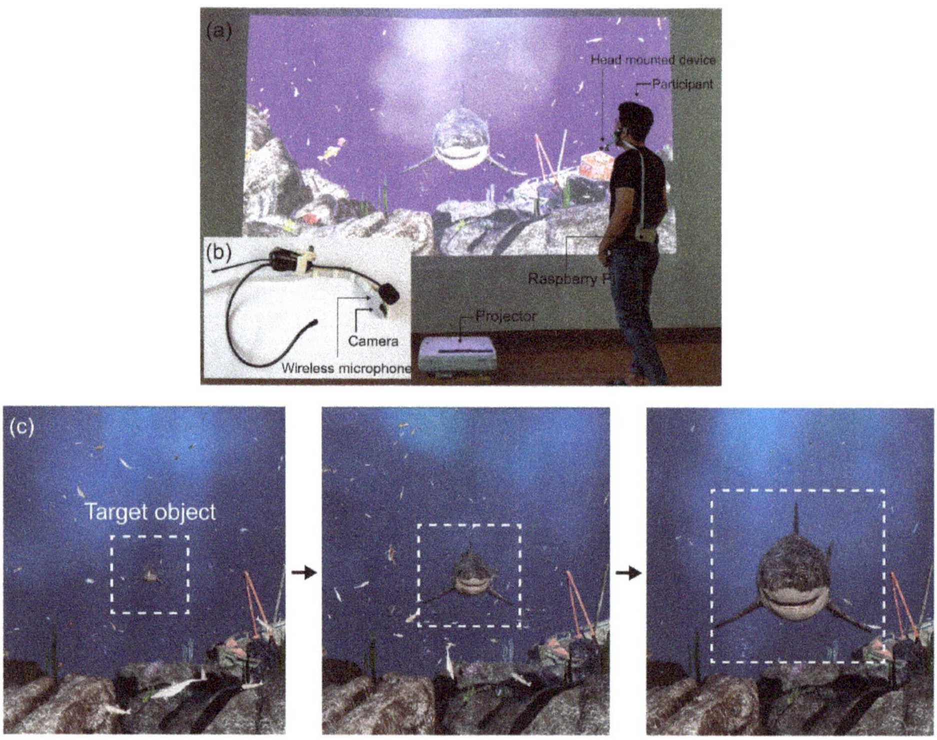

Figure 8. (**a**) Participant in the real-world interaction environment; (**b**) head-mounted device design; (**c**) target object approaching as a result of speech interaction.

We trained the proposed model on data fired 100 times per class and divided the data into training and validation sets in a 7:3 ratio. During the data collection process, there were participants whose pronunciation became weaker as the number of experiments increased because of the repetitive speech. Therefore, to prevent training overfitting from the problem of poor pronunciation, we used the most focused early (20–30) utterances, middle (50–60) utterances, and finally some of the utterances when concentration was lowest among the 100 utterances for validation. The utterances were selected and divided into training and validation sets in a 7:3 ratio. Specifically, the training dataset comprised utterances 1–9, 10–20, 31–50, 61–80, and 91–100, and the validation dataset comprised utterances 21–30, 51–60, and 81–90.

We used the Dlib [54] face detector with a HoG feature-based linear classifier to search the target region as a preliminary step for extracting human lip information in the data preprocessing step. The Dlib library, which can utilize image processing and different machine learning methods, is a general-purpose cross-platform software library developed in C++ that can use HoG features or a trained CNN model for face recognition. To create a bounding box around the mouth, the detected output was presented as (x, y) diagonal edge coordinates. Then, 68 landmarks and the same lip point as the data obtained from the training dataset with the online Kalman filter iBug [55] program were extracted. Using affine transformation, we extracted pictures with a size of 100×50 pixels from the target face region that was extracted in relation to the mouth's center region. We then normalized the RGB channels of the entire training set so that the mean and unit variance were both zero. Furthermore, because—unlike other image classification data—overfitting may occur owing to the small amount of data, data augmentation was performed to prevent overfitting [41]. All models were trained and assessed using the same dataset pretreatment and augmentation approaches, and normal and horizontally mirrored image sequences were used for data augmentation throughout the training phase.

4.2. Implementation

The diagram on the left side of Figure 8 is a schematic of the overall system for processing the respective audio and visual information. Figure 8a shows a participant interacting with a virtual aquarium through speech commands. The participant wears a device that combines a wireless microphone and a small camera, as shown in Figure 8b, and controls the target object (e.g., a shark) via speech commands. A single-channel wireless head microphone is used to acquire the experimenter's voice information, which is wirelessly transmitted to a remote computer. Simultaneously, visual information is acquired using a camera module attached to the wireless head microphone, and the connected single-board computer, Raspberry Pi, uses the robot operating system (ROS) to wirelessly transmit the visual information to the remote computer.

We used the CTC decoder with the beam search method to evaluate the character accuracy rate (CAR) of the proposed model. The evaluation environment consisted of an Intel® CoreTM i7-7700K CPU running the Linux Ubuntu 18.04 LTS operating system, with 32 GB RAM, and an NVIDIA GeForce RTX 2080-Ti GPU. In addition, Keras based on the TensorFlow backend was used to evaluate the performance of the CTC decoder. Table A1 provides information on the architecture utilized for the evaluation. The specific requirements for each layer of the proposed model are listed as hyperparameters.

For model optimization during training, three evaluations were performed: optimization, batch size, and learning rate. To perform model optimization, AdaDelta [56], AdaGrad [57], adaptive moment estimation (Adam) [58], stochastic gradient descent (SGD) [59], RMSprop [55], AdaMax, and Nadam [59] were evaluated with a mini-batch of four and a learning rate of 0.0001. After performing model optimization, the model was executed to determine the optimal mini-batch size (0.1, 0.001, 0.000.1, 0.00001, or 0.000001) and learning rate (4, 8, 16, 32, or 64). A maximum batch size of 64 was used because of the limitations of our GPU memory.

We performed data collection and performance evaluation while considering factors such as lighting that could have a detrimental effect on the proposed system. The performance evaluation was conducted in an environment different from the conventional data collection environment, and for the evaluation, participants used a head-mounted webcam device (Figure 8). The performance evaluation environment was a natural environment without any controls, i.e., an environment with natural light and noise. The examination was separated into three sections: audio only, visual only, and combined auditory and visual information. Furthermore, none of the individuals that engaged in data collection during the performance evaluation stage participated in the generalized performance evaluation.

While taking into account external elements impacting the accuracy, such as illumination, the proposed system was assessed in an environment that differed from the data gathering setting. The participants used a head-mounted device with a webcam and did the evaluation while standing 1.5 m away from the virtual aquarium screen during the performance evaluation stage (Figure 8). In the experiment room, it was performed with normal lighting and noise, and only participants who had not taken part in the data collection were selected. The multimodal technique used in this investigation was divided into three categories: audio-only, visual-only, and audiovisual speech.

4.3. Performance Evaluation Metrics

We examined the learning loss, batch size, and optimization to assess the learning state throughout the proposed model's training process, and we utilized the character error rate to assess the accuracy. Specifically, we converted the error rate measure to a percentage by calculating the total edit distance and compared the predicted text with the original text. In addition, we used a confusion matrix approach for visualization of the predicted data. The CAR for accuracy evaluation consists of five variables (C, S, D, I, N). They signify the characters (C), false prediction characters (S), number of deleted characters (D), unselected characters (I), and total number of correct characters (N). The proposed model is a maximum-probability prediction method that performs the CTC beam search technique. The CAR equation is expressed as follows:

$$\mathrm{CAR}\ (\%) = 100 - \left(\frac{C_S + C_D + C_I}{C_N}\right) \times 100 \tag{2}$$

To visualize and analyze the predicted data after learning, we used a confusion matrix, which is commonly used to summarize the performance of a classification model. The matrix compares the real findings with the test data to the number of properly and incorrectly categorized samples. Using the confusion matrix as a tool for evaluation provides the benefit of allowing a more thorough analysis in the event that the dataset is imbalanced, as opposed to depending solely on the fraction of correctly recognized samples (which could cause misleading conclusions). In Table 2, the confusion matrix for two classes is presented.

Table 2. Confusion matrix for two classes.

		Predicted Class	
		Actual Positive	**Actual Negative**
Actual	Predictive Positive	True Positive (TP)	False Positive (FP)
Class	Predictive Negative	False Negative (FN)	True Negative (TN)

The true-positive, false-positive, true-negative, and false-negative values in the confusion matrix served as the basis for the algorithms' performance parameters. Precision, recall, and F1-score were among the metrics computed. Classwise accuracy, recall, and F1-score were used to assess the classification models. These performance metrics were calculated using the following formulas.

Precision denotes the outcome of a scenario that is thought to be positive:

$$\text{Precision} = \frac{\text{TP}}{(\text{TP} + \text{FP})} \tag{3}$$

Recall reveals the estimation of the success of positive situations:

$$\text{Recall} = \frac{\text{TP}}{(\text{TP} + \text{FN})} \tag{4}$$

The F1-score, which represents the overall classification accuracy, is calculated as the harmonic average of recall and precision:

$$\text{F1} - \text{score} = \frac{2 \times \text{Precision} \times \text{Recall}}{(\text{Precision} + \text{Recall})} \tag{5}$$

Additionally, noise was generated with a signal-to-noise ratio (SNR) approach using the gathered source data to assess the trained model's quantitative performance. Both the commonplace noise found in daily life and the multichannel acoustic noise database (DEMAND) served as the source of noise data for the noise-generation process. DEMAND comprises different types of noise from eight environments (parks, corridors, restaurants, stations, cafés, plazas, cars, living) and ambient noise (Table 3).

Table 3. Noise database structure: categories and recordings conducted in each category.

	Category	Place	Environment
(a)	Office	Hallway	Hallway inside an office building, with individuals and groups passing by occasionally
(b)	Public	Cafeteria	Busy office cafeteria
(c)		Station	Main transfer area of a busy subway station
(d)	Street	Cafe	Terrace of a cafe at a public square

5. Results

5.1. Training Procedure and Convergence Rate

Figure 9 compares the learning loss and convergence rates for different batch sizes and learning rates, respectively. Batch size and learning rate are important hyperparameters in model training, and various studies related to the effect of batch size and learning rate on model training have been conducted [60–62]. Masters and Luschi [60] showed that higher test accuracy can be obtained with a small batch size by changing the batch size while fixing the learning rate. In addition, the results of fixing the batch size and changing the learning rate showed that when a small batch is used, stable learning is possible over a wider range of learning rates. In addition, Keskar et al. [63] showed that the use of a large batch size increases the likelihood of convergence to a sharp minimum of the training function, which lowers the generalization performance. Therefore, we evaluated the optimal batch size and learning rate to optimize the proposed model.

Figure 9a–c show the evaluation of different batch sizes, and Figure 9d–f show the evaluation of different learning rates. In training and validation loss, as the batch size increased, the convergence speed became slower, and when the batch size was four, it showed the fastest convergence speed. A moving average strategy was used to better discern the visualization as smooth. Figure 9 illustrates how the smoothed value was displayed as a curve and the real value was expressed as the shadow portion of the image for the proposed model's training. The uneven fluctuation of the real value caused by the small batch size was also addressed, and smoothing was performed to improve the understanding of the curve. In the case of learning rate, there was no training at 0.1 and

0.001, and 0.1 showed a tendency to diverge. On the other hand, among the remaining three learning rates, the 0.0001 value resulted in the fastest convergence speed. Thus, the proposed model exhibits the fastest convergence rate for the collected dataset, with a batch size of four and a learning rate of 0.0001.

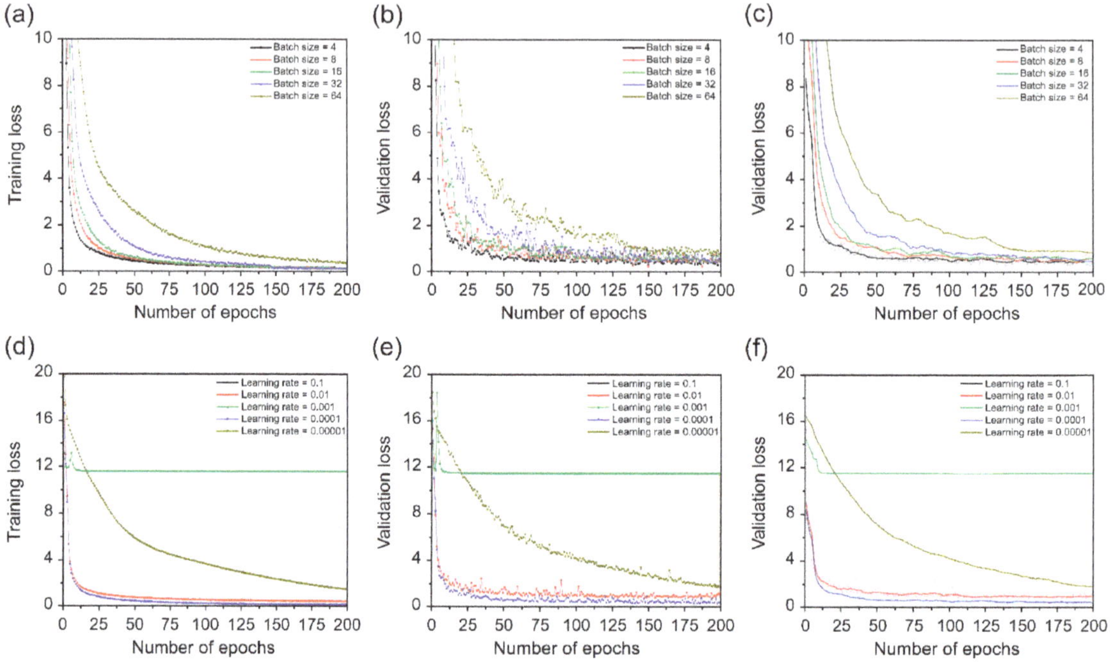

Figure 9. Training and validation loss of the collected dataset. (**a**) Different training batch sizes; (**b**) different validation batch sizes; (**c**) different validation batch sizes using moving average; (**d**) different training learning rates; (**e**) different validation learning rates; and (**f**) different validation learning rates using moving average.

5.2. Optimization

Hyperparameters that affect training are typically used to determine optimal model updates (e.g., batch size and learning rate). Prior to decreasing the error or loss function caused by the difference between the actual and predicted values, the optimizer must update the weight parameters repeatedly with different weights. However, selecting the right optimizer for the best model training might be challenging. To increase the prediction accuracy and learning rate, training an ideal model is crucial. To determine the model that fits the data the best, we employed the optimization models SGD, RMSprop, Adam, Nesterov-accelerated Adam (Nadam), AdaMax, AdaGrad, and AdaDelta, which are the models most often used for deep learning neural network training.

In comparison to other deep learning neural networks, the SGD [59] optimization strategy is quicker and simpler to train because it eliminates duplication by performing one update at a time. The objective function has considerable fluctuations when the frequent update method with high variance is used, and these fluctuations then have the ability to shift the parameters to new and improved local minima. Because of the ongoing overshooting of SGD, convergence to an accurate minimum is challenging. AdaGrad [57] is a gradient-based optimization approach that adjusts the learning rate of parameters to perform larger updates on repeatedly occurring parameters, while performing fewer updates on less frequently occurring parameters. This approach is suitable for processing sparse data and significantly improves the robustness of SDG optimization [59]. The

AdaDelta [56] optimization approach is an extended version of AdaGrad optimization that reduces the learning rate aggressively and monotonically, with parameters varying the learning rates and the learning process stopping after a particular point. The RMSprop [55] optimization approach was developed to overcome the rapidly decreasing learning rate of AdaGrad. It is an adaptive learning rate method, and uses variable learning rates that vary with the results for each sample of each iteration. The Adam [58] optimization approach was developed based on SDG, AdaDelta, and RMSprop. It dynamically calculates the learning rate for each sample of the dataset based on parameters to be used as adaptive optimization approaches with limited memory. The optimization strategy used by Nadam is almost identical to that used by Adam, with the slight difference being that Adam's and Nesterov's momentums are combined in Nadam to replace the flat momentum, which greatly improves its performance. The AdaMax [59] optimization approach is an extended approach to Adam optimization that consists of a simpler constraint than the parameter update size of Adam optimization, resulting in stable weight update rules.

We compared the training results using a Bi-GRU classifier for the optimization approach of the proposed model. Figure 10 shows the loss curve of learning and validation of the optimization approach. Among the seven optimization techniques, Adam shows the fastest learning process and convergence rate, which means that Bi-GRU classifiers were trained more successfully than other optimization approaches. Conversely, the AdaDelta optimization approach exhibits the lowest learning rate. Consequently, in the optimization approach following batch size and learning rate, Adam was adopted as the optimal approach best suited for training lip-based classification by the proposed model.

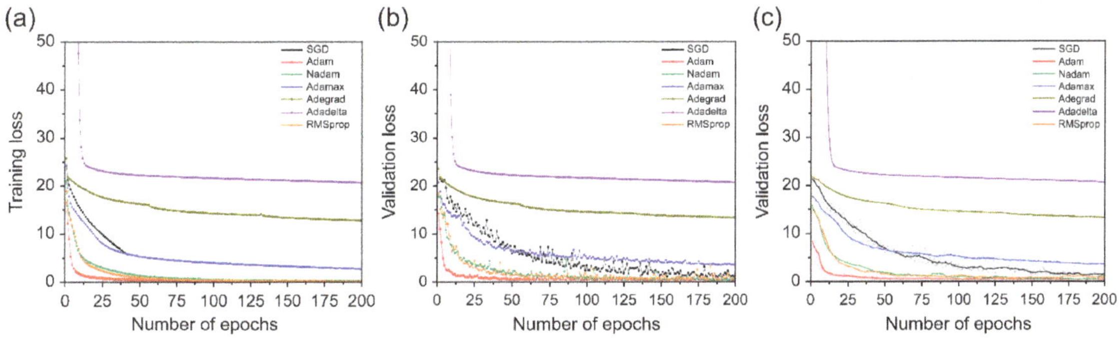

Figure 10. Loss curves comparing various optimizers. (**a**) Training loss; (**b**) validation loss; (**c**) validation loss using moving average.

5.3. Performance and Accuracy

The performance evaluation results of the proposed model are presented in Figure 11 and Table 4. The proposed model obtained the best results, with a CAR of 98.698% at four, the smallest size among different batch sizes, and the CAR decreased as the batch size increased (Figure 11a,b and Table 4). At different learning rates, 0.0001 yielded the best performance, and the remaining 0.01 and 0.0001 yielded similar performances (Figure 11c,d). However, at 0.1 and 0.001, they were excluded from the performance evaluation owing to divergence in the process of training the model. Among the seven optimization approaches, Adam optimization yielded the highest results, followed by Nadam and RMSprop with similar performance (Figure 11e,f). In addition, SGD and AdaMax showed good performance in that order, and AdaGrad showed low performance at 46.392%. AdaDelta was excluded from the performance evaluation because it diverged during the training process of the model. Thus, the proposed model exhibited the best performance when trained with a batch size of four, a learning rate of 0.0001, and the Adam optimization approach.

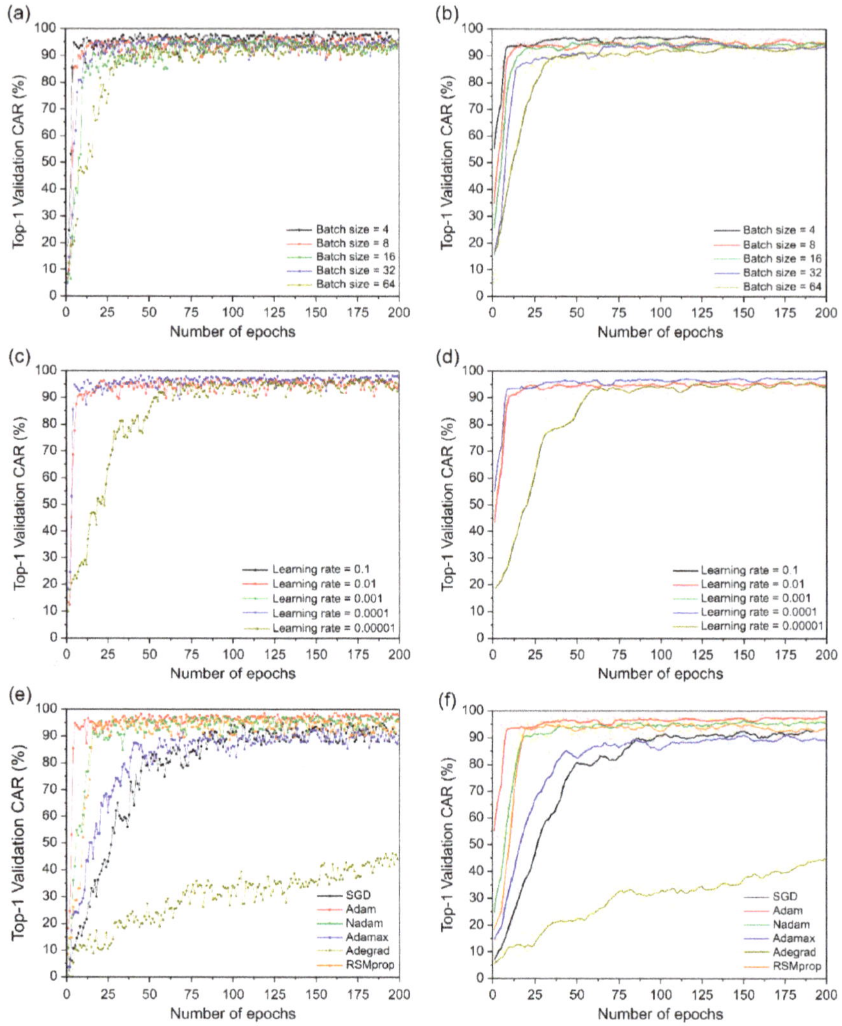

Figure 11. Training steps for character accuracy rate (CAR) comparing different batch sizes, learning rates, and optimizers: (**a**) Different batch sizes; (**b**) different batch sizes using moving average; (**c**) different learning rates; (**d**) different learning rates using moving average; (**e**) different optimizers; (**f**) different optimizers using moving average.

Table 4. Performance of different batch sizes, learning rates, and optimizers on the collected dataset.

Batch Size	Top-1 CAR (%)	Learning Rate	Top-1 CAR (%)	Optimizer	Top-1 CAR (%)
4	98.698	0.1	-	SGD	94.867
8	97.663	0.01	97.301	Adam	98.698
16	97.657	0.001	-	Nadam	97.500
32	96.972	0.0001	98.698	AdaMax	92.953
64	96.126	0.00001	97.095	AdaGrad	46.392
				AdaDelta	-
				RMSprop	97.323

5.4. Confusion Matrix

The performance of the proposed model was evaluated using 30% of the dataset as a validation sample. The results are shown in Figures 12 and A1 as confusion matrices and Figure 13 as a classification report. The average (mean) precision of each of the proposed models was 0.9870%, recall was 0.9869%, and F1-score was 0.9869% (shown in Table A2). In the classified recognition results, misclassification occurred in "go", "on", and "off", where the mouth was opened only to a small degree and the duration of utterance was shortest. Further, a small misclassification occurred in "center" and "under", and "fish", "jellyfish", and "starfish", which have similar utterance endings. As reported by Kaburagi et al. [1], this is a problem that occurs because of the similar lip shape at the beginning and end of speech, but it is not a factor that has a great influence on the overall performance evaluation. Based on our evaluation of the proposed model, it is clear that it can help overcome technical obstacles to practical implementation.

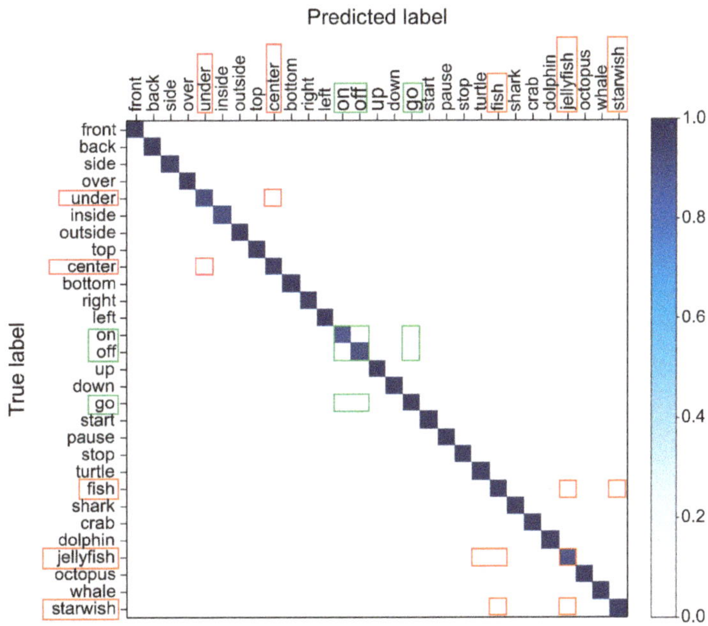

Figure 12. Confusion matrix of the model that proposed many misclassifications out of 53 classes.

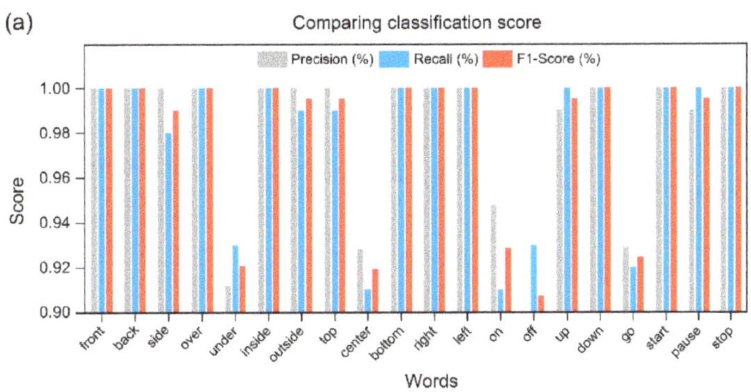

Figure 13. *Cont.*

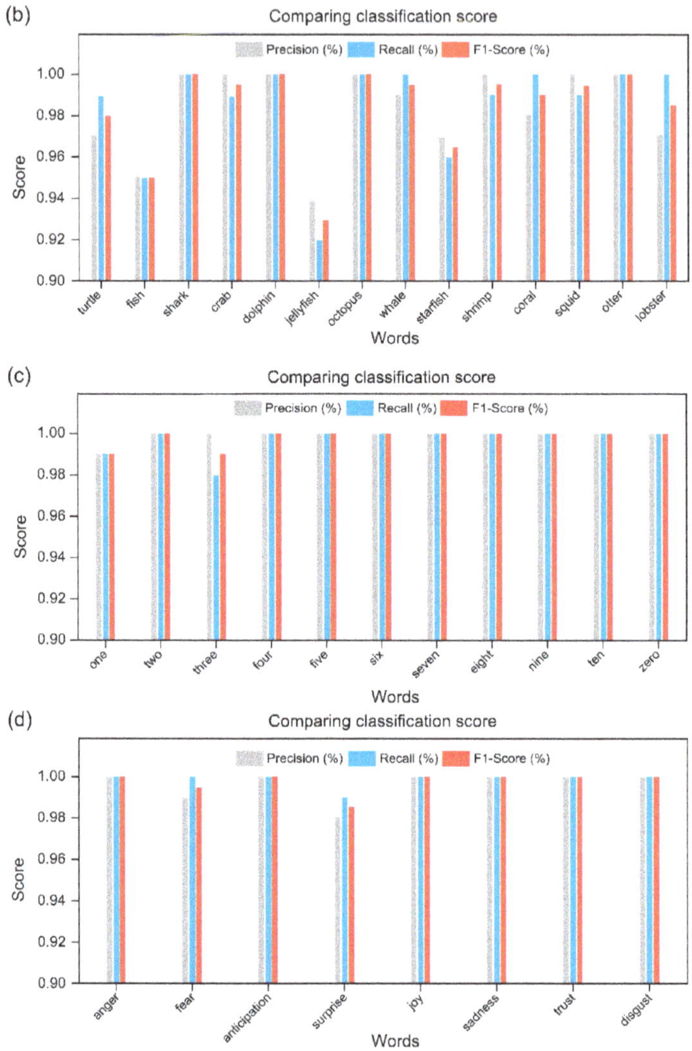

Figure 13. Comparing classification scores of the proposed model trained on 53 classes. (**a**) Control commands; (**b**) marine life; (**c**) number; (**d**) emotion.

5.5. Performance and Accuracy

Figure 8 shows the actual experimental scene where participants interacted using their speech in a virtual aquarium environment, and Figure 14 and Table A3 show the performance of each recognizer at different SNR (dB) levels considering the case of four noises. In Figure 14a, which was evaluated at different SNR levels considering the case of corridor noise, it was 91.14% ± 1.24% in clean, and the highest accuracy was 90.88% ± 1.08% in 40 dB. The visual-only accuracy for various SNR levels was 88.79% ± 0.73%, and because this value depends only on visual information, it is not affected by the clearance of the audio signal or the SNR level. Participants who participated in the data collection stage did not participate in the performance evaluation of the proposed model, and there were no external factors such as lighting for clear data collection during data collection. In addition, in order not to control external factors, the experiment was carried out naturally without limitations of variables (such as mouth shape during pronunciation by participants or mouth shadow according to lighting). The multimodal approach using both auditory and visual signals was 97.87% ± 0.62%, which improved recognition rates by 6.73% and 9.08%, respectively, compared to recognition using only auditory and visual signals. As a result, speech may be inferred even in the presence of background noise by utilizing a mix of sound and visual information.

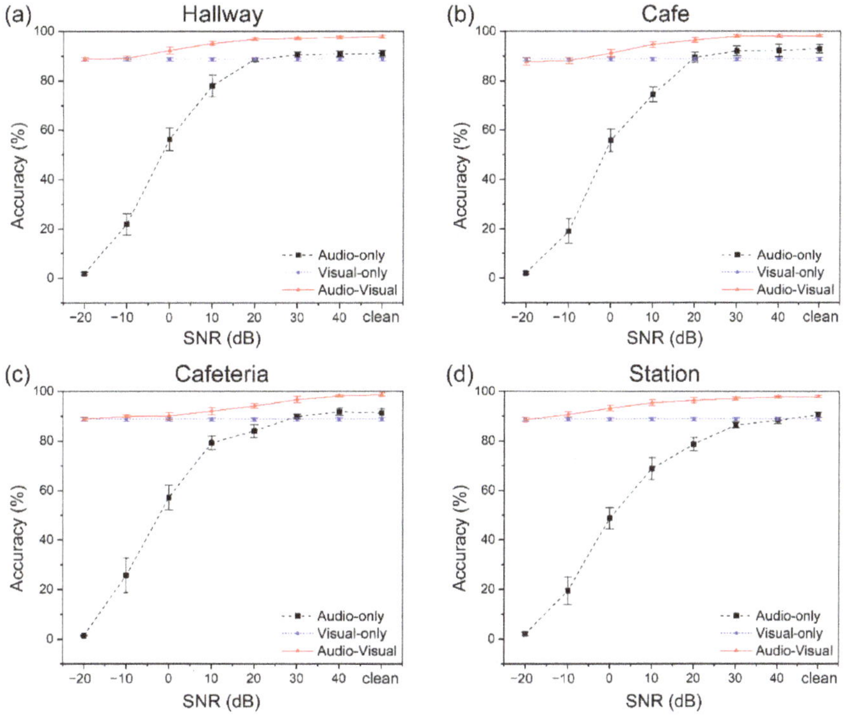

Figure 14. Standard variation of the identification accuracy rate across four noise situations. Standard deviation is shown by error bars. The audio recognition result is shown by the black (single-modal recognition) line, the visual recognition result is represented by the blue (single-modal recognition) line, and the audiovisual recognition result is represented by the red (multimodal recognition) line. The identification outcome in a true experimental setting is called clean SNR.

Based on the previous results, different noises were synthesized and compared for the three components. For the results in Figure 14b and Table A3b, the performance evaluation was performed by synthesizing cafe noise at various SNR levels, and only audio was used for recognition at the clean level in 92.90% ± 1.64% of cases. On the other hand, when audio and visual information were used together for recognition, it was 98.17%± 0.52%, an improvement of 5.27% compared to when only audio was used. Figure 14c,d consist of noise from two different environments: a cafeteria and a subway station. The cafeteria is a busy office cafeteria environment, and the subway is the noise of the main transit area of a crowded subway station. When only audio is used, the recognition rates are 91.26% ± 1.69% and 90.37% ± 1.10%, respectively; and when audio and visual are used together, the recognition rates are 98.60% ± 0.70% and 97.85% ± 0.51%, respectively. The recognition rates were improved by 7.34% and 7.48%, respectively, compared to the case where only audio was used.

The statistical significance of the three-group t-test for four noise settings is shown in Figure 15 for each noise environment. In all the noise environments, the recognition rate was improved by 6.04% on average in the case of using both audio and visual compared to the case of using only audio. On the other hand, the recognition rate was improved by 9.33% compared to the case where only vision was used. Additionally, the standard deviation was reduced when audio and visual information were used together, compared to when only audio was used, and a similar recognition rate for repeated experiments was also observed.

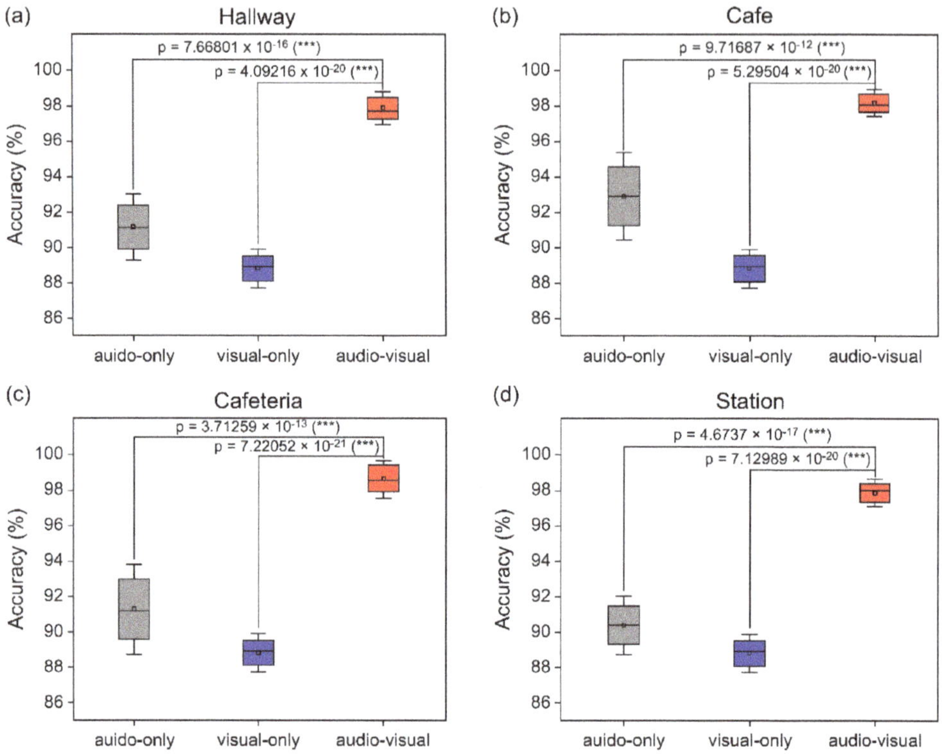

Figure 15. Average recognition accuracy rates in four noisy scenarios. The error bars show the standard deviation. *t*-tests with statistical significance between each group are indicated by asterisks (* for $p < 0.05$, ** for $p < 0.01$, *** for $p < 0.001$). The gray (single-modal recognition), blue (single-modal recognition), and red (multimodal recognition) boxes show the results of auditory-only, visual-only, and audiovisual recognition, respectively.

6. Discussion

Recently, owing to the exponential increase in large-capacity data-processing capability, speech recognition-based interactive edutainment systems have become more widespread. Along with gesture interaction, speech interaction is used in various applications, and more difficult acoustic scenarios have to be considered than in the past for extensive practical applications, taking into account the challenging issue of adequate noise management for varied settings. We conducted a study on speech interaction in a virtual aquarium considering various scenarios.

In this study, an end-to-end visual speech recognition-based interaction system for speech interaction in a virtual aquarium environment was proposed. Recognized words were vectored by combining pretrained word embedding with Microsoft API, which showed the best performance and dense dispersion in previous studies [45]. Feature vectorization was performed on the image sequence input through video via the visual processing module, and word vectorization was combined to output the predicted word. For the optimization of the proposed model, the performance was evaluated using different batch sizes, learning rates, and optimization approaches. In addition, performance evaluation was performed by synthesizing the data used in the actual evaluation and four noise types using different SNR levels.

The system performance in terms of SNR was evaluated using four sets of synthesized data associated with four noise environments along with other evaluation data. Compared to the speech recognition system using only audio, which has a large standard deviation in the four noise environments, the system using visual information showed small standard deviation and superior performance. Furthermore, in the case of clean, the average recognition rate was 91.42% when only speech was used in the four noise environments, whereas when audio and visual information were used together, the recognition rate was improved by 6.7% to 98.12%.

The utilization of multimodal interactions based on visual information is necessary to develop antinoise automated speech recognition (ASR) systems. The proposed system could help patients who have difficulty with noise during conversation. It can also provide an opportunity for people with hearing problems to have a conversation. However, it is difficult to apply this technique to real-world conversation recognition. Therefore, in future studies, we will consider expanding the system's ability to identify non-keyword-centric phrases. Additionally, by displaying an actual virtual aquarium in a science museum, we will assess the efficiency of the proposed model in a real environment and create a reliable portable gadget for real-world usage.

7. Conclusions

By combining visual information with the existing speech interaction-based edutainment system, a new visual information-based speech recognition system that is robust to noise was applied to a virtual aquarium and demonstrated. The proposed system combines audio and visual information, and for optimization, the performance was evaluated considering three factors (batch size, learning rate, and optimization approach), and the performance was further evaluated in four noise environments. It was shown that visual information contributed to the improvement of speech recognition by using visual information, unlike the existing method that used only speech to interact. To achieve consistent and high performance in diverse noise conditions, our technique blends visual speech recognition, a technology that can enhance speech recognition systems, with current cloud-based speech recognition systems. This method offers the potential for real-world speech recognition applications in noisy locations. It can be utilized for speech interaction in venues such as museums and scientific centers that have significant amounts of interior noise and noise from visitors, and it also has the potential for use in diverse applications that employ voice recognition in noise, such as IoT and robot applications.

Author Contributions: Methodology, S.J.; software, S.J.; validation, S.J.; formal analysis, S.J.; investigation, S.J.; data curation, S.J.; writing—original draft preparation, S.J.; writing—review and editing, S.J.; visualization, S.J.; supervision, M.S.K.; funding acquisition, M.S.K. All authors have read and agreed to the published version of the manuscript.

Funding: This research was funded by National Research Foundation of Korea (NRF) grant funded by the Korea government (MSIT), grant number NRF-2018X1A3A1069795.

Institutional Review Board Statement: As a reason for indemnification, in accordance with Article 13 of the Enforcement Rules of "Research on Human Subjects", "A study that does not collect and record personal identification information, and one or more of the following studies are exempt from deliberation, and in detail, it is a study that uses only simple contact measurement equipment or observation equipment that does not cause physical changes". Only human voice and lip information were collected and used; thus, there is no personal identification information and only observation equipment is used. Therefore, there is no IRB registration number owing to the indemnification reasons.

Informed Consent Statement: Not applicable.

Data Availability Statement: The data presented in this study are available on request from the corresponding author. The data are not publicly available due to ethical restrictions.

Conflicts of Interest: The authors declare no conflict of interest. The funders had no role in the design of the study; in the collection, analyses, or interpretation of data; in the writing of the manuscript; or in the decision to publish the results.

Appendix A

The following Tables A1–A3 and Figure A1 provide (1) the details of the hyperparameters associated with the proposed architecture, (2) the confusion matrix of the proposed model trained on 53 classes, (3) the average word accuracy and standard deviation of the proposed system in four noise environments, and (4) the performance in terms of precision, recall, and F1-score, respectively.

Table A1. Hyperparameters associated with the proposed architecture.

Layers	Size/Stride/Pad	Visual	Audio
		Output Size	
Input Layer	-	40 × 100 × 50 × 3	1500
3D Conv Layer 3D Max Pooling	[3 × 5 × 5]/(1, 2, 2)/(1, 2, 2) [1 × 2 × 2]/(1, 2, 2)	40 × 50 × 25 × 64 40 × 50 × 13 × 64	
Densely Connected 3D CNN	[3 × 1 × 1] 3D Conv [3 × 3 × 3] 3D Conv (×6)	40 × 25 × 13 × 96	
	[3 × 1 × 1] 3D Conv [1 × 2 × 2] average pool/(1 × 2 × 2)	40 × 12 × 6 × 6	
	[3 × 1 × 1] 3D Conv [3 × 3 × 3] 3D Conv (×12)	40 × 12 × 6 × 38	
	[3 × 1 × 1] 3D Conv [1 × 2 × 2] average pool/(1 × 2 × 2)	40 × 6 × 3 × 3	
	[3 × 1 × 1] 3D Conv [3 × 3 × 3] 3D Conv (×24)	40 × 12 × 6 × 38	
	[3 × 1 × 1] 3D Conv [1 × 2 × 2] average pool/(1 × 2 × 2)	40 × 3 × 1 × 1	
	[3 × 1 × 1] 3D Conv [3 × 3 × 3] 3D Conv (×16)	40 × 3 × 1 × 33	
Multiscale 3D CNN (1)	[3 × 5 × 5]/(1, 2, 2)/(1, 2, 2)	40 × 3 × 1 × 32	
Multiscale 3D CNN (2)	[3 × 5 × 5]/(1, 2, 2)/(1, 2, 2)	40 × 3 × 1 × 64	
Multiscale 3D CNN (3)	[3 × 5 × 5]/(1, 2, 2)/(1, 2, 2)	40 × 3 × 1 × 92	
Spatial Attention (1)	[1 × 2 × 2] max pool/(1 × 2 × 2) [1 × 2 × 2] average pool/(1 × 2 × 2) [3 × 7 × 7]/(1, 2, 2)/(1, 2, 2)	40 × 3 × 1 × 32	
Spatial Attention (2)	[1 × 2 × 2] max pool/(1 × 2 × 2) [1 × 2 × 2] average pool/(1 × 2 × 2) [3 × 7 × 7]/(1, 2, 2)/(1, 2, 2)	40 × 3 × 1 × 64	
Spatial Attention (3)	[1 × 2 × 2] max pool/(1 × 2 × 2) [1 × 2 × 2] average pool/(1 × 2 × 2) [3 × 7 × 7]/(1, 2, 2)/(1, 2, 2)	40 × 3 × 1 × 96	
Bi-GRU (1)	256	40 × 512	
Bi-GRU (2)	256	40 × 512	

Table A1. *Cont.*

Layers	Size/Stride/Pad	Visual	Audio
		Output Size	
Concatenation	-	40 × 2012	
Dense Layer	27 + blank	40 × 2012	
Softmax		40 × 28	

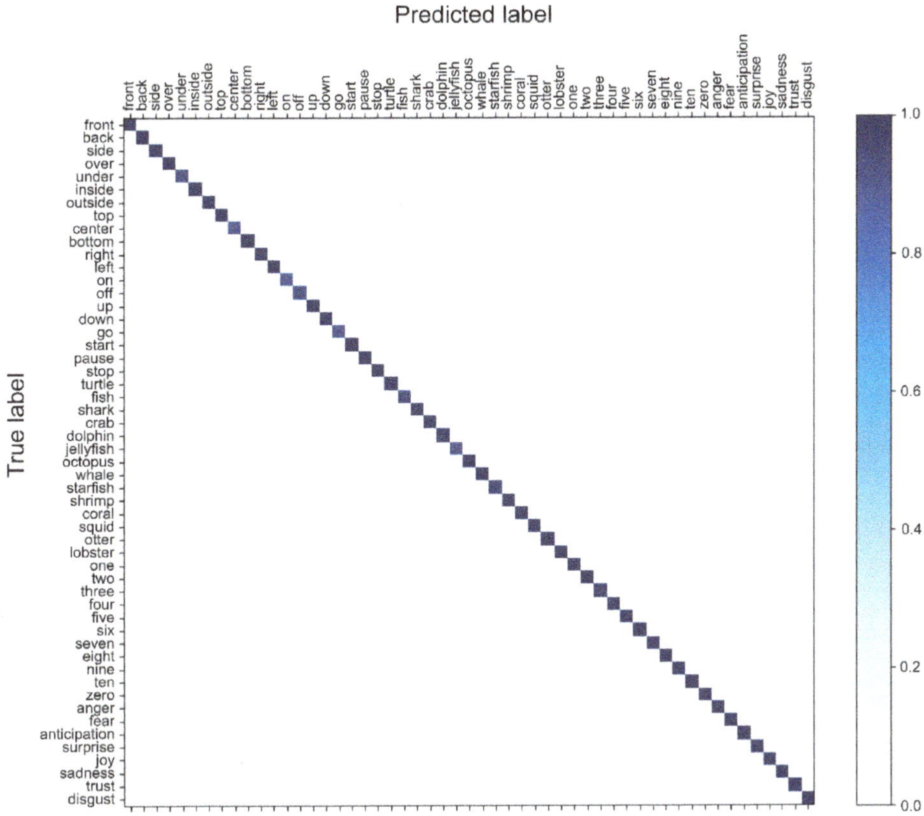

Figure A1. Confusion matrix of the proposed model trained on 53 classes.

Table A2. Average word accuracy and standard deviation of the proposed system in four noise environments.

SNR (dB)		−20	−10	0	10	20	30	40	Clean
(a) Hallway	A	1.84% ± 0.62%	21.98% ± 4.30%	56.26% ± 4.63%	78.01% ± 4.42%	88.60% ± 0.88%	90.59% ± 1.07%	90.88% ± 1.08%	91.14% ± 1.24%
	V	88.79% ± 0.73%	88.79% ± 0.73%	88.79% ± 0.73%	88.79% ± 0.73%	88.79% ± 0.73%	88.79% ± 0.73%	88.79% ± 0.73%	88.79% ± 0.73%
	AV	88.84% ± 0.65%	89.27 ± 0.70%	92.33 ± 1.41%	95.19 ± 0.96%	96.84 ± 0.58%	97.20% ± 0.44%	97.58% ± 0.54%	97.87% ± 0.62%

Table A2. *Cont.*

SNR (dB)		−20	−10	0	10	20	30	40	Clean
(b) Cafe	A	1.99% ± 0.28%	19.05% ± 4.99%	55.77% ± 4.54%	74.42% ± 3.05%	89.34% ± 2.04%	92.04% ± 2.01%	92.23% ± 2.50%	92.90% ± 1.64%
	V	88.79% ± 0.73%	88.79% ± 0.73%	88.79% ± 0.73%	88.79% ± 0.73%	88.79% ± 0.73%	88.79% ± 0.73%	88.79% ± 0.73%	88.79% ± 0.73%
	AV	87.70% ± 1.51%	88.17% ± 1.29%	91.16% ± 1.43%	94.70% ± 1.15%	96.48% ± 0.93%	98.02% ± 0.61%	98.10% ± 0.65%	98.17% ± 0.52%
(c) Cafeteria	A	1.44% ± 0.31%	25.74% ± 6.92%	57.30% ± 5.00%	79.23% ± 2.81%	84.06% ± 2.57%	89.90% ± 1.05%	91.76% ± 1.54%	91.26% ± 1.69%
	V	88.79% ± 0.73%	88.79% ± 0.73%	88.79% ± 0.73%	88.79% ± 0.73%	88.79% ± 0.73%	88.79% ± 0.73%	88.79% ± 0.73%	88.79% ± 0.73%
	AV	88.92% ± 0.79%	89.94% ± 0.57%	90.07% ± 1.36%	92.05% ± 1.43%	94.13% ± 1.04%	96.63% ± 1.30%	98.17% ± 0.46%	98.60% ± 0.70%
(d) Station	A	2.08% ± 0.69%	19.49% ± 5.64%	48.84% ± 4.30%	68.79% ± 4.46%	78.61% ± 2.67%	86.31% ± 1.17%	88.23% ± 1.34%	90.37% ± 1.10%
	V	88.79% ± 0.73%	88.79% ± 0.73%	88.79% ± 0.73%	88.79% ± 0.73%	88.79% ± 0.73%	88.79% ± 0.73%	88.79% ± 0.73%	88.79% ± 0.73%
	AV	88.57% ± 0.99%	90.64% ± 0.96%	93.10% ± 1.12%	95.34% ± 1.20%	96.34% ± 1.05%	97.06% ± 0.63%	97.67% ± 0.47%	97.85% ± 0.51%

Table A3. Precision, recall, and F1-score performance of in four noise environments. (a) control commands; (b) marine life; (c) numbers; (d) emotions.

(a) Control Commands				(b) Marine Life			
Word	Precision	Recall	F1-Score	Word	Precision	Recall	F1-Score
front	1.0000	1.0000	1.0000	turtle	0.9706	0.9900	0.9802
back	1.0000	1.0000	1.0000	fish	0.9500	0.9500	0.9500
side	1.0000	0.9800	0.9899	shark	1.0000	1.0000	1.0000
over	1.0000	1.0000	1.0000	crab	1.0000	0.9900	0.9950
under	0.9118	0.9300	0.9208	dolphin	1.0000	1.0000	1.0000
inside	1.0000	1.0000	1.0000	jellyfish	0.9388	0.9200	0.9293
outside	1.0000	0.9900	0.9950	octopus	1.0000	1.0000	1.0000
top	1.0000	0.9900	0.9950	whale	0.9901	1.0000	0.9950
center	0.9286	0.9100	0.9192	starfish	0.9697	0.9600	0.9648
bottom	1.0000	1.0000	1.0000	shrimp	1.0000	0.9900	0.9950
right	1.0000	1.0000	1.0000	coral	0.9804	1.0000	0.9901
left	1.0000	1.0000	1.0000	squid	1.0000	0.9900	0.9950
on	0.9479	0.9100	0.9286	otter	1.0000	1.0000	1.0000
off	0.8857	0.9300	0.9073	lobster	0.9709	1.0000	0.9852
up	0.9901	1.0000	0.9950				
down	1.0000	1.0000	1.0000				
go	0.9293	0.9200	0.9246				
start	1.0000	1.0000	1.0000				
pause	0.9901	1.0000	0.9950				
stop	1.0000	1.0000	1.0000				

Table A3. *Cont.*

(c) Numbers				(d) Emotions			
Word	**Precision**	**Recall**	**F1-Score**	**Word**	**Precision**	**Recall**	**F1-Score**
one	0.9900	0.9900	0.9900	anger	1.0000	1.0000	1.0000
two	1.0000	1.0000	1.0000	fear	0.9901	1.0000	0.9950
three	1.0000	0.9800	0.9899	anticipation	1.0000	1.0000	1.0000
four	1.0000	1.0000	1.0000	surprise	0.9802	0.9900	0.9851
five	1.0000	1.0000	1.0000	joy	1.0000	1.0000	1.0000
six	1.0000	1.0000	1.0000	sadness	1.0000	1.0000	1.0000
seven	1.0000	1.0000	1.0000	trust	1.0000	1.0000	1.0000
eight	1.0000	1.0000	1.0000	disgust	1.0000	1.0000	1.0000
nine	1.0000	1.0000	1.0000				
ten	1.0000	1.0000	1.0000				
zero	1.0000	1.0000	1.0000				

References

1. Mich, O.; Schiavo, G.; Ferron, M.; Mana, N. Framing the design space of multimodal mid-air gesture and speech-based interaction with mobile devices for older people. *Int. J. Mob. Hum. Comput. Interact.* **2020**, *12*, 22–41. [CrossRef]
2. Kaburagi, R.; Ishimaru, Y.; Chin, W.H.; Yorita, A.; Kubota, N.; Egerton, S. Lifelong robot edutainment based on self-efficacy. In Proceedings of the 2021 5th IEEE International Conference on Cybernetics (CYBCONF), Sendai, Japan, 8–10 June 2021; IEEE: New York, NY, USA, 2021; pp. 79–84.
3. Soo, V.-W.; Huang, C.-F.; Su, Y.-H.; Su, M.-J. AI applications on music technology for edutainment. In Proceedings of the International Conference on Innovative Technologies and Learning, Portoroz, Slovenia, 27–30 August 2018; Springer: Cham, Switzerland, 2019; pp. 594–599.
4. Tsai, T.-H.; Chi, P.-T.; Cheng, K.-H. A sketch classifier technique with deep learning models realized in an embedded system. In Proceedings of the 2019 IEEE 22nd International Symposium on Design and Diagnostics of Electronic Circuits & Systems (DDECS), Cluj-Napoca, Romania, 24–26 April 2019; IEEE: New York, NY, USA, 2019; pp. 1–4.
5. Disney, W. Educational values in factual nature pictures. *Educ. Horiz.* **1954**, *33*, 82–84.
6. Rapeepisarn, K.; Wong, K.W.; Fung, C.C.; Depickere, A. Similarities and differences between "learn through play" and "edutainment". In Proceedings of the 3rd Australasian Conference on Interactive Entertainment, Perth, Australia, 4–6 December 2006; pp. 28–32.
7. Bellotti, F.; Kapralos, B.; Lee, K.; Moreno-Ger, P.; Berta, R. Assessment in and of serious games: An overview. *Adv. Hum.-Comput. Interact.* **2013**, *2013*, 136864. [CrossRef]
8. Zin, H.M.; Zain, N.Z.M. The effects of edutainment towards students' achievements. In Proceedings of the Regional Conference on Knowledge Integration in ICT, Putrajaya, Malaysia, 1 June 2010; Volume 129, p. 2865.
9. Kara, Y.; Yeşilyurt, S. Comparing the impacts of tutorial and edutainment software programs on students' achievements, misconceptions, and attitudes towards biology. *J. Sci. Educ. Technol.* **2008**, *17*, 32–41. [CrossRef]
10. Efthymiou, N.; Filntisis, P.; Potamianos, G.; Maragos, P. A robotic edutainment framework for designing child-robot interaction scenarios. In Proceedings of the 14th Pervasive Technologies Related to Assistive Environments Conference, Corfu, Greece, 29 June–2 July 2021; pp. 160–166.
11. Matulík, M.; Vavrečka, M.; Vidovićová, L. Edutainment software for the Pepper robot. In Proceedings of the 2020 4th International Symposium on Computer Science and Intelligent Control, Newcastle Upon Tyne, UK, 17–19 November 2020; pp. 1–5.
12. Che Hashim, N.; Abd Majid, N.A.; Arshad, H.; Khalid Obeidy, W. User satisfaction for an augmented reality application to support productive vocabulary using speech recognition. *Adv. Multimed.* **2018**, *2018*, 9753979. [CrossRef]
13. Yum, M.S. Istanbul Aquarium edutainment project. *Online J. Art Des.* **2022**, *10*, 207–228.
14. Hepperle, D.; Weiß, Y.; Siess, A.; Wölfel, M. 2D, 3D or speech? A case study on which user interface is preferable for what kind of object interaction in immersive virtual reality. *Comput. Graph.* **2019**, *82*, 321–331. [CrossRef]
15. Janowski, K.; Kistler, F.; André, E. Gestures or speech? Comparing modality selection for different interaction tasks in a virtual environment. In Proceedings of the Tilburg Gesture Research Meeting, Tilburg, The Netherlands, 19–21 June 2013.
16. Venezia, J.; Matchin, W.; Hickok, G. Multisensory integration and audiovisual speech perception. *Brain Mapp. Encycl. Ref.* **2015**, *2*, 565–572.
17. Campbell, R. The processing of audio-visual speech: Empirical and neural bases. *Philos. Trans. R. Soc. B Biol. Sci.* **2008**, *363*, 1001–1010. [CrossRef]
18. Sumby, W.H.; Pollack, I. Visual contribution to speech intelligibility in noise. *J. Acoust. Soc. Am.* **1954**, *26*, 212–215. [CrossRef]
19. Dodd, B. The role of vision in the perception of speech. *Perception* **1977**, *6*, 31–40. [CrossRef] [PubMed]
20. Jones, J.A.; Callan, D.E. Brain activity during audiovisual speech perception: An fMRI study of the McGurk effect. *Neuroreport* **2003**, *14*, 1129–1133. [CrossRef] [PubMed]

21. Risberg, A. The importance of prosodic speech elements for the lipreader. *Scand. Audiol.* **1974**, *4*, 153–164.
22. Grant, K.W.; Ardell, L.H.; Kuhl, P.K.; Sparks, D.W. The contribution of fundamental frequency, amplitude envelope, and voicing duration cues to speechreading in normal-hearing subjects. *J. Acoust. Soc. Am.* **1985**, *77*, 671–677. [CrossRef] [PubMed]
23. Bernstein, L.E.; Eberhardt, S.P.; Demorest, M.E. Single-channel vibrotactile supplements to visual perception of intonation and stress. *J. Acoust. Soc. Am.* **1989**, *85*, 397–405. [CrossRef] [PubMed]
24. McGurk, H.; MacDonald, J. Hearing lips and seeing voices. *Nature* **1976**, *264*, 746–748. [CrossRef] [PubMed]
25. Huang, G.; Liu, Z.; Van Der Maaten, L.; Weinberger, K.Q. Densely connected convolutional networks. In Proceedings of the IEEE Conference on Computer Vision and Pattern Recognition, Honolulu, HI, USA, 21–26 July 2017; pp. 4700–4708.
26. Graves, A.; Fernández, S.; Gomez, F.; Schmidhuber, J. Connectionist temporal classification: Labelling unsegmented sequence data with recurrent neural networks. In Proceedings of the 23rd International Conference on Machine Learning, Pittsburgh, PA, USA, 25–29 June 2006; pp. 369–376.
27. Google Cloud Speech to Text. Available online: https://cloud.google.com/speech-to-text (accessed on 7 July 2022).
28. Watson Speech to Text. Available online: https://www.ibm.com/kr-ko/cloud/watson-speech-to-text (accessed on 7 July 2022).
29. Microsoft Azure Cognitive Services. Available online: https://azure.microsoft.com/en-us/services/cognitive-services/ (accessed on 7 July 2022).
30. Xiong, W.; Wu, L.; Alleva, F.; Droppo, J.; Huang, X.; Stolcke, A. The Microsoft 2017 conversational speech recognition system. In Proceedings of the 2018 IEEE International Conference on Acoustics, Speech and Signal Processing (ICASSP), Calgary, AB, Canada, 15–20 April 2018; IEEE: New York, NY, USA, 2018; pp. 5934–5938.
31. Amazon Alexa. Available online: https://developer.amazon.com/en-US/alexa (accessed on 7 July 2022).
32. Ji, S.; Xu, W.; Yang, M.; Yu, K. 3D convolutional neural networks for human action recognition. *IEEE Trans. Pattern Anal. Mach. Intell.* **2013**, *35*, 221–231. [CrossRef] [PubMed]
33. Petridis, S.; Pantic, M. Deep complementary bottleneck features for visual speech recognition. In Proceedings of the 2016 IEEE International Conference on Acoustics, Speech and Signal Processing (ICASSP), Shanghai, China, 20–25 March 2016; IEEE: New York, NY, USA, 2016; pp. 2304–2308.
34. Wand, M.; Koutník, J.; Schmidhuber, J. Lipreading with long short-term memory. In Proceedings of the 2016 IEEE International Conference on Acoustics, Speech and Signal Processing (ICASSP), Shanghai, China, 20–25 March 2016; IEEE: New York, NY, USA, 2016; pp. 6115–6119.
35. Cooke, M.; Barker, J.; Cunningham, S.; Shao, X. An audio-visual corpus for speech perception and automatic speech recognition. *J. Acoust. Soc. Am.* **2006**, *120*, 2421–2424. [CrossRef] [PubMed]
36. Noda, K.; Yamaguchi, Y.; Nakadai, K.; Okuno, H.G.; Ogata, T. Lipreading using convolutional neural network. In Proceedings of the Fifteenth Annual Conference of the International Speech Communication Association, Singapore, 14–18 September 2014.
37. Assael, Y.M.; Shillingford, B.; Whiteson, S.; De Freitas, N. Lipnet: End-to-end sentence-level lipreading. *arXiv* **2016**, arXiv:1611.01599.
38. Fenghour, S.; Chen, D.; Guo, K.; Li, B.; Xiao, P. An effective conversion of visemes to words for high-performance automatic lipreading. *Sensors* **2021**, *21*, 7890. [CrossRef] [PubMed]
39. Li, H.; Yadikar, N.; Zhu, Y.; Mamut, M.; Ubul, K. Learning the relative dynamic features for word-level lipreading. *Sensors* **2022**, *22*, 3732. [CrossRef] [PubMed]
40. Xu, K.; Li, D.; Cassimatis, N.; Wang, X. LCANet: End-to-end lipreading with cascaded attention-CTC. In Proceedings of the 2018 13th IEEE International Conference on Automatic Face & Gesture Recognition (FG 2018), Xi'an, China, 15–19 May 2018; IEEE: New York, NY, USA, 2018; pp. 548–555.
41. Këpuska, V.; Bohouta, G. Comparing speech recognition systems (Microsoft API, Google API and CMU Sphinx). *Int. J. Eng. Res. Appl.* **2017**, *7*, 20–24. [CrossRef]
42. Yoo, H.J.; Seo, S.; Im, S.W.; Gim, G.Y. The performance evaluation of continuous speech recognition based on Korean phonological rules of cloud-based speech recognition open API. *Int. J. Netw. Distrib. Comput.* **2021**, *9*, 10–18. [CrossRef]
43. Alibegović, B.; Prljača, N.; Kimmel, M.; Schultalbers, M. Speech recognition system for a service robot—A performance evaluation. In Proceedings of the 2020 16th International Conference on Control, Automation, Robotics and Vision (ICARCV), Shenzhen, China, 13–15 December 2020; IEEE: New York, NY, USA, 2020; pp. 1171–1176.
44. Caute, A.; Woolf, C. Using voice recognition software to improve communicative writing and social participation in an individual with severe acquired dysgraphia: An experimental single-case therapy study. *Aphasiology* **2016**, *30*, 245–268. [CrossRef]
45. Jeon, S.; Kim, M.S. End-to-end lip-reading open cloud-based speech architecture. *Sensors* **2022**, *22*, 2938. [CrossRef]
46. Mikolov, T.; Sutskever, I.; Chen, K.; Corrado, G.; Dean, J. Distributed Representations of Words and Phrases and Their Compositionality. In Proceedings of the 26th International Conference on Neural Information Processing Systems, Lake Tahoe, NV, USA, 5–10 December 2013; Ser. NIPS'13. Curran Associates Inc.: Red Hook, NY, USA, 2013; pp. 3111–3119.
47. Mikolov, T.; Chen, K.; Corrado, G.; Dean, J. Efficient estimation of word representations in vector space. *arXiv* **2013**, arXiv:1301.3781.
48. Hinton, G.E.; Srivastava, N.; Krizhevsky, A.; Sutskever, I.; Salakhutdinov, R.R. Improving neural networks by preventing co-adaptation of feature detectors. *arXiv* **2012**, arXiv:1207.0580.
49. Tompson, J.; Goroshin, R.; Jain, A.; LeCun, Y.; Bregler, C. Efficient object localization using convolutional networks. In Proceedings of the IEEE Conference on Computer Vision and Pattern Recognition, Boston, MA, USA, 7–12 June 2015; pp. 648–656.

50. Lee, S.; Lee, C. Revisiting spatial dropout for regularizing convolutional neural networks. *Multimed. Tools Appl.* **2020**, *79*, 34195–34207. [CrossRef]
51. Woo, S.; Park, J.; Lee, J.-Y.; Kweon, I.S. Cbam: Convolutional block attention module. In Proceedings of the European Conference on Computer Vision (ECCV), Munich, Germany, 8–14 September 2018; pp. 3–19.
52. Chung, J.; Gulcehre, C.; Cho, K.; Bengio, Y. Empirical evaluation of gated recurrent neural networks on sequence modeling. *arXiv* **2014**, arXiv:1412.3555.
53. Warden, P. Speech commands: A dataset for limited-vocabulary speech recognition. *arXiv* **2018**, arXiv:1804.03209.
54. King, D.E. Dlib-ml: A machine learning toolkit. *J. Mach. Learn. Res.* **2009**, *10*, 1755–1758.
55. Tieleman, T.; Hinton, G. Lecture 6.5-rmsprop: Divide the gradient by a running average of its recent magnitude. *COURSERA Neural Netw. Mach. Learn.* **2012**, *4*, 26–31.
56. Zeiler, M.D. Adadelta: An adaptive learning rate method. *arXiv* **2012**, arXiv:1212.5701.
57. Duchi, J.; Hazan, E.; Singer, Y. Adaptive subgradient methods for online learning and stochastic optimization. *J. Mach. Learn. Res.* **2011**, *12*, 2121–2159.
58. Kingma, D.P.; Ba, J. Adam: A method for stochastic optimization. *arXiv* **2014**, arXiv:1412.6980.
59. Bottou, L. Stochastic gradient descent tricks. In *Neural Networks: Tricks of the Trade*; Springer: Berlin/Heidelberg, Germany, 2012; pp. 421–436.
60. Masters, D.; Luschi, C. Revisiting small batch training for deep neural networks. *arXiv* **2018**, arXiv:1804.07612.
61. Kandel, I.; Castelli, M. The effect of batch size on the generalizability of the convolutional neural networks on a histopathology dataset. *ICT Express* **2020**, *6*, 312–315. [CrossRef]
62. You, Y.; Gitman, I.; Ginsburg, B. Large batch training of convolutional networks. *arXiv* **2017**, arXiv:1708.03888.
63. Keskar, N.S.; Mudigere, D.; Nocedal, J.; Smelyanskiy, M.; Tang, P.T.P. On large-batch training for deep learning: Generalization gap and sharp minima. *arXiv* **2016**, arXiv:1609.04836.

sensors

Article

End-to-End Lip-Reading Open Cloud-Based Speech Architecture

Sanghun Jeon and Mun Sang Kim *

Center for Healthcare Robotics, Gwangju Institute of Science and Technology (GIST), School of Integrated Technology, Gwangju 61005, Korea; jeon7887@gist.ac.kr

* Correspondence: munsang@gist.ac.kr; Tel.: +82-10-9126-4628

Abstract: Deep learning technology has encouraged research on noise-robust automatic speech recognition (ASR). The combination of cloud computing technologies and artificial intelligence has significantly improved the performance of open cloud-based speech recognition application programming interfaces (OCSR APIs). Noise-robust ASRs for application in different environments are being developed. This study proposes noise-robust OCSR APIs based on an end-to-end lip-reading architecture for practical applications in various environments. Several OCSR APIs, including Google, Microsoft, Amazon, and Naver, were evaluated using the Google Voice Command Dataset v2 to obtain the optimum performance. Based on performance, the Microsoft API was integrated with Google's trained word2vec model to enhance the keywords with more complete semantic information. The extracted word vector was integrated with the proposed lip-reading architecture for audio-visual speech recognition. Three forms of convolutional neural networks (3D CNN, 3D dense connection CNN, and multilayer 3D CNN) were used in the proposed lip-reading architecture. Vectors extracted from API and vision were classified after concatenation. The proposed architecture enhanced the OCSR API average accuracy rate by 14.42% using standard ASR evaluation measures along with the signal-to-noise ratio. The proposed model exhibits improved performance in various noise settings, increasing the dependability of OCSR APIs for practical applications.

Keywords: audio-visual speech recognition; lip-reading; application programming interface; multi-modal interaction; deep neural networks

Citation: Jeon, S.; Kim, M.S. End-to-End Lip-Reading Open Cloud-Based Speech Architecture. *Sensors* **2022**, *22*, 2938. https://doi.org/10.3390/s22082938

Academic Editors: Bruce Denby, Tamás Gábor Csapó, Michael Wand and Nikolaos Doulamis

Received: 17 February 2022
Accepted: 8 April 2022
Published: 12 April 2022

1. Introduction

Automatic speech recognition (ASR) uses algorithms implemented in devices such as computers or computer clusters to convert voice signals into a sequence of words or other linguistic entities [1,2]. Previous ASR applications were based on interactive voice response, device control by voice, content-based voice audio search, and robotics. However, ASR technology has improved significantly in recent years owing to the exponential increase in data and processing power, which makes it possible to perform difficult applications. Voice search using mobile devices, voice control in home, and numerous speech-centric information processing applications that benefit from the downstream processing of ASR outputs are some of the examples of advancements in ASR technology [3]. Thus, noise robustness has become an essential core technology for large-scale, real-world applications given that OCSR APIs must exhibit improved functionality because of the significantly demanding acoustic scenarios.

In this study, we propose a noise enhancement system that uses a multimodal interaction approach based on multisensory integration that refers to the interplay of information from several senses. Multisensory integration often influences the perception of human speech. The human nervous system comprises several specialized sensory organs, each of which conveys certain sensory information [4]. The organization of sensory organs in the human body is advantageous considering that each organ serves as a non-redundant source of information, which allows organisms to detect critical sensory events with higher certainty by separately examining the input received from each sense. However, when several sources of information are merged, information from different senses can be linked [5], and this can synergistically influence the capacity to notice, assess, and start reactions to sensory events (Figure 1) [6]. The brain is divided into four lobes; Figure 1 shows the audio (red, temporal lobe) and visual (green, occipital lobe) information in multisensory integration.

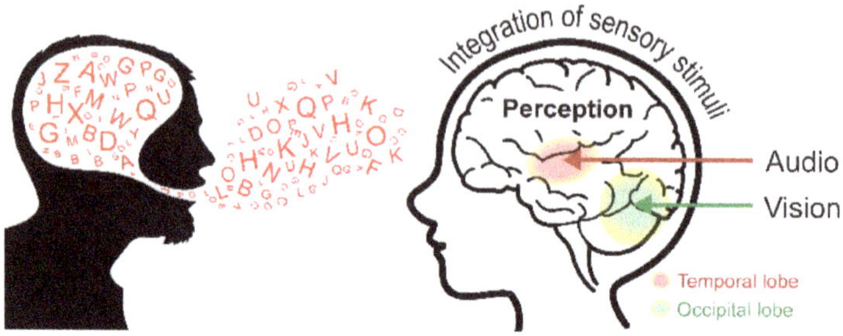

Figure 1. Multisensory integration comprising both auditory and visual information.

Several studies have shown that the contemporaneous observation of visual speech, such as the movement around the speaker's lips, significantly affects speech perception. Visual speech information improves the ability to understand speech in scenarios when words are spoken with an accent, or when the surrounding environment is noisy [7–9]. For example, lip-reading can substantially improve the understanding of speech if the audio signal is unclear [10–12]. The McGurk effect illustrates how mismatched auditory and visual speech information affects speech perception [13]. For example, when we hear the sound "ba" while seeing a person's face express "ga", many people hear "da", a third sound that is a combination of the two. This fusion approach contributes to the robustness of speech detection in a variety of real-world applications, such as human–machine interaction, by overcoming the problems of noise, auditory ambiguity, and visual ambiguity.

To the best of our knowledge, this is the first study that proposes a noise-robust OCSR API system based on an end-to-end lipreading architecture for practical applications in various environments. This system exhibits performance superior to those systems that comprise only audio or visual speech recognition technology. For auditory-based speech recognition, we evaluated the performance of four OCSR APIs (Google, Microsoft, Amazon, and Naver) using Google Speech Commands Dataset v2 and the collected dataset prepared by us to select the best API for our design. The word lists recognized in the highest-performance API were expressed as word vectors using the Google Word2Vec model, which was trained using the dataset of 1,791,232-word sentences. Similarly, we developed a new end-to-end lipreading architecture comprising two end-to-end neural subnetworks for visual-based speech recognition. The feature extraction method consists of the following components: a 3D convolutional neural network (CNN), 3D dense connection CNN for each time step to reduce the number of model parameters and avoid overfitting, and multilayer 3D CNN to capture multichannel information in the temporal dimension of the entire video to overcome insufficient visual information and obtain specific image features. A bi-directional gated recurrent unit (GRU) with two layers, followed by a linear layer, was used in the sequence processing module. Therefore, the integrated values of word vectors obtained from speech API and the integrated values of vectors obtained from the lip-reading model were concatenated to form vector matrix. After introducing a SoftMax layer at each time step with a concatenated vector matrix, the entire network was trained using the connectionist temporal classification (CTC) loss function to obtain probabilities. Furthermore, we compared the proposed system's accuracy and efficiency with those of existing standalone techniques that extract the visual features and several OCSR APIs for the collected datasets. An extensive assessment revealed that the proposed system achieved an excellent performance and efficiency. Thus, we propose a noise-robust open cloud-based speech recognition API system based on an end-to-end lip-reading architecture for practical applications.

The remainder of this paper is organized as follows. Section 2 examines relevant research on OCSR APIs and visual speech recognition VSR systems. Section 3 introduces the architecture of the proposed system. Section 4 presents information on the benchmark datasets, custom collected datasets, audio-visual information processing, augmentation technique, experimental setup, and evaluation. Finally, Section 5 presents the discussion and conclusions.

2. Related Work

Google has published "Google Assistant" (an AI voice recognition assistant) and various speech recognition features for autos and consumer electronics as an open API. The technology, which supports more than 120 languages, includes various features, such as automatic punctuation, speaker distinction, automatic language identification, and enhanced voice adaptability. As part of Watson's AI service package, which currently handles 11 languages, the company offers an ASR service. However, the custom acoustic and language model desired by the user must be initially trained using user data. In other words, to use a personalized acoustic model, the user's own audio must be used, and a new corpus must be added to expand the language model. Further, Microsoft offers a cloud service package known as Azure, and speech-to-text is an API for speech recognition provided by Azure's cognitive services. According to the official website, Azure employs "breakthrough voice technology" based on decades of study. Furthermore, their website alludes to a 2017 publication where Microsoft achieved the first-ever human-level accuracy on the switchboard test [14]. Alexa by Amazon is a voice-activated artificial intelligence (AI) smart personal assistant that includes features such as voice interaction and the ability to ask and answer questions. Further, Alexa can control smart gadgets featured in intelligent home technology. Alexa is easily available on Amazon Echo, Echo Dot, Echo Plus, and other smart speakers. In South Korea, the Naver collaboration aggressively developed speech recognition by launching Clova speech recognition (CSR) on 12 May 2017 [15]. The CSR now supports Korean, English, Japanese, and Chinese languages, although Korean has a better recognition rate.

Deep learning technology has recently demonstrated remarkable performance in a variety of applications, including VSR systems. Deep learning algorithms can achieve higher accuracy compared to older approaches with traditional predictions. For example, when a CNN is used in combination with conventional approaches, the CNN architecture can differentiate different visemes, and temporal information is added after obtaining CNN output using an HMM framework [16,17]. Furthermore, other studies [18,19] have integrated long short-term memory (LSTM) with histograms of oriented gradients (HOGs) and used the GRID dataset to input recognized short words. Similarly, word predictions were generated using an LSTM classifier with a discrete cosine transformation (DCT) trained with the OuluVS and AVLetters datasets [17]. The sequence-to-sequence model (seq2seq) is a deep speech recognition architecture that can read and predict the output of an entire input sequence. For longer sequences, it takes advantage of global information. These studies [20,21] demonstrated the recognition of audio-visual speech in a dataset based on real words using the first seq2seq model that incorporates both audio and visual information. The initial model is LipNet to use the end-to-end model and be trained using sentence-level datasets (GRID corpus) for performance evaluation [22]. The overlapped and unseen-speaker databases had word error rates of 4.8% and 11.4%, respectively, whereas human lip-readers had a success rate of 47.7% for the same database. In [23–25], digit sequence prediction using 18 phonemes and 11 terms, and other similar architectures such as the CTC cascade model were implemented to evaluate the convergence of audio-visual features. Therefore, deep learning techniques can extract more detailed information from experimental data, which demonstrates their high level of resiliency against big data and visual ambiguity.

Several previous studies [26–28] focused on evaluating the recognition performance of existing disclosed OCSR APIs and aimed to apply them to applications such as robots. However, this study focused on improving the system using visual information in the existing OCSR API system and showed a low error rate in a noisy environment; as a result, a high recognition performance was demonstrated. This study presents a model that achieves a low error rate even in noisy environments using deep learning-based audio and visual information compared to the existing OCSR APIs that rely solely on audio information.

3. Architecture of the Proposed Model

Our lipreading system is combined with existing cloud-based speech recognition systems, and the proposed audio-visual speech recognition system is shown in Figure 2. The audio architecture among Microsoft, Google, Amazon, and Naver was compared to select and combine the best performing OCSR API. The vision architecture of the proposed model was combined with the following feature extraction methods: LipNet (used as the baseline method), LeNet-5, Autoencoder, ResNet-50, DenseNet-121, and multi-layer CNN, all of which exhibit exceptional feature performance.

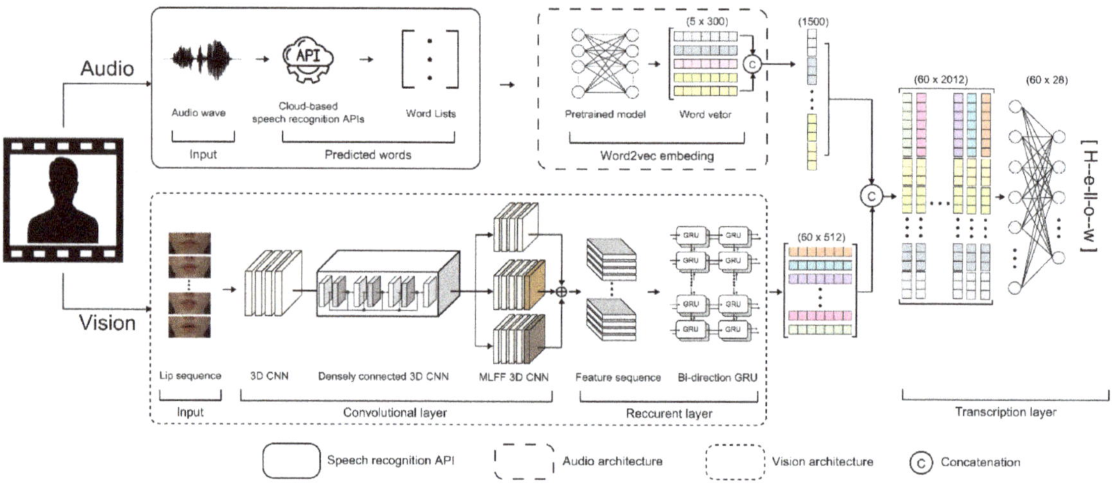

Figure 2. Block diagram of proposed audio-visual speech recognition system.

In all speech recognition engines, the user's voice is transmitted to the recognition system using a microphone (Figure 2). To this end, we used two generic algorithms. The voice was processed on a local device, and the recorded voice was forwarded to a cloud server provided by Google or Microsoft for additional processing. Microsoft Cortana and Google are commercial engines that simultaneously separate speech recognition systems into closed and open-source code systems [29]. Speech recognition in OCSR APIs, which is a type of closed source, allows the rapid and easy construction of application speech recognition systems. Application speech recognition systems can be developed easily using OCSR APIs and are therefore gaining traction in a variety of sectors. Thus, developers of application speech recognition systems need to select suitable OCSR APIs based on the function and performance of the system. Furthermore, the performance of OCSR APIs varies depending on the date of the research and the type of learning data. Words output from the OCSR API are represented by word vectors using Google's pre-trained Google's Word2Vec model. Word2Vec is a set of shallow neural network models developed by Mikolov et al. to build "high-quality distributed vector representations that capture a large number of exact syntactic and semantic word associations" [30,31]. The dimensions of these word vector representations, also known as word embeddings, can be in the hundreds.

To represent a document, word embeddings can be concatenated. Google has provided a Word2Vec model that has been pre-trained on the 100 billion words in the Google News corpus, resulting in 3 million 300-dimension word embeddings for academics. Therefore, five-word list were outputted and converted into 300-dimensional vectors and summed into one single vector (Figure 2).

As mentioned above, we present a deep-learning-based VSR architecture and propose a new feature extraction method (Figure 2). Figure 3 and Supplementary Table S1 shows the detailed hyperparameters of the proposed architecture.

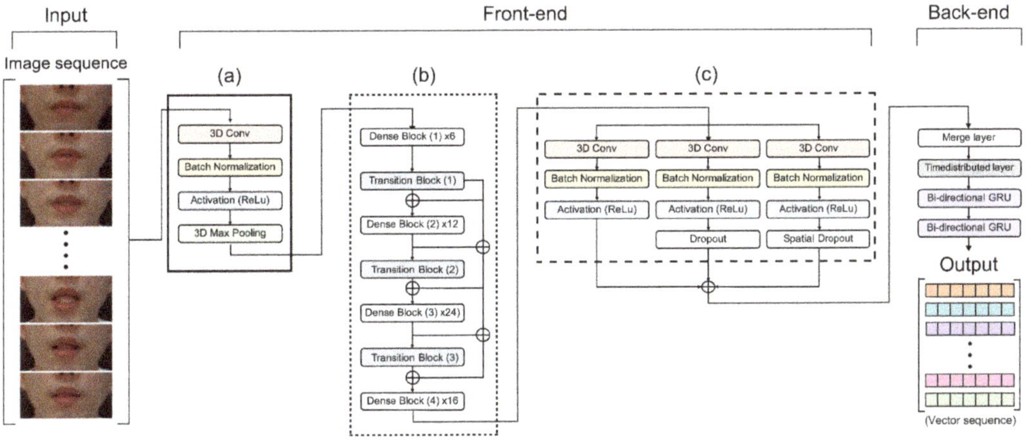

Figure 3. Proposed VSR system architecture. (**a**) 3D CNN; (**b**) 3D dense connection CNN; (**c**) multi-layer 3D CNN.

3.1. Convolutional Layer

CNNs use raw input data directly, which results in the automation of the feature development process. For image recognition, a 2D CNN is used to collect encoded information for a single picture dataset and to convert that information to 2D feature maps for computing features from spatial dimensions. However, the motion information contained in numerous contiguous frames fails when utilizing a 2D CNN for video recognition (Figure 4a). We used a spatial-temporal 3D CNN to calculate spatial and temporal features to capture distinct lip-reading actuations around the lips, such as tongue and teeth movements. When spatial and temporal information from following frames is considered, 3D CNNs have been found to be effective in extracting attributes from video frames in several experiments [16,22] (Figure 4b).

In this experiment, all consecutive frames input to encode the visual information of the lips were transmitted to the CNN layers in 64 3D kernels with a size of $3 \times 7 \times 7$ to obtain feature information, as shown in Figure 3a. We reduced the internal covariate transformation using a batch normalization (BN) layer and ReLU to accelerate the training process. Additionally, a max-pooling 3D layer was added to reduce the spatial scale of the 3D feature maps (Supplementary Figure S1a).

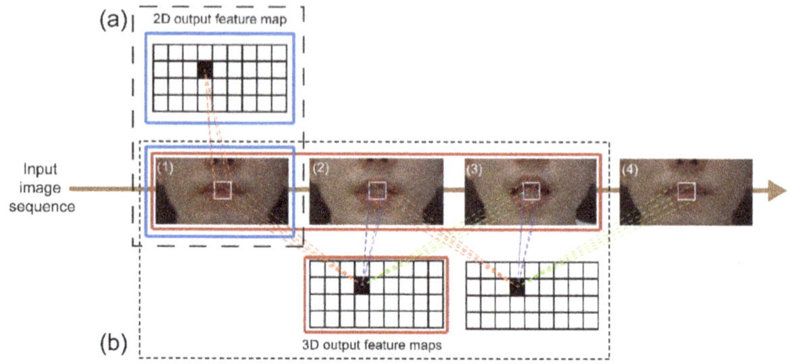

Figure 4. Comparison of convolutions in (**a**) 2D; (**b**) 3D.

A dense connection CNN generates relationships between multiple connected layers, allowing for full feature usage, vanishing gradient, and network depth. The input features are decreased by the bottleneck layer placed prior to the convolution layer. As a result, following the bottleneck layer operation, multichannel feature volumes are fused. Because the previous features are still present, the subsequent layer is only applied to a small number of feature volumes. Transition layers are also incorporated to increase model compactness due to the hyperparameter theta that controls the degree of compression. A decreased growth rate was achieved by using bottleneck and transition layers, resulting in a narrower network, reduced model parameters, efficiently controlled overfitting, and reduced processing resources.

We implemented the 3D dense connection CNN architecture comprising transition layers and dense blocks, as shown in Figure 5. The transition layers (Figure 5a) are connected in the following order: BN layer, ReLU, 3D convolution layer ($3 \times 1 \times 1$), and average pooling 3D layer ($2 \times 2 \times 2$). The dense blocks (Figure 5b) are organized in the following order: BN layer, ReLU, 3D convolution layer ($3 \times 1 \times 1$), BN layer, ReLU, 3D convolution layer, and 3D convolution layers ($3 \times 3 \times 3$).

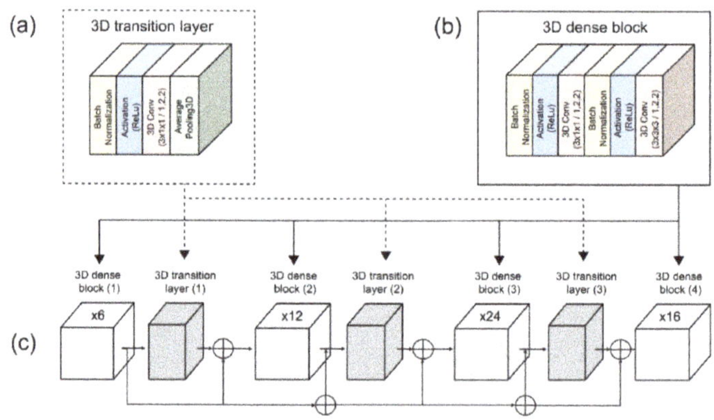

Figure 5. 3D dense connection CNN architecture. (**a**) 3D transition layer structure; (**b**) 3D dense block structure; (**c**) detailed 3D dense connection CNN.

To date, image classification tasks handled with various CNN models have demonstrated exceptional performance. For example, using the fusion of several CNNs for feature aggregation, it is feasible to extract diverse spatial and temporal information by building different scales and depths [32]. In addition, a different convolutional layer can extract different features for the multilayer 3D CNN training phase to obtain more diverse feature information. Furthermore, by using different depths and filters of varying sizes, multiple features may be created from this training process. Certain associated qualities that were lost in the layered design can be chosen using this strategy, resulting in a richer final feature. The suggested multilayer 3D CNN architecture is shown in Figure 3c. The first module follows the 3D dense connection convolution layer output feature in the order of a 3D convolution layer (64 3D kernels of size $3 \times 5 \times 5$) and then a BN-ReLU layer (Figure 6a). The second module (Figure 6b) includes a dropout layer to prevent overtraining and overfitting that improves and generalizes the CNN's performance by preventing strongly correlated activations. This is important because of the small size of the benchmark dataset compared to other image datasets [24]. The structure of the third module drops the entire feature map by adding a spatial dropout layer to the structure of the first module (Figure 6c). Unlike the traditional dropout method that removes pixels at random, this method uses CNN models with significant spatial correlation to provide superior picture categorization [33]. As a result, we used a spatial dropout layer to efficiently extract the shape of the lips, teeth, and tongue, which are fine movements around the mouth, with a significant spatial correlation.

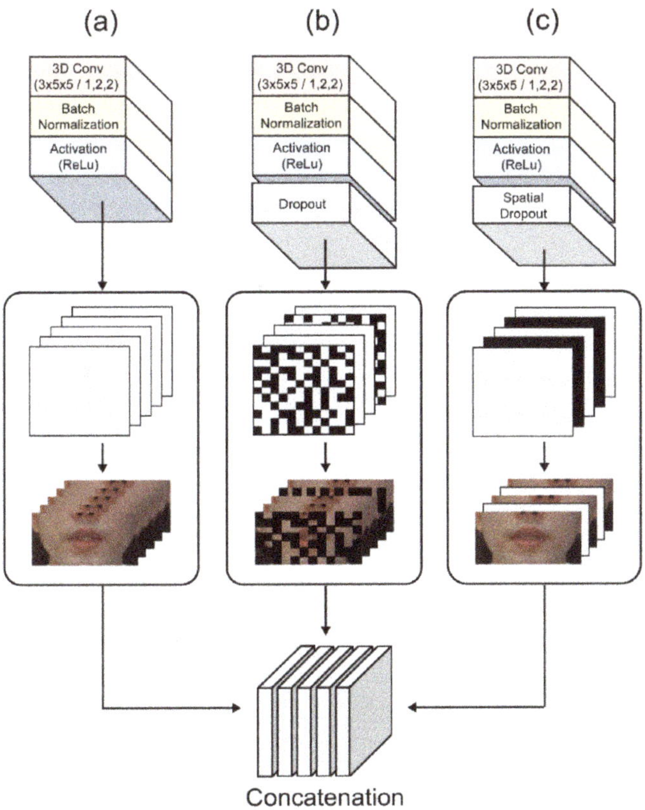

Figure 6. Multilayer 3D CNN architectures: (**a**) first architecture; (**b**) second design with the dropout layer; (**c**) third architecture with the spatial dropout layer.

3.2. *Structure of Comparative Feature Extraction Methods*

We compared the proposed method to other feature extraction methods, such as LipNet, LeNet-5, CNN Autoencoder, and ResNet-50, which all exhibit an outstanding feature extraction performance. The feature extraction method of LipNet as a baseline comprises 3 × (spatiotemporal convolutions, channel-wise dropout, and spatial max-pooling) [22]. LeNet-5 is the earliest model of deep learning and uses a gradient-based CNN structure for handwritten digital recognition [34]. The input layer of a typical LeNet-5 structure diagram is a handwritten digital image of 0 with a size of 32 × 32, whereas the output layer comprises 10 nodes corresponding to 0. LeNet-5 comprises six layers in total, namely the input and output layers: three convolutional layers, two pooling levels, and one fully connected layer. Convolutional core sizes in the convolutional and pooling layers were set to 5 × 5 and 2 × 2, respectively. However, the training parameters were reduced when the connection layer decreased the number of neurons from 120 to 84. Thus, an unsupervised model that learns to rebuild the input is used as a typical autoencoder [35].

In several domains, such as speech recognition and computer vision, deep learning models can learn intricate hierarchical nonlinear features that can provide superior representations of original data [36]. Encoder, hidden, and decoder layers comprise the autoencoder. The hidden layer's input is the encoder layer's output, and the decoder layer's input is the encoder layer's output. We created an autoencoder model using LipNet's feature extraction method for experimental comparison. ResNet-50, a convolutional neural network with 50 layers, is a ResNet [37] version comprising 48 convolution layers, a MaxPool layer, and an average pool layer. The deep residual learning architecture lies at the heart of ResNet. ResNet-50 is substantially smaller than other current designs, with 50 layers and over 23 million trainable parameters; extremely deep neural networks can be employed to circumvent the vanishing gradient problem.

3.3. Recurrent Layer

The GRU is one of the recurrent neural networks and is a method of governing and propagating information flow across many time stages [25]. GRUs are derived from LSTM units that determine what information should be carried forward and which should be disregarded. Given that the 3D CNN only captures brief viseme-level data, it may be able to comprehend wider temporal contexts, which is beneficial for ambiguity detection. Because the GRU uses update and reset gates, the gradient vanishing problem can also be overcome. A bi-directional GRU is used as a sequence processing module in the proposed architecture. Compared to typical GRU deployment, a bi-directional GRU provides information in both forward and backward directions to two distinct neural network topologies coupled to the same output layer, allowing both networks to gain full knowledge of the input.

3.4. Transcription Layer

We used the CTC method, which employs a loss function to parameterize the distribution of a label token sequence without requiring the alignment of the input sequence to an end-to-end deep neural network. CTC is conditionally independent of the marginal distributions established at each time step of the temporal module as it restricts the usage of autoregressive connections to manage the inter-time-step dependencies of the label sequence. Therefore, CTC models are decoded using a beam search procedure to restore label temporal dependence, and the language model's probabilities are mixed.

4. Experiment

4.1. Dataset

We utilized Google Speech Command Dataset v2 and gathered a dataset to analyze the performances of the five OCSR APIs and the proposed model (Supplementary Figure S1a) [38]. Google Speech Command Dataset v2 was released in April 2018, and it contained 105,829,35 word utterances in one second or less. Several experiments have been conducted using this dataset to evaluate the performance of speech recognizers [39–41]. In addition, we employed the most frequently used speech recognition command in IoT or real life from the Word Choice part of Google Voice Command Data Set v2 to assess the proposed model [38]. In real-life applications, because an important unit of speech recognition is not the entire sentence but words or short phrases, we selected words useful as commands in automobiles, robot applications, etc. Data were gathered by enlisting people acquainted with speech recognition technology. A total of

40 people (20 males and 20 females with a mean age of 29.14 years) participated, and they were compensated with a $20 voucher. The participants were provided pin microphones to wear, as shown in Supplementary Table S2a. The participants stood at a distance of 1.2 m from the front camera and lighting, and each participant received a list of 20 keywords. The participants repeatedly uttered the same word 100 times at a rate of 2 s per keyword for 2 h; we stored the audio and video information and generated a total of 80,000 videos (Supplementary Table S2b).

Using a Dlib face detector, the targeted area used as the input for the end-to-end lip-reading was detected in the data pre-processing phase by employing a HoG-feature-based linear classifier [23]. The output was subsequently supplied in the form of (x, y) diagonal edge coordinates, which were then used to construct the bounding box around the mouth. Then, using a facial landmark predictor to detect movement around the lips and extract points of lips identical to those obtained from the training dataset, 68 land-marks and online Kalman filters were used in the iBug program [42]. Using an affine transformation, a mouth-centered region with dimensions of 100×50 pixels per frame was extracted, and the RGB channels throughout the full training set were normalized to provide zero mean and unit variance. To avoid overfitting, we adopted the data augmentation approach from [22] for the training data. In the training procedure, regular and horizontally mirrored picture sequences were employed. We used individual words as additional training cases with a decay rate of 0.925, considering that the dataset contained the starting and ending terms serving as timers for each "clip" sample to enhance training data at the word level. With a probability of 0.05/frame to eliminate variation, we identified the movement speed and duplicate frames. The same dataset pre-processing and augmentation approaches were used to train and assess all models.

4.2. Implementation

We used custom Python scripts to build the OCSR API methods (Python3.6; Rossum, 2019). These scripts were used as wrappers for loading and submitting audio recordings to OCSR API providers and for saving the resulting transcripts. All four providers (Microsoft Azure, Naver Clova, AWS, and Google Cloud) adopted OCSR APIs. While an entire chapter could be transcribed in one instance using Microsoft Azure, Naver Clova, and AWS, only 60 s of the audio per file could be transcribed using Google Cloud. Before analyzing the text, all capitalization and punctuation were removed, and all numbers were converted to text. Our metric for assessing the OCSR API performance was the percentage of correctly transcribed phrases for each recording. In addition, to compare the predicted words to the actual words, the recognition performance was evaluated. If the two words were found to be the same, the recognized word would be true; otherwise, it would be false. The five-word lists following the OSCR API were converted into vectors using the pre-trained Word2Vec model and concatenated into one single vector.

All models of the end-to-end lip-reading architecture were constructed using Keras with a TensorFlow backend and TensorFlow-CTC decoder to evaluate the character accuracy rate (CAR) using the CTC beam search. The complete configuration and parameters utilized for the layers in our proposed architecture are shown in Figure 3 and listed in Supplementary Table S1. All model network parameters were initialized through He initialization, with the exception of square GRU matrices with orthogonal initialization and default hyperparameters. The proposed lip-reading model was trained in the multilayer 3D CNN using channel-wise dropped pixels and the dropped channel utilizing spatial dropout, where the proposed lip-reading architecture was the baseline that was trained on the collected dataset until overfitting. Therefore, our combined proposed system learned using the Adam Optimizer [43] with a learning rate of 0.0001 and mini batches of size 8 by combining the audio and visual vector. The combined proposed system was evaluated in an environment different from that of the data collection, while considering factors affecting accuracy, such as illumination. For the evaluation phase, a person stood 1.5 m away from the KIOSK screen, and a webcam was installed above the screen (Supplementary Figure S1b). The room had normal lightning conditions and controlled noise levels. The evaluation process was divided into three categories to test whether the different components of the system not based on the participants contributed to model training. In the audio-only (A) and visual-only (V) categories, only auditory or visual information was used for recognition, whereas both auditory and visual information were used simultaneously for recognition (multimodal information) in the AV category. None of the participants participated in the data collection phase.

4.3. Performance Evaluation Metrics

We employed standard automated voice recognition assessment measures to evaluate the proposed deep-learning model. The learning loss of all models was examined to assess their learning status during the training process. In addition, we evaluated the parameters, training time, and character accuracy rate (CAR) for each model to compare their performances and computational efficiencies. The total edit distance was calculated to convert the error rate measurements, which is used to evaluate accuracy, into percentages. It was important to compare the decoded and original texts for analyzing misclassifications. The CAR percentage is given by

$$\mathrm{CAR}(\%) = 100 - \left(\frac{C_S + C_D + C_I}{C_N}\right) \times 100, \tag{1}$$

where N, S, I, and D represent the total number of characters in the ground truth, number of characters substituted for incorrect classifications, number of characters inserted for non-picked characters, and number of deletions that should not be present for decoded characters, respectively. As a result, CAR is computed using (N), with C denoting the words. We generated an approximate maximum-probability prediction for all experimental models with the CTC beam search using a Tensor-Flow-CTC decoder. We also analyzed the CAR over the training period in terms of the number of parameters and computational efficiency. To visualize the data, we employed the phoneme-to-viseme mapping approach reported in [44]. The signal-to-noise ratio (SNR) was evaluated by synthesizing noise into the acquired source data. The diverse environments multichannel acoustic noise database (DEMAND) was collected using 16-channel array microphones [45]. For ambient noise, we used different types of noises divided into eight environments (park, hallway, cafeteria, station, café, square, car, and living), and we synthesized noise to evaluate the SNR (Supplementary Table S3). In each sample, the target speech was mixed with eight noises.

5. Results

5.1. Performance of OCSR APIs

Figure 7 shows the mean and distribution of the individual and overall performances for all 35 words of the Google Speech Commands Dataset V2 using 5 speech recognizer APIs. Each word has a different number of datasets, and the results for the individual performance are listed in Supplementary Table S4. The performance evaluation was considered to be correct if the word was the same as the prediction result of the recognized data. The Google API shows a low performance for certain words (e.g., forward, off, and up) and a wide variance. Naver shows a low overall performance and has a wide distribution, as shown in Figure 7b. Amazon shows better performance than the other two APIs; however, it demonstrates lower recognition rates for certain words (e.g., off, tree, and up) as well as the Google API. However, among all APIs, Microsoft shows the highest performance for all words, with excellent average and dense distribution results. Thus, Microsoft Azure was selected as the main API.

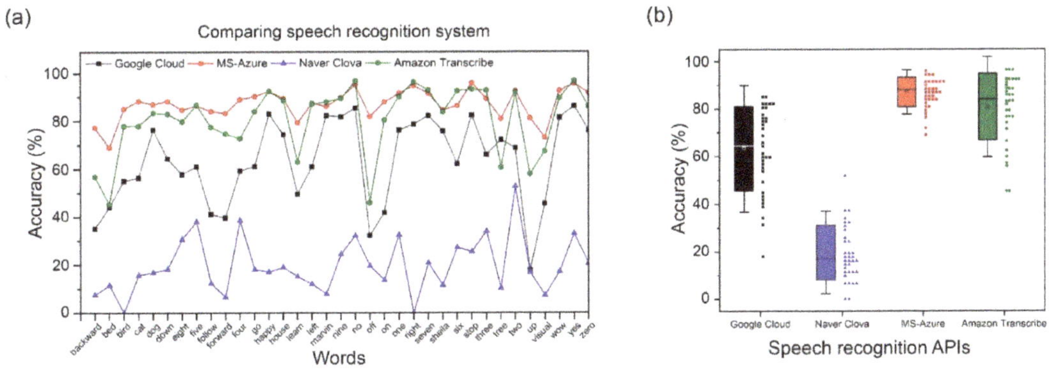

Figure 7. (**a**) Comparison of the performance of speech recognition APIs for 35 individual words in Google dataset V2; (**b**) Comparison of the mean and distribution of speech recognition APIs for Google dataset V2. Error bars represent standard deviation. The squares on the left represent the mean and distribution, and the 35 small structures on the right represent the respective accuracies for 35 words of Google dataset V2.

5.2. Training Procedure and Learning Loss

Figure 8 shows the training and validation losses that occurred when training the collected dataset with the audio and visual information. For audio, words output from the OCSR API are represented by word vectors using the pre-trained Google's Word2Vec model. For visual, the seven models have different visual feature extraction modules at the front end and the same sequence processing modules at the back end (Table 1). Model A contained the same architecture as LipNet, the baseline model, whereas Models B, C, D, E, and F used the feature extraction methods of LeNet-5, Autoencoder, ResNet-50, DenseNet-121, and Multilayer CNN, respectively (Section 3.2). The training and validation losses of all seven models were in good agreement. However, the gap between the training and validation losses was the largest in Model C, and its overfitting phenomenon was higher compared to those of other models. Further, while Model F showed lower overfitting results (smallest among all models), it exhibited a lower convergence speed rate than those of Models A, B, D, and E. Therefore, the learning and convergence speeds of Model G (proposed model) were high, and the gap was small. These findings show that the suggested model for the gathered dataset had the smallest difference between training and validation losses, hence preventing overfitting.

Table 1. Performance, number of parameters, and training time of proposed model compared to those of the baseline and other models.

Model	Audio	Vision		Parameters	Average Epoch Time (s)	CAR (%)
		Front-End	Back-End			
A		3D CNN		4,571,388	914.74	89.654
B		3D CNN + 3D LeNet-5		3,651,038	847.18	88.948
C		3D CNN + Autoencoder		10,576,269	1242.89	89.189
D	Google's Pertained model	3D CNN + 3D ResNet-50	Bi-direction GRU + CTC	66,705,692	1957.81	92.365
E		3D CNN + 3D DenseNet-121		2,247,537	798.77	91.237
F		3D CNN + Multilayer 3D CNN		44,327,404	1372.30	92.459
G		Proposed architecture		3,455,857	849.44	95.893

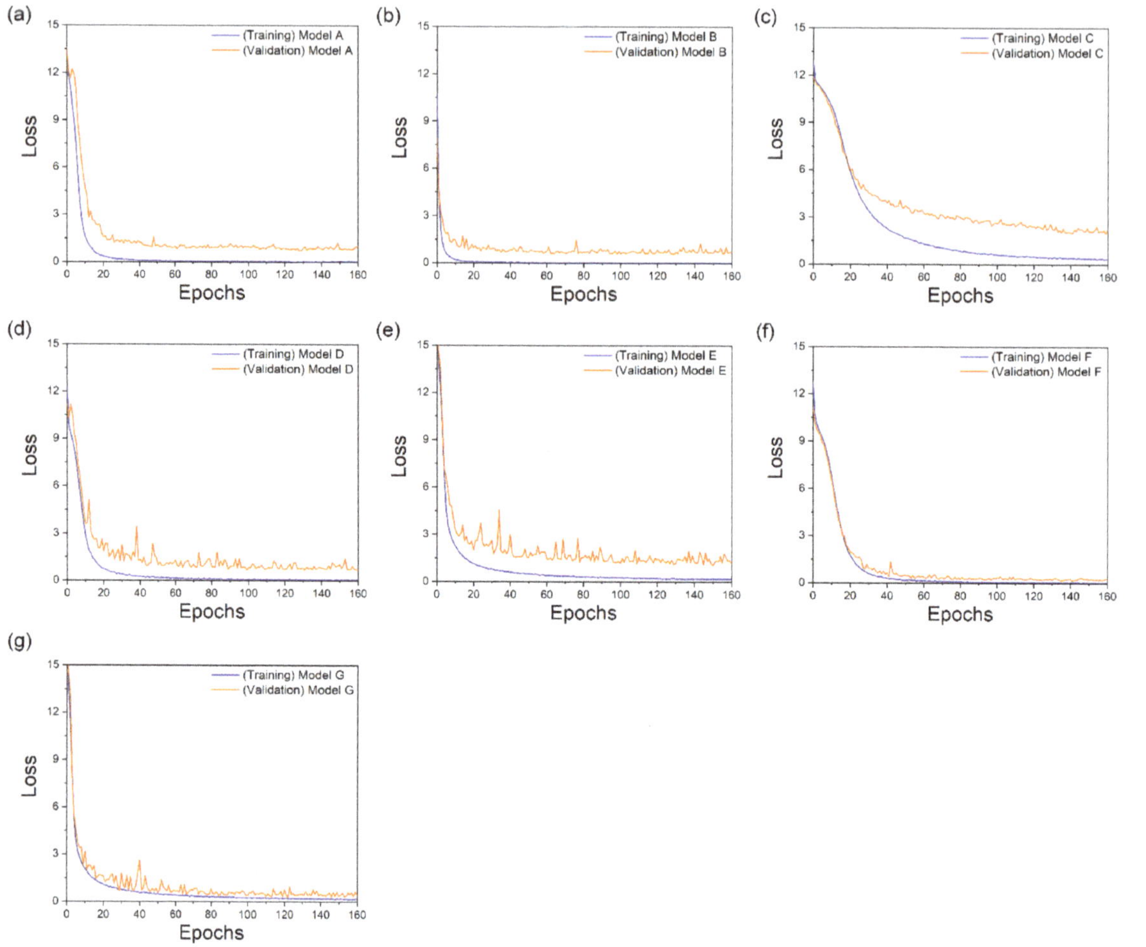

Figure 8. Training and validation loss of the collected dataset. Models (**a**) A; (**b**) B; (**c**) C; (**d**) D; (**e**) E; (**f**) F; and (**g**) G.

5.3. Characteristic Accuracy Rate

The results of the comparison between the proposed model and current deep-learning models are listed in Table 1. The suggested model obtained the best results, with a CAR of 95.893%, which was higher than those of the other models and baseline values in all cases. Despite the increase in accuracy of Models (C), (E), (D), and (F) over the baseline, no significant differences were detected (Table 1). Therefore, the proposed model outperformed the existing models, including the baseline model, in terms of accuracy; this could be owing to the combined use of various 3D CNN architectures. Figure 9 compares the proposed model to all other models after training with CAR on the obtained dataset. In addition, a dataset available from [46] was used to assess the performance of the suggested model. The performance of the suggested model remained unchanged, proving its superiority for both open and collected datasets.

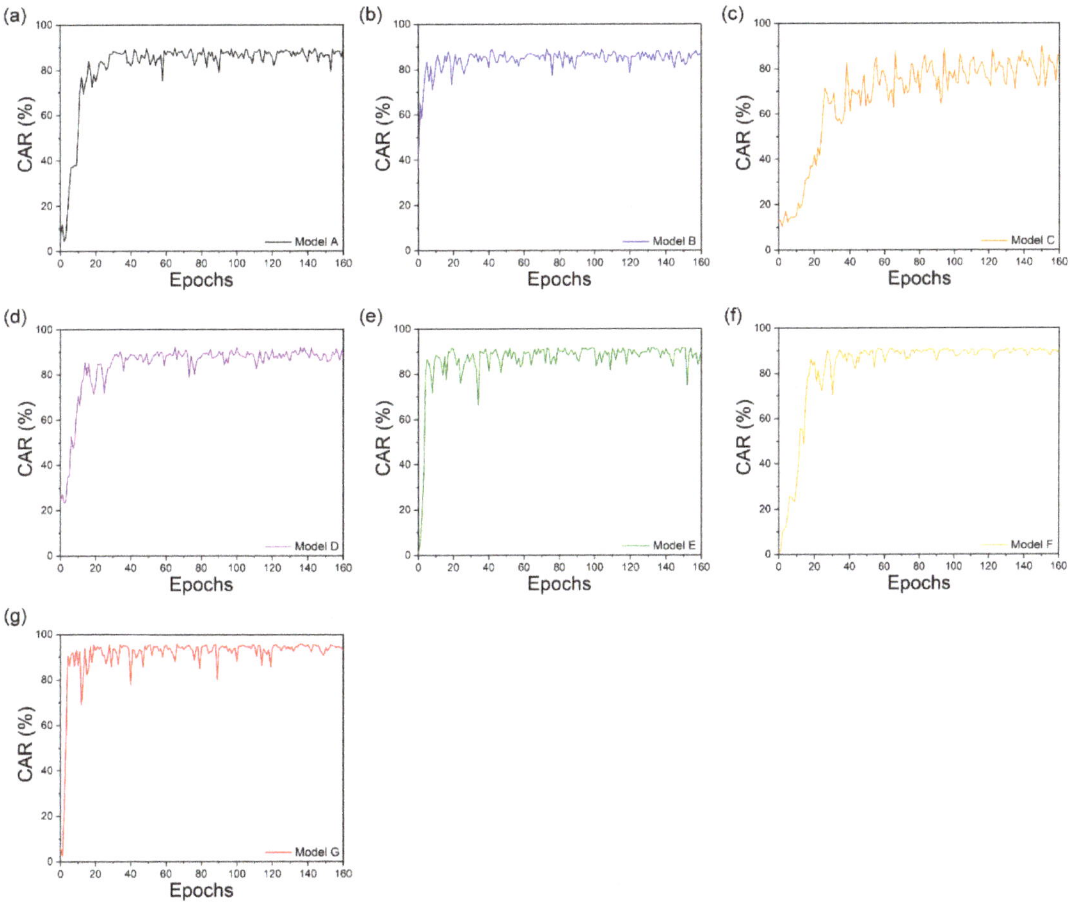

Figure 9. Training steps for CAR comparing the proposed model toother models: Models (**a**) A; (**b**) B; (**c**) C; (**d**) D; (**e**) E; (**f**) F; and (**g**) G.

5.4. Model and Computational Efficiency

The model size and computational efficiency of the proposed systems are the main limitations of real-time applications. We examined the accuracy and computational efficiency of the models using varying numbers of trained parameters and training times (Figure 10). Figure 10a shows the performance according to the number of parameters; Figure 10b shows the results of the average training time comparison of the seven models for 160 epochs. Although the proposed model is similar to the baseline model (approximately 50 s), it showed a high CAR while using approximately 11.2 M fewer parameters compared to the other six models that used the collected dataset listed in Table 1. Compared with the baseline model performance, we were able to improve the accuracy while lowering the number of training parameters by approximately 11.2 M and achieving a comparable training time when using our database.

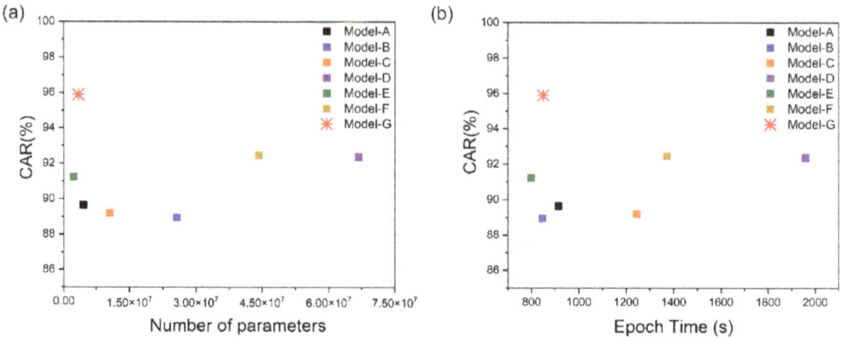

Figure 10. CAR of the baseline and other models according to the (**a**) number of parameters; (**b**) average epoch time.

5.5. Confusion Matrix

Visual analysis was conducted using IBM ViaVoice database mapping [44]. A confusion matrix was created for the most confusing phoneme in the bilabial viseme class (Figure 11), which included lip-rounding-based vowels, intra-visemes, and bilabial visemes. The experimental results showed that {/AA/, /AO/} is frequently misclassified during the text decoding process (Figure 11a). To produce the vowel sound /AA/, as in "bat", the mid-back portion of the tongue must be raised, followed by the front and back portions of the tongue stretching in opposite directions. For producing the sound /AO/, as in "orange" and "port", the tongue and mouth become tighter than that when making the /AA/ sound. The recognized experiment results showed that misclassification generated from "on" and "off" with the shortest duration time with a small mouth opening and incorrect recognition results were obtained from the "er" portion of "center" and "under". The intra-viseme categorical confusion matrix is shown in Figure 11b. As illustrated in the experimental results for [p], [b], and [m] in Figure 11c, distinguishing homophones was difficult. Based on our assessment of the proposed model considering different perspectives, this model can assist in overcoming technological impediments to practical implementation.

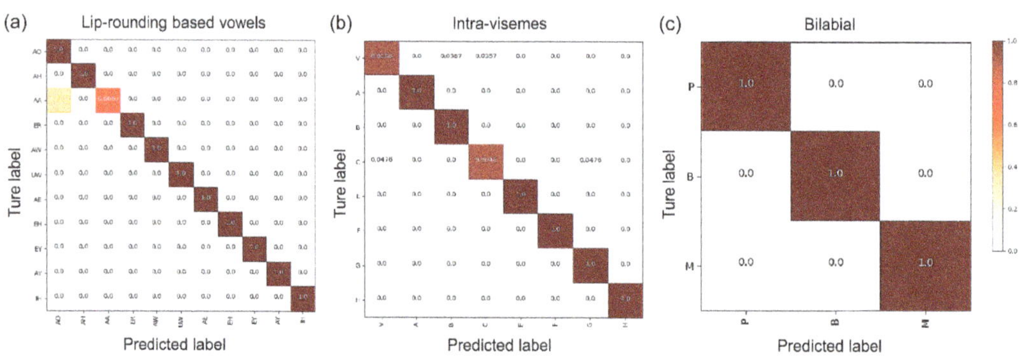

Figure 11. Detailed proposed architecture confusion matrices for the (**a**) lip-rounding based vowels; (**b**) intra-visemes; (**c**) bilabial groups. The three groups with the greatest confusions and confusions inside viseme clusters were selected.

5.6. Performance of Combined System

Considering the case of park noise in Supplementary Table S5a, we described the WAR for each recognizer (e.g., A, V, and AV) at different SNR levels. For an auditory-only (A) recognizer, the highest word accuracy rate was 78.28% ± 4.21%, which was obtained at an SNR of 35 dB, as shown in Figure 12a. The WAR of visual-only (V) for various SNR values were calculated as 74.54% ± 1.96%. The V recognition cases relied only on visual information and remained unaffected by the clearance of the audio signal or SNR level. Participants in the A, V, and AV recognition cases did not participate in the data collection phase. The process of evaluating A, V, and AV recognition using the KIOSK for performance was different from the data collection environment because the experiment was conducted naturally without limiting variables such as the height of the speaker against the fixed cameras, the shape of the mouth during pronunciation, and lighting. Therefore, different factors could contribute to the difference in the WAR between learning outcomes and evaluation in some real-world applications.

The AV recognition cases, in which both auditory and visual signals were used simultaneously for recognition, exhibited WAR scores of 90.94% ± 1.62%, which showed a 12.66% improvement in the recognition rate compared to A recognition. Therefore, combining sound and images can effectively infer spoken words in the presence of ambient noise.

Based on previous results, we compared the performance of multimodal recognition cases (AV) and single-modal recognition (A) while changing the noise environment. We switched the applied noise environment to a hallway-like noise environment, as listed in Supplementary Table S5b. Thus, the WAR of AV was calculated as 92.09% ± 1.18% with enhancement rates of up to 4.44% compared to the WAR of A.

Supplementary Table S5c,d show the results obtained when the noise environment is switched to the cafeteria and station, respectively. For the cafeteria-like noise environment, the (A) recognition cases scored the lowest value with a WAR of 74.53% ± 5.14%. However, the recognition rate of the AV was 89.76% ± 1.96%, with a 15.23% improvement compared to case A (Figure 12c). In the station noise environment, the AV recognition rate was 93.14% ± 1.51% with an improvement of 3.77% compared to that of A (89.37% ± 2.61%), as shown in Figure 12d.

The street category comprised two different environments: a café and square. The terrace of the café in a public square was considered, and the square was a public town square with many tourists, as listed in Supplementary Table S5e,f, respectively. For the café noise environment, the recognition rates for A and AV were 90.12% ± 3.21%, and 92.78% ± 1.35%, respectively, with an improvement in recognition accuracy up to 2.66% for AV (Figure 12e). For the square noise environment, the recognition rates of A and AV were 89.96% ± 3.18%, and 92.93% ± 1.73%, respectively, with an improvement in recognition accuracy of up to 2.97% for AV.

The environment in which audio-only recognition (A) had the highest recognition rate was cars (Figure 12g and Supplementary Table S5g). The recognition rates for A and AV were 93.68 ± 2.03%, and 95.01% ± 1.18%, respectively. The recognition accuracy for AV was up to 1.33%.

For the living room noise listed in Supplementary Table S5h, the recognition rates of A and AV were 77.71% ± 3.94% and 92.13% ± 1.35%, respectively. For AV, the recognition rates were improved by 14.42% (Figure 12h).

Figure 13 shows the statistical significance of the t-test between each group. In general, AV showed an average improvement in the recognition rate of 11.05% and 7.19%, respectively, compared to A. Supplementary Table S6 lists a significant difference of 20.48% between the highest (95.01%) and lowest (74.53%) recognition rates across all eight environments. However, the difference between the maximum and minimum was significantly reduced from 19.15% (A (difference of cafeteria and car)) to 5.251.93% (AV, (cafeteria and car)) by combining multimodal inputs (audio and visual) to recognize spoken words. Thus, unlike A, which showed a good recognition rate in a specific environment (e.g., car noise environment), AV exhibited a superior recognition rate in multiple noise environments (eight environments).

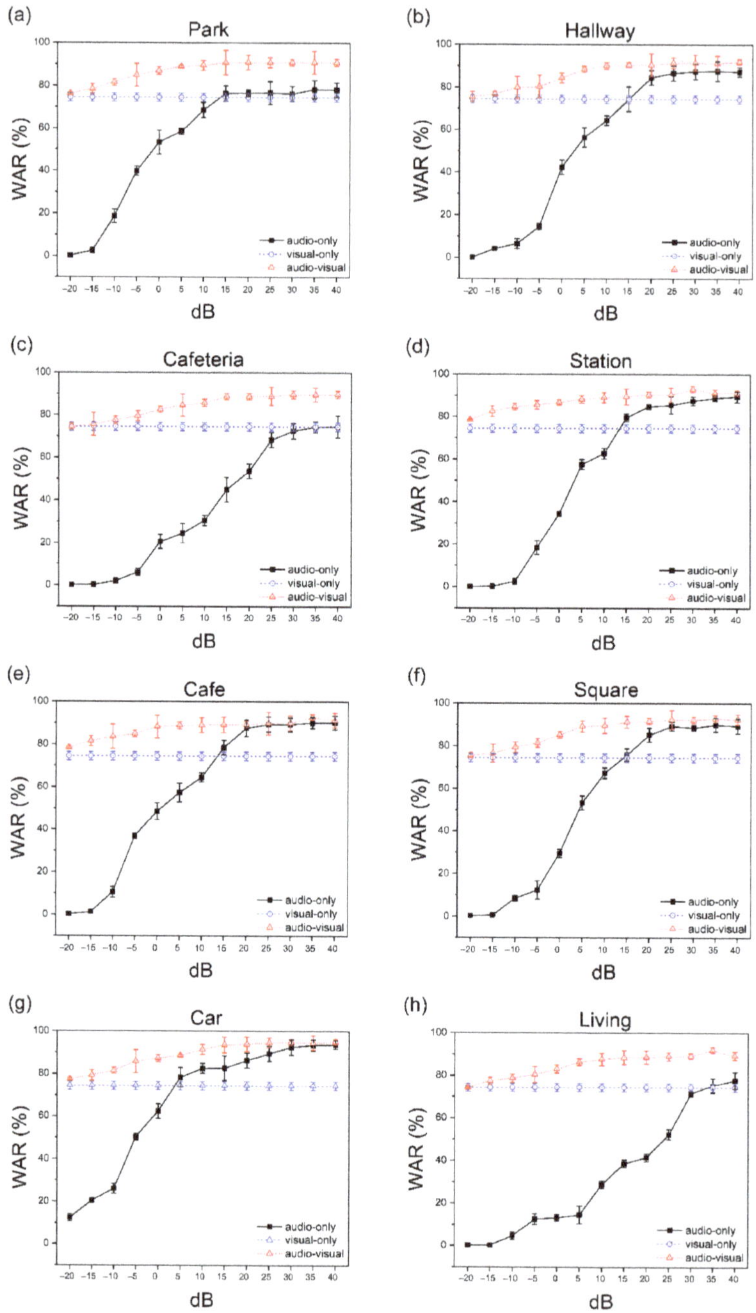

Figure 12. Average recognition accuracy rate with a standard deviation under eight noise environments. Error bars represents standard deviation. The black (single-modal recognition) line represents the audio recognition result, the blue (single-modal recognition) line represents the visual recognition result, and the red (multimodal recognition) line represents the audio-visual recognition result.

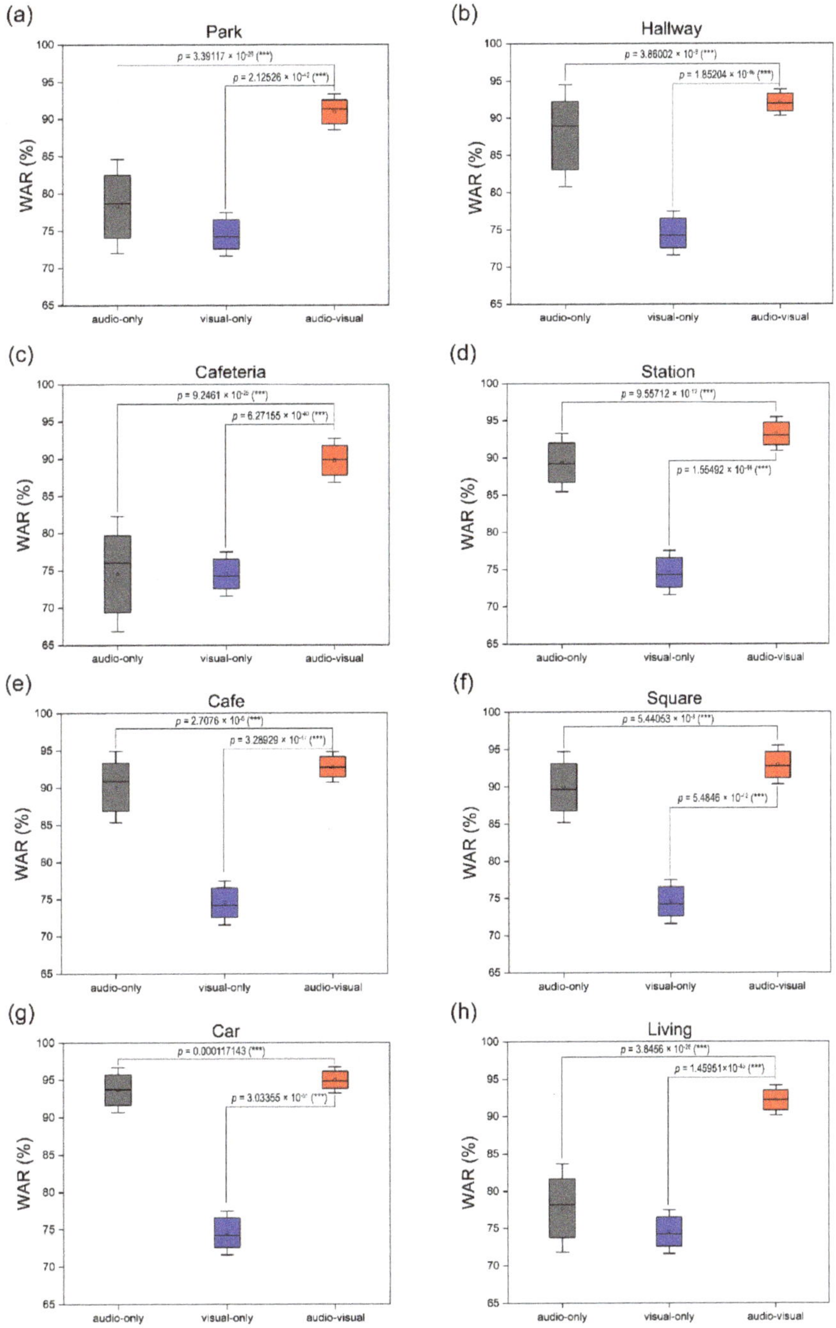

Figure 13. Best average recognition accuracy rates under eight noise environments. Error bars represent standard deviation. Asterisks represent statistical significance-based *t*-tests between each group (* for $p < 0.05$, ** for $p < 0.01$, *** for $p < 0.001$). The gray (single-modal recognition) line represents the audio recognition result, the blue (single-modal recognition) line represents the visual recognition result, and the red (multimodal recognition) line represents the audio-visual recognition result.

6. Discussion

In recent years, the ASR technology has significantly improved because of the exponential increase in large data and processing power, which has made it possible to create complex applications such as voice search and interactions with mobile devices, voice control in home entertainment systems, and various speech-centric information processing applications that benefit from the downstream processing of ASR outputs [3]. Considering that OCSR APIs must function appropriately in demanding acoustic scenarios compared to those in the past, noise robustness has become an essential core technology for large-scale, real-world applications.

This study proposes noise-robust OCSR APIs based on an end-to-end lip-reading architecture for practical applications in various environments. We compared the performance of five OCSR APIs with excellent performance ability. Among all OCSR APIs, Microsoft's API achieved the best performance on the Google Speech Command Dataset V2. Further, we evaluated the performance of several deep-learning models that analyzed visual information to predict keyword sequences. The results show that the proposed architecture achieves the best performance. Moreover, the proposed system requires fewer parameters and provides faster training times than those of the existing models. Compared to the baseline model, the proposed model decreased the number of parameters by 11.2 M and increased the accuracy by 6.239%.

We measured the SNR of the combined proposed system by synthesizing eight noise data and OCSR API outputs to compare the performance for various noise environments. Audio-based speech recognition systems, which showed excellent performance in only specific environment such as a car, demonstrated stable and excellent performance in all environments using visual information. Supplementary Table S6 shows the highest word accuracy and standard deviation values for each of the eight environments. The lowest and highest recognition rates of audio-based speech recognition were calculated as 74.53% and 95.01%, respectively, with a difference of 19.15%, which indicates a significant performance difference based on the specific environment. However, the difference between the two performances was reduced from 19.15% to 5.25% by adding visual information using multimodal interaction methods, and the same performance was achieved in several environments. To solve problems based on the type of environment, two sets of experiments (A, V, and AV) were conducted; stable performances were observed in all environments. The proposed system showed consistent performance in various environments compared to the performance of conventional audio-based speech recognition, which showed excellent performance in only specific environments.

7. Conclusions

We demonstrated a speech recognition system robust to noise using multimodal interaction based on visual information. Our system consisted of an architecture that combines audio and visual information, and its performance was evaluated under eight noise environments. Unlike conventional speech recognition, which shows high performance only in specific environments, we showed the same stable high performance in various noise environments, and simultaneously showed that visual information contributed to improving speech recognition. Therefore, our method showed a stable and high performance in various noise environments by combining lip-reading, a technology that can enhance the speech recognition system, with existing cloud-based speech recognition systems. This system has potential in various applications, such as IoT and robot applications, that use speech recognition in noise and can be useful in various real-life applications where speech recognition is frequently used, particularly indoors, including hallways, cars, and stations, and outdoors, such as parks, cafés, and squares.

Future Work

Multimodal interactions based on visual information must be used to produce noise-resistant ASR. The proposed system may be helpful for patients who have difficulty in conversation owing to problems with speech recognition in noisy environments. However, applying this technology to conversation recognition is problematic. Therefore, we will seek to expand the system's capabilities to identify phrases rather than individual words in the future. We will also examine the performance of the suggested system in real-world settings involving humans and machines. Despite the intense effort invested into the development of an accurate speech recognition system, the development of a lightweight system that is robust with respect to real-life circumstances while accounting for all uncertainties is still challenging.

Supplementary Materials: The following supporting information can be downloaded at: https://www.mdpi.com/article/10.3390/s22082938/s1, Figure S1. (a) Data recording environment for audiovisual data and (b) evaluation of auditory-visual speech recognition system; Table S1. Hyperparameters of the proposed CNN architecture; Table S2. (a) Google Speech Commands Dataset v2 and (b) Collected Speech Commands Dataset; Table S3. Noise database structure: Categories and recordings conducted in each category; Table S4. Performance evaluation of speech recognition systems on Google Speech Commands Dataset V2; Table S5. Average word accuracy and standard deviation of proposed system in eight environments; Table S6. Best word accuracy and standard deviation of proposed system in eight environments.

Author Contributions: Conceptualization, S.J.; methodology, S.J.; software, S.J.; validation, S.J.; formal analysis, S.J.; investigation, S.J.; resources, S.J.; data curation, S.J.; writing—original draft preparation, S.J.; writing—review and editing, M.S.K.; visualization, S.J.; supervision, M.S.K.; project administration, S.J.; funding acquisition, M.S.K. All authors have read and agreed to the published version of the manuscript.

Funding: This research was funded by National Research Foundation of Korea (NRF) grant funded by the Korea government (MSIT), grant number NRF-2018X1A3A1069795.

Institutional Review Board Statement: As a reason for indemnification, in accordance with Article 13 of the Enforcement Rules of "Research on Human Subjects", "A study that does not collect and record personal identification information, and one or more of the following studies are exempt from deliberation, and in detail, it is a study that uses only simple contact measurement equipment or observation equipment that does not cause physical changes". Only human voice and lip information were collected and used; thus, there is no personal identification information and only observation equipment is used. Therefore, there is no IRB registration number owing to the indemnification reasons.

Informed Consent Statement: Informed consent was obtained from all subjects involved in the study.

Data Availability Statement: The data presented in this study are available on request from the corresponding author. The data are not publicly available due to ethical restrictions. The database used in this article is Google Speech Commands Dataset V2. For details, please refer to [38].

Acknowledgments: We would like to thank Ahmed Elsharkawy and Dongwoo Koo for their help with data collection, analytic review, editing, and experimentation.

Conflicts of Interest: The authors declare no conflict of interest. The funder had no role in the design of the study; in the collection, analyses, or interpretation of data; in the writing of the manuscript, or in the decision to publish the results.

References

1. Huang, X.; Acero, A.; Hon, H. *Spoken Language Processing*; Prentice Hall: Hoboken, NJ, USA, 2001.
2. Deng, L.; O'Shaughnessy, D. *Speech Processing: A Dynamic and Optimization-Oriented Approach*; CRC Press: London, UK, 2003.
3. He, X.; Deng, L. Speech-Centric Information Processing: An Optimization-Oriented Approach. *Proc. IEEE* **2013**, *101*, 1116–1135. [CrossRef]
4. Venezia, J.; Matchin, W.; Hickok, G. Multisensory Integration and Audiovisual Speech Perception. *Brain Mapp. Encycl. Ref.* **2015**, *2*, 565–572.
5. Campbell, R. The Processing of Audio-Visual Speech: Empirical and Neural Bases. *Philos. Trans. R. Soc. Lond. B Biol. Sci.* **2008**, *363*, 1001–1010. [CrossRef] [PubMed]
6. Calvert, G.; Spence, C.; Stein, B.E. *The Handbook of Multisensory Processes*; MIT Press: London, UK, 2004.
7. Sumby, W.H.; Pollack, I. Visual Contribution to Speech Intelligibility in Noise. *J. Acoust. Soc. Am.* **1954**, *26*, 212–215. [CrossRef]
8. Dodd, B. The Role of Vision in the Perception of Speech. *Perception* **1977**, *6*, 31–40. [CrossRef] [PubMed]

9. Jones, J.A.; Callan, D.E. Brain Activity During Audiovisual Speech Perception: An fMRI Study of the McGurk Effect. *Neuroreport* **2003**, *14*, 1129–1133. [CrossRef]
10. Risberg, A. The Importance of Prosodic Speech Elements for the Lipreader. *Scand. Audiol.* **1974**, *4*, 153–164.
11. Grant, K.W.; Ardell, L.H.; Kuhl, P.K.; Sparks, D.W. The Contribution of Fundamental Frequency, Amplitude Envelope, and Voicing Duration Cues to Speechreading in Normal-Hearing Subjects. *J. Acoust. Soc. Am.* **1985**, *77*, 671–677. [CrossRef]
12. Bernstein, L.E.; Eberhardt, S.P.; Demorest, M.E. Single-Channel Vibrotactile Supplements to Visual Perception of Intonation and Stress. *J. Acoust. Soc. Am.* **1989**, *85*, 397–405. [CrossRef]
13. McGurk, H.; MacDonald, J. Hearing Lips and Seeing Voices. *Nature* **1976**, *264*, 746–748. [CrossRef]
14. Xiong, W.; Wu, L.; Alleva, F.; Droppo, J.; Huang, X.; Stolcke, A. The Microsoft 2017 Conversational Speech Recognition System. In Proceedings of the 2018 IEEE International Conference on Acoustics, Speech and Signal Processing, Calgary, AB, Canada, 15–20 April 2018; pp. 5934–5938.
15. Kim, J.-B.; Kweon, H.-J. The Analysis on Commercial and Open Source Software Speech Recognition Technology. In *International Conference Computability Science Intellettuale Appliance Informatics*; Studies in Computational Intelligence; Springer: Cham, Switzerland, 2020; pp. 1–15. [CrossRef]
16. Ji, S.; Xu, W.; Yang, M.; Yu, K. 3D Convolutional Neural Networks for Human Action Recognition. *IEEE Trans. Pattern Anal. Mach. Intell.* **2012**, *35*, 221–231. [CrossRef] [PubMed]
17. Petridis, S.; Pantic, M. Deep Complementary Bottleneck Features for Visual Speech Recognition. In Proceedings of the IEEE International Conference on Acoustics, Speech and Signal Processing, Shanghai, China, 20–25 March 2016; Volume 2016, pp. 2304–2308.
18. Wand, M.; Koutník, J.; Schmidhuber, J. Lipreading with Long Short-Term Memory. In Proceedings of the 2016 IEEE International Conference on Acoustics, Speech and Signal Processing, Shanghai, China, 20–25 March 2016; pp. 6115–6119.
19. Cooke, M.; Barker, J.; Cunningham, S.; Shao, X. An Audio-Visual Corpus for Speech Perception and Automatic Speech Recognition. *J. Acoust. Soc. Am.* **2006**, *120*, 2421–2424. [CrossRef] [PubMed]
20. Noda, K.; Yamaguchi, Y.; Nakadai, K.; Okuno, H.G.; Ogata, T. Lipreading Using Convolutional Neural Network. In Proceedings of the Fifteenth Annual Conference Interna Speech Commentata Associação, Singapore, 14–18 September 2014.
21. Zhou, Z.; Zhao, G.; Hong, X.; Pietikäinen, M. A Review of Recent Advances in Visual Speech Decoding. *Image Vis. Comput.* **2014**, *32*, 590–605. [CrossRef]
22. Assael, Y.M.; Shillingford, B.; Whiteson, S.; De Freitas, N. Lipnet: End-to-End Sentence-Level Lipreading. *arXiv* **2016**, arXiv:1611.01599.
23. Zhang, P.; Wang, D.; Lu, H.; Wang, H.; Ruan, X. Amulet: Aggregating Multi-Level Convolutional Features for Salient Object Detection. In Proceedings of the IEEE International Conference Computability Vision, Venice, Italy, 22–29 October 2017; pp. 202–211.
24. Graves, A.; Fernández, S.; Gomez, F.; Schmidhuber, J. Connectionist Temporal Classification: Labelling Unsegmented Sequence Data with Recurrent Neural Networks. In Proceedings of the 23rd International Conference Machine Learning, Pittsburgh, PA, USA, 25–29 June 2006; pp. 369–376.
25. Chung, J.S.; Zisserman, A. Learning to Lip Read Words by Watching Videos. *Comput. Vis. Image Understand* **2018**, *173*, 76–85. [CrossRef]
26. Këpuska, V.; Bohouta, G. Comparing Speech Recognition Systems (Microsoft API, Google API and CMU Sphinx). *Int. J. Eng. Res. Appl.* **2017**, *7*, 20–24. [CrossRef]
27. Yoo, H.J.; Seo, S.; Im, S.W.; Gim, G.Y. The Performance Evaluation of Continuous Speech Recognition Based on Korean Phonological Rules of Cloud-Based Speech Recognition Open API. *Int. J. Network Distr Comput.* **2021**, *9*, 10–18. [CrossRef]
28. Alibegović, B.; Prljača, N.; Kimmel, M.; Schultalbers, M. Speech Recognition System for a Service Robot-A Performance Evaluation. In Proceedings of the International Conference on Control, Automation, Robotics and Vision, Shenzhen, China, 13–15 December 2020; Volume 2020, pp. 1171–1176.
29. Caute, A.; Woolf, C. Using Voice Recognition Software to Improve Communicative Writing and Social Participation in an Individual with Severe Acquired Dysgraphia: An Experimental Single-Case Therapy Study. *Aphasiology* **2016**, *30*, 245–268. [CrossRef]
30. Mikolov, T.; Sutskever, I.; Chen, K.; Corrado, G.; Dean, J. Distributed Representations of Words and Phrases and Their Compositionality. In Proceedings of the 26th International Conference on Neural Information Processing Systems, Lake Tahoe, NV, USA, 5–10 December 2013; Ser. NIPS'13; Curran Associates Inc.: Red Hook, NY, USA, 2013; pp. 3111–3119.
31. Mikolov, T.; Chen, K.; Corrado, G.; Dean, J. Efficient Estimation of Word Representations in Vector Space. *arXiv* **2013**, arXiv:1301.3781.
32. Chung, J.; Gulcehre, C.; Cho, K.; Bengio, Y. Empirical Evaluation of Gated Recurrent Neural Networks on Sequence Modeling. *arXiv* **2014**, arXiv:1412.3555.
33. King, D.E. Dlib-ml: A Machine Learning Toolkit. *J. Mach. Learn. Res.* **2009**, *10*, 1755–1758.
34. Tivive, F.H.C.; Bouzerdoum, A. An Eye Feature Detector Based on Convolutional Neural Network. 2005. Available online: https://ro.uow.edu.au/infopapers/2860/ (accessed on 17 February 2022).
35. Hinton, G.E.; Zemel, R.S. Autoencoders, Minimum Description Length, and Helmholtz Free Energy. *Adv. Neural Inf. Process. Syst.* **1994**, *6*, 3–10.

36. Nix, R.; Zhang, J. Classification of Android Apps and Malware Using Deep Neural Networks. In Proceedings of the 2017 International Joint Conference on Neural Networks, Anchorage, AK, USA, 14–19 May 2017; pp. 1871–1878.
37. He, K.; Zhang, X.; Ren, S.; Sun, J. Deep Residual Learning for Image Recognition. In Proceedings of the IEEE Conference Computer Vision Pattern Recognition, Las Vegas, NV, USA, 30 June 2016; pp. 770–778.
38. Warden, P. Speech Commands: A Dataset for Limited-Vocabulary Speech Recognition. *arXiv* **2018**, arXiv:1804.03209.
39. Majumdar, S.; Ginsburg, B. MatchboxNet: 1-d Time-Channel Separable Convolutional Neural Network Architecture for Speech Commands Recognition. *arXiv* **2020**, arXiv:2004.08531.
40. Vygon, R.; Mikhaylovskiy, N. Learning Efficient Representations for Keyword Spotting with Triplet Loss. *arXiv* **2021**, arXiv:2101.04792.
41. Mo, T.; Liu, B. Encoder-Decoder Neural Architecture Optimization for Keyword Spotting. *arXiv* **2021**, arXiv:2106.02738.
42. Sagonas, C.; Tzimiropoulos, G.; Zafeiriou, S.; Pantic, M. Volume 300 faces in-the-wild challenge: The first facial landmark localization challenge. In Proceedings of the IEEE International Conference Computability Vision Workshops 2013, Sydney, Australia, 1–8 December 2013; pp. 397–403.
43. Kingma, D.P.; Ba, J. Adam: A Method for Stochastic Optimization. *arXiv* **2014**, arXiv:1412.6980.
44. Neti, C.; Potamianos, G.; Luettin, J.; Matthews, I.; Glotin, H.; Vergyri, D.; Sison, J.; Mashari, A. *Audio Visual Speech Recognition (No, R.E.P. Work)*; IDIAP: Martigny, Switzerland, 2000.
45. Thiemann, J.; Ito, N.; Vincent, E. DEMAND: Diverse Environments Multichannel Acoustic Noise Database. *Proc. Mtgs. Acoust.* **2013**, *19*, 035081.
46. Jeon, S.; Elsharkawy, A.; Kim, M.S. Lipreading Architecture Based on Multiple Convolutional Neural Networks for Sentence-Level Visual Speech Recognition. *Sensors* **2021**, *22*, 72. [CrossRef]

sensors

MDPI

Article

End-to-End Sentence-Level Multi-View Lipreading Architecture with Spatial Attention Module Integrated Multiple CNNs and Cascaded Local Self-Attention-CTC

Sanghun Jeon and Mun Sang Kim *

Center for Healthcare Robotics, Gwangju Institute of Science and Technology (GIST), School of Integrated Technology, Gwangju 61005, Korea; jeon7887@gist.ac.kr
* Correspondence: munsang@gist.ac.kr; Tel.: +82-10-9126-4628

Abstract: Concomitant with the recent advances in deep learning, automatic speech recognition and visual speech recognition (VSR) have received considerable attention. However, although VSR systems must identify speech from both frontal and profile faces in real-world scenarios, most VSR studies have focused solely on frontal face pictures. To address this issue, we propose an end-to-end sentence-level multi-view VSR architecture for faces captured from four different perspectives (frontal, 30°, 45°, and 60°). The encoder uses multiple convolutional neural networks with a spatial attention module to detect minor changes in the mouth patterns of similarly pronounced words, and the decoder uses cascaded local self-attention connectionist temporal classification to collect the details of local contextual information in the immediate vicinity, which results in a substantial performance boost and speedy convergence. To compare the performance of the proposed model for experiments on the OuluVS2 dataset, the dataset was divided into four different perspectives, and the obtained performance improvement was 3.31% (0°), 4.79% (30°), 5.51% (45°), 6.18% (60°), and 4.95% (mean), respectively, compared with the existing state-of-the-art performance, and the average performance improved by 9.1% compared with the baseline. Thus, the suggested design enhances the performance of multi-view VSR and boosts its usefulness in real-world applications.

Keywords: lipreading; visual speech recognition; multi-view VSR; deep learning; attention mechanism; spatial attention module; convolutional neural network; local self-attention; connectionist temporal classification

Citation: Jeon, S.; Kim, M.S. End-to-End Sentence-Level Multi-View Lipreading Architecture with Spatial Attention Module Integrated Multiple CNNs and Cascaded Local Self-Attention-CTC. *Sensors* **2022**, *22*, 3597. https://doi.org/10.3390/s22093597

Academic Editors: Bruce Denby, Tamás Gábor Csapó and Michael Wand

Received: 11 April 2022
Accepted: 7 May 2022
Published: 9 May 2022

1. Introduction

Hearing and vision, sometimes known as verbal and visual signals, are widely employed in communication. Because audio signals typically include more information than visual signals, various experiments on automatic speech recognition (ASR) have been performed. Consequently, ASR has attained a very high recognition rate without causing significant signal deterioration. Moreover, it has been used in numerous applications. In contrast, visual speech recognition (VSR) recognizes speech content based on the speaker's lip-movement features in the absence of speech signals, that is, the speech information is inferred from the movement of the lips. In particular, the visual channel receives two-dimensional visual information, which typically contains more redundant information than that contained in the one-dimensional spoken information received via the auditory channel. Overcoming these VSR limitations is challenging.

People with hearing loss frequently communicate using sign language or by reading the movement of the person's lips. However, sign language has limitations, such as learning and comprehension difficulties, as well as insufficient expression skills. In this regard, VSR can help people with hearing loss interact effectively with others [1,2]. In noisy environments, interference from ambient noise can reduce audio recognition rates. By contrast, the visual information required for VSR does not change; consequently, VSR can increase speech-recognition performance in noisy contexts [3,4]. In particular, owing to the dominance of facial recognition technology in the field of security, including the use of photographs, video playback, and 3D modeling, VSR technology has been subjected to a large number of attacks. In this approach, including lip movement into a security system might improve its reliability [5]. Additionally, conventional speech synthesis can only generate a single voice in the primary domain of visual synthesis, whereas lipreading technology may generate high-resolution speeches of several characters in a video [6]. Furthermore, lip gestures can be employed to increase sign-language identification accuracy or comprehension [7,8].

Recent research has predominantly focused on lipreading from a frontal perspective [9–15]. This approach contradicts previous findings in the literature showing that human lipreaders prefer non-frontal views [16,17], owing to noticeable lip protrusion and lip rounding at these angles. Therefore, it might be practical to improve frontal-view lipreading abilities using non-frontal lip view information. This information can also be helpful when a frontal view of the mouth, which is the region of interest (ROI), is unavailable. This is true in real-life situations in which the subject's face is not visible [11,18,19]. In other words, in an audio VSR or VSR system, the speaker is not continually facing the smart device, kiosk, or camera.

Recently, several VSR systems have been proposed [20–26]. However, most VSR studies focus on frontal facial images because of the shortage of published datasets that include facial images from different angles. These investigations include lipreading studies, in which the emphasis is on frontal, diagonal, and profile images. The OuluVS2 [27] dataset, a publicly accessible multi-view VSR dataset, is typically used as a research corpus for evaluating novel approaches.

Estellers and Thiranin [28] trained a system using both frontal (0°) and profile (90°) faces and performed exploratory research on multi-view lipreading. Their study demonstrated that the frontal perspective exhibited a lower word error rate (WER) than the profile view. Isobe et al. [29] examined the frontal (0°), left profile (90°), and right profile (90°) viewpoints using a multi-angle approach. When the frontal perspective was used instead of the other perspectives, the system performance improved. As a breakthrough sequence-picture encoding approach, Saitoh et al. [21] proposed concatenated frame image encoding (CFI). They developed a framework for a convolutional neural network (CNN) based on CFI and compared two data augmentation methodologies for CFI.

Bauman et al. [16] observed that AI lipreaders perform better when human faces are slightly inclined because of lip protrusion and rounding. They used the active appearance model (AAM) to extract features from five distinct angles. Using a regression technique in feature space to assess lipreading on both view-dependent and view-independent systems, they reported that the view-dependent system outperformed benchmark models in all tests, receiving a perfect score of 30. Aiming at blending diverse views, Zimmermann et al. [22] coupled principal component analysis-based convolutional networks with long short-term memory (LSTM), a deep learning model, a conventional voice recognition model, hidden Markov models, and Gaussian mixture models. They found that a 30° face inclination produced the best effects. Anina et al. [27] recorded the best accuracy at 60°. Lipreading with a profile view produces lower WERs than lipreading with a frontal viewpoint, according to Kumar et al. [20].

Deep learning has also been used to blend multiple view angles and edit photographs. In particular, Komai et al. [30] implemented AAMs to transform frontal faces viewed from various angles. Their results suggested that identification accuracy increased even when the face orientation was rotated roughly 30° from the frontal perspective. The "View2View" system developed by Koumparoulis and Potamianos [23] relies on a CNN-based encoder–decoder paradigm. The technique converts non-frontal mouth photographs into frontal mouth images. Their view-mapping method for VSR and audio-visual speech recognition (AVSR) was reported to be successful.

By synthesizing virtual frontal views from non-frontal images, Estellers et al. [28] devised a position normalization technique and accomplished multi-view speech recognition. Petridis et al. [24] proposed a multi-view bidirectional LSTM-based lipreading model. The proposed approach considers data directly from pixels while simultaneously performing VSR from various perspectives. They discovered that combining the frontal and profile images boosted the accuracy when compared to using only the frontal view. Zimmermann et al. [25] implemented a PCA-based CNN, LSTM network, and GMM–HMM model to extract features in a decision fusion-based lipreading model. They reported that the decision fusion was effective because Viterbi pathways were included. In addition, to perform multi-angle lipreading, Sahrawat et al. [26] employed view-temporal attention to expand a hybrid attention-based connectionist temporal classification (CTC) system. Finally, Lee et al. [31] trained a CNN–LSTM model from beginning to end.

Evidently, numerous studies have been conducted based on deep learning. However, fewer studies have been conducted on multi-view lipreading than existing speech recognition and front lipreading studies.

Therefore, considering the above-mentioned limitations, we propose a multi-view VSR architecture that supports VSR when both frontal and non-frontal lip pictures are identified. In particular, for non-frontal views, we developed an end-to-end sentence-level multi-view lipreading neural-network architecture that outperforms traditional and current deep learning VSR systems. Convolutional, recurrent, and transcriptional layers were sequentially applied to develop the multi-view VSR architecture.

The remainder of this paper is structured as follows: Section 2 delves into the details on the proposed architecture, Section 3 discusses the experiments, and Section 4 discusses the results. Finally, Section 5 provides the concluding remarks of this study.

2. Proposed Architecture

In this section, we propose a novel feature-extraction approach. In particular, the proposed architecture is divided into three layers (convolutional layer, recurrent layer, transcription layer) based on an end-to-end neural network with four different perspective inputs, as shown in Figure 1. The three layers are compared against various modules for their performance evaluation. In the convolutional layer, based on the visual extraction module proposed in a previous study [32], the model was modified to improve the feature extraction performance and convergence speed. To compare the modules of the proposed architecture, three current equivalent designs were implemented: multi-scale 3D CNN, spatial attention module (SAM), and integrated multi-scale 3D CNN (Figure 1a). In addition, the recurrent layer was compared as a sequence-processing module with other modules, such as residual neural network (RNN), LSTM, gated recurrent unit (GRU), Bi-LSTM, and Bi-GRU (Figure 1b). The transcription layer was compared as a process for decoding the output features with other components, such as standard CTC, global self-attention-CTC, and local self-attention-CTC (Figure 1c).

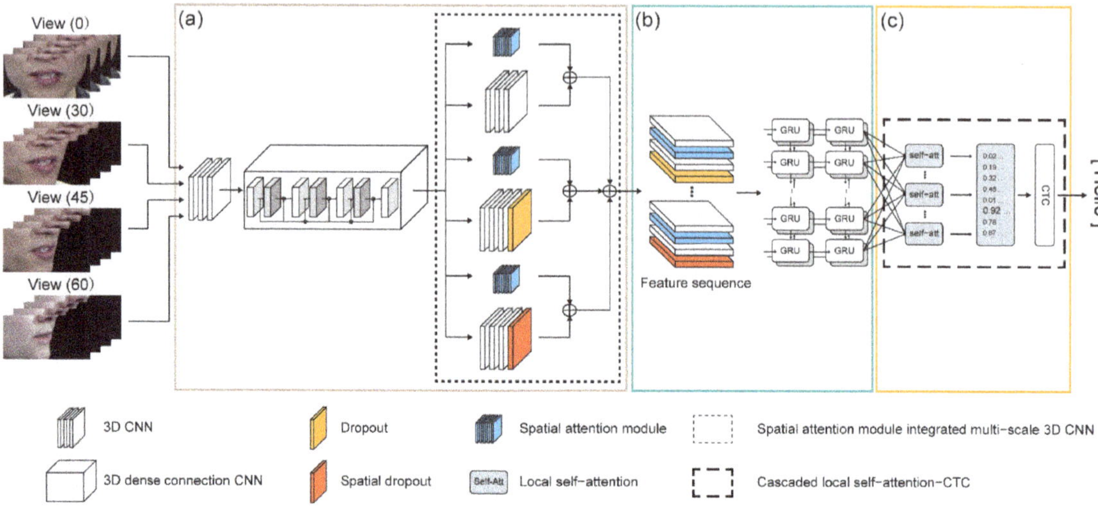

Figure 1. Block diagram of the proposed multi-view lipreading architecture; (**a**) convolutional layer; (**b**) recurrent layer; and (**c**) transcription layer.

2.1. Convolutional Layer

To encode visual information from the extracted lips, all input-image sequences were loaded into a spatiotemporal CNN. We extracted spatiotemporal information from an input image composed of numerous continuous frames using a three-dimensional convolutional layer with 64 kernels; $3 \times 5 \times 5$, $(1, 2, 2)$, and $(1, 2, 2)$ are the sizes, strides, and pads, respectively. To minimize the transformation of internal variables, we used a batch normalization (BN) layer and a rectified linear unit (ReLU) layer to accelerate the training process. Subsequently, a max-pooling 3D layer was used to decrease the spatial size of the 3D feature maps. Thus, the output form was observed utilizing $40 \times 50 \times 25 \times 64$ tensors with an input sequence of $40 \times 100 \times 50 \times 3$ frames.

A densely linked connection contains several connections. In this regard, CNN connects numerous layers of a connection, allowing for efficient feature usage, decreased gradient disappearance, and increased network depth. The input-feature volumes are reduced by the bottleneck layer, which comes before the convolutional layer. The multichannel feature volumes are merged using the bottleneck layer approach. The second layer is applied to only a fraction of the volume of the previous features because the prior features remain visible. Additionally, transition layers are utilized to increase the model's compactness, with the hyperparameter theta controlling the degree of compression. A bottleneck layer, transition layer, and slower growth rate are used to create a tight network. This strategy saves computing power while minimizing model parameters and preventing overfitting.

Dense connection CNN is an architecture that focuses on making deep learning networks go even deeper, while simultaneously making them more efficient to train by using shorter connections between the layers (Figure 2). Figure 2a displays a CNN, where each layer is connected to all of the other layers that are deeper in the network, and it consists of two important blocks other than the basic convolutional and pooling layers, that is, the dense blocks and the transition layers. Dense block (1) was built using the following layers in order: BN, ReLU, 3D convolutional, BN, ReLU, and 3D convolutional layers (see Figure 2b). Dense blocks (2), (3), and (4) have the same structure as dense block (1). The transition layer is depicted in Figure 2c, which comprises a BN layer, ReLU layer, three 3D convolutional layers, and two 2D pooling layers.

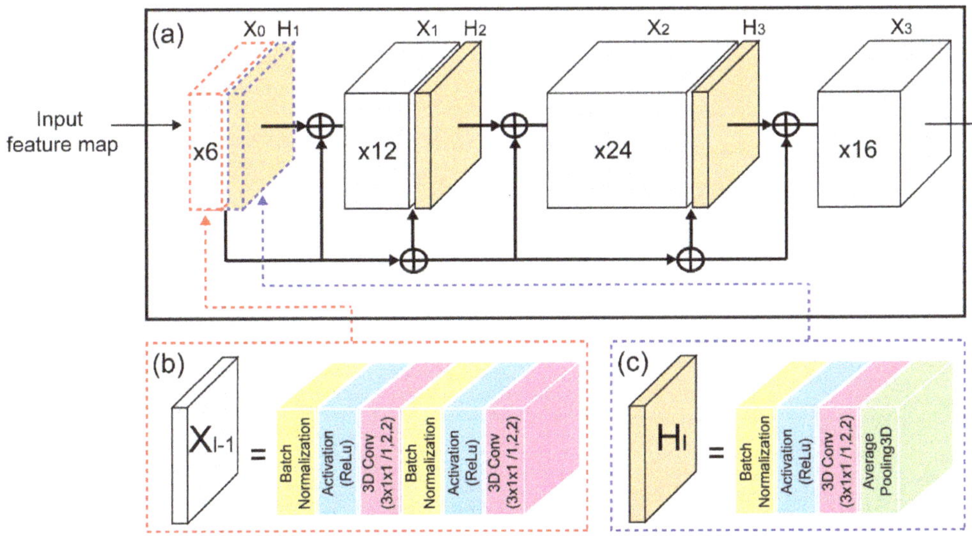

Figure 2. Details of 3D dense connection CNN architecture: (**a**) dense connection CNN; (**b**) dense block layer structure; and (**c**) transition layer structure.

Different CNN models have yielded outstanding results in picture classification tasks. One such example is feature aggregation using numerous CNNs, which allows the extraction of diverse spatial and temporal information by creating separate structures and depths [33]. Several convolutional layers with varying degrees of abstraction can be extracted during the multi-scale 3D CNN training phase. This training technique can also produce a range of features with various depths and filter sizes. Some of the essential characteristics lost in the layered design can be selected using this strategy, resulting in a more feature-rich final product.

The attention mechanism can boost the feature representation strength of our interests by telling us "what" and "where" to focus our attention. Attention weighting is used in computer vision to boost the feature representation capacity by emphasizing relevant characteristics and limiting inconsequential characteristics. Moreover, attention can be regarded as a strategy for allocating a finite computational force to more informative areas [34–36]. Hu et al. [37] proposed the "Squeeze-and-Excitation" module to describe the channel-wise correlation of convolutional features without considering the spatial information. The convolutional block attention module [38] empirically demonstrated that both max-pooling and average-pooling operations contribute to the attention mechanism. Additionally, the inter-spatial interactions feature may be utilized to produce a map of spatial attention. Spatial attention, in contrast to channel attention, focuses on the locations of informative sections and serves as a supplement to channel attention. As a result, the weights associated with attention are distributed over two separate dimensions in this model: channel and space.

The model initially executes average-pooling and max-pooling operations along the channel axis before concatenating them to build an efficient feature descriptor to compute spatial attention. To construct a spatial attention map $M_s(F) \in \mathcal{R}^{H \times W}$, a convolutional layer is applied to the concatenated feature descriptor. Subsequently, two pooling processes are used to aggregate the channel information of a feature map, resulting in two 3D maps: $F^s_{avg} \in \mathbb{R}^{H \times W}$ and $F^s_{max} \in \mathbb{R}^{H \times W}$, each representing the average- and max-pooled features over the channel. A 3D spatial attention map is created by concatenating and convolving them with a conventional convolutional layer. In brief, spatial attention is calculated using the following formula:

$$M_s(F) = \sigma(f^{7\times7}([AvgPool(F); MaxPool(F)])), \quad (1)$$

$$M_s(F) = \sigma(f^{7\times7}(\left[F^s_{avg}; F^s_{max} \right])), \quad (2)$$

where σ denotes the sigmoid function, and $f^{7\times7}$ represents a convolution operation with a filter size of 7 × 7 (Figure 3b).

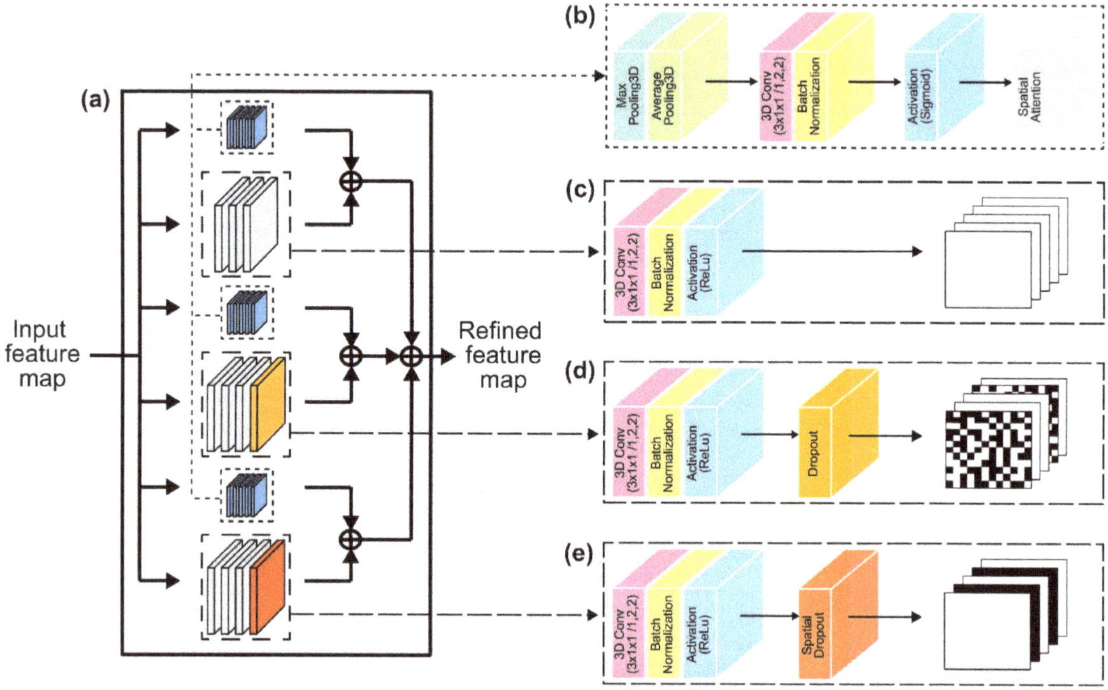

Figure 3. Details of the spatial attention module-integrated MLFF 3D CNN: (**a**) Block diagram of the proposed module; (**b**) spatial attention module; (**c**) first module's architecture; (**d**) second module's architecture with dropout layer; and (**e**) third module's architecture with spatial dropout layer.

Because several existing studies implement learning approaches based on sentence front-view datasets [32,39–41], it is difficult to expect high accuracy using the same model for multiple viewpoints. Therefore, we propose an SAM-integrated-MLFF 3D CNN, which is a network module focusing on spatial attention with different neighborhoods in the feature maps (Figure 3a). The first module (Figure 3c) comprises a 3D convolutional layer on a 3D dense connection convolutional layer output feature with 32 kernels, followed by a BN layer and a ReLU layer. The second module (Figure 3d) is structured similarly to the benchmark dataset, with a 3D convolutional layer with 64 kernels, followed by a dropout layer to prevent overfitting. By inhibiting the formation of highly correlated activations, the dropout layer enhances and generalizes the performance by avoiding overtraining and overfitting [42].

The third module, which contains a 3D convolutional layer with 96 kernels, is similar to the second module, except for the absence of a dropout layer (Figure 3e). In particular, this method drops the entire feature map. Moreover, in contrast to the traditional dropout method, which removes pixels at random, this method employs CNN models with substantial spatial correlation to improve image classification [43]. Consequently, we employed a spatial dropout layer to extract lips, teeth, and tongue morphologies, which have strong spatial connectivity and contain few movements. Each SAM multi-scale 3D CNN module consists of 3D average-pooling, 3D max-pooling, and 3D convolutional layers, with 32, 64, and 96 3D kernel operations, respectively, along the channel axis and a concatenated BN layer (Figure 3b). Therefore, the output of each multi-scale 3D CNN and SAM is merged and concatenated. As a result, SAM exploits the inter-spatial interaction of the characteristics to better select and focus on the most identifiable and helpful portions of an tinput picture [38].

2.2. Recurrent Layer

Traditional recurrent neural networks (RNNs), LSTM, and GRU are examples of previously implemented RNN algorithms. Owing to the gradient vanishing issue, a typical RNN has difficulties in learning long-range dependent input and output data, owing to the backpropagation technique's inability to perform adequately with an increase in input data. To overcome this issue, Hochreiter and Schmidhuber [44] created the LSTM network, which is currently widely used in time-series-data processing [45–47]. By efficiently overcoming the gradient vanishing issue through effective learning, LSTM and GRU achieve higher levels of validation and prediction accuracy than traditional RNNs, particularly for long-range dependent input and output data [45,47].

A GRU is an RNN that, through multiple stages, learns to manage and transmit information flow [48]. GRUs are constructed using LSTM units that can decide which data to retain and discard. While the 3D CNN only gathers data at the viseme level, GRUs can differentiate across greater temporal contexts, which is crucial for resolving ambiguity. GRU, which consists of an update gate and a reset gate, can also be used to address the gradient vanishing issue.

A two-layer bidirectional GRU is implemented in the proposed architecture, providing a faster convergence speed than a sequence processing module. The two-layer bidirectional GRU is used to transfer information both ways to two distinct neural network topologies coupled to the same output layer, enabling both networks to acquire substantial knowledge of the input. The SAM-integrated-multi-scale 3D CNN provides the input to the two-layer bidirectional GRU layer. For instance, to obtain an output containing 40×512 tensors, we submitted a bidirectional GRU $40 \times 3 \times 1 \times 384$ frame sequence into the merging layer.

2.3. Transcription Layer

Assael et al. [18] used "LipNet" (their neural network, which had outperformed experienced human lip readers) to train a network of end-to-end deep neurons on a benchmark dataset, using the effective CTC loss function [49] for acoustic-based speech recognition. The CTC loss function parameterizes the distribution of the label token sequence without having to align the input sequence; it is conditionally independent of the surrounding distribution generated at each time step. Therefore, the CTC model is a decoding method that uses a beam search technique to detect the temporal dependence of labels.

It is worth noting that the CTC loss function assumes conditional independence of independent labels (i.e., individual character symbols). Each output unit corresponds to the probability of seeing one label at a time. As a result, although CTC is built on RNNs, it is primarily concerned with local data (nearby frames) [50]. While this strategy is effective for forecasting acoustic phonemes, it is not effective for predicting visemes, which require additional background information to discern tiny variations.

Figure 4 illustrates that the self-attention mechanism [36,51] is a technique to better encode the word at the target location by looking at the word at another location and taking hints from each word in the input full-sequence sentence. Figure 4a depicts the processing process of the self-attention mechanism, with the global area enclosed by a blue-line square and the local area by a red dotted line. Furthermore, Figure 4b shows an example of the mechanism processing process presented in Figure 4a for the sentence "Nice to meet you".

The multi-head self-attention modules that transformers are known for constitute their distinguishing feature [36]. Given an input $X \in \mathbb{R}^{T \times n}$, where T is the number of time steps and n is the hidden state dimension, a set comprising query, key, and value matrices is generated using the weight matrices W_h^Q, W_h^K, and $W_h^V \in \mathbb{R}^{n \times d_k}$, respectively, where d_k is the dimension of the heads of the attention module. There is one embedding per head, denoted by the subscript h.

$$Q_h = XW_h^Q, \tag{3}$$

$$K_h = XW_h^K, \tag{4}$$

$$V_h = XW_h^V. \tag{5}$$

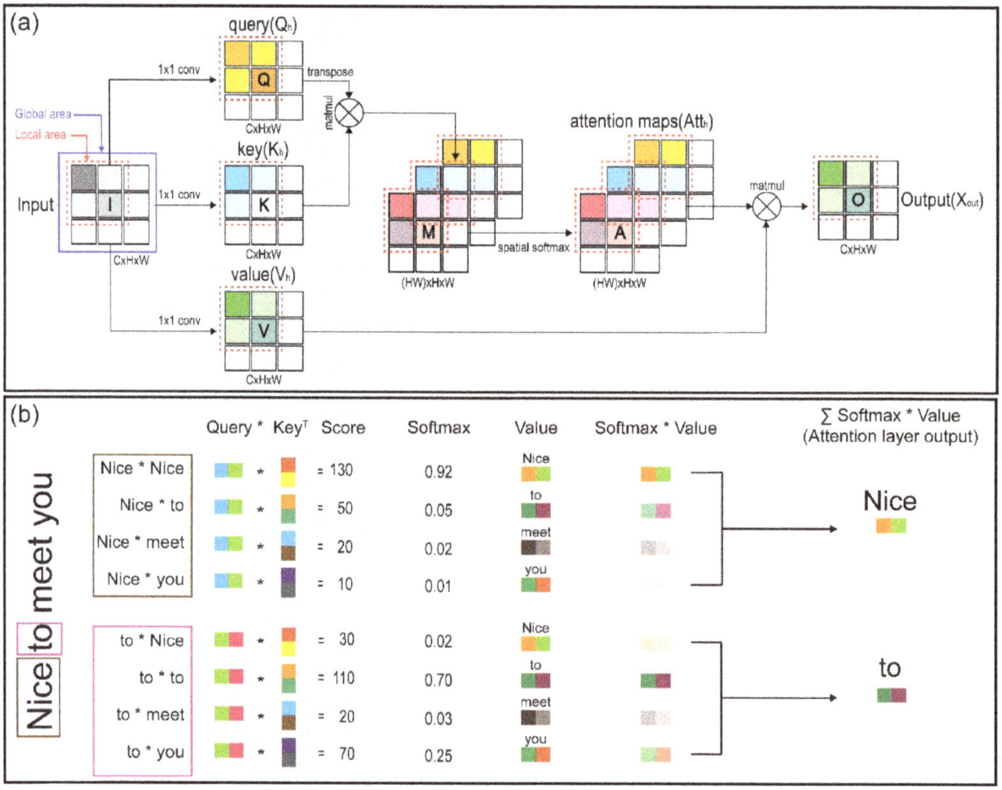

Figure 4. (**a**) Details regarding the global and local self-attention process: the blue line square encloses the global area, and red dotted line square encloses the local area; and (**b**) self-attention mechanism processing process presented for the sentence "Nice to meet you". (* for dot product).

The keys and queries are multiplied to obtain a $T \times T$ attention matrix A. This matrix encodes the relative relevance of each time step, that is, how much attention each time step receives, by assigning a scalar to each pair of time steps. A SoftMax function with temperature $\sqrt{d_k}$ is applied to convert this into a normalized distribution. The value matrix is subsequently multiplied by the normalized attention matrix. Consequently, each time step has a linear combination of value embeddings, with the most significant embedding receiving the largest weights as follows:

$$\mathrm{Att_h} = \mathrm{Softmax}\left(\frac{Q_h K_h^T}{\sqrt{d_k}}\right) V_h. \tag{6}$$

The heads are then concatenated and transformed back to the original dimension n using the weight matrix $W^{out} \in \mathbb{R}^{d_k \cdot n_h \times n}$, where n_h is the number of heads. Moreover, a residual connection connecting the output to the input is added as follows:

$$X^{out} = \mathrm{Concat_h}(\mathrm{Att_h}) W^{out} + X. \tag{7}$$

Subsequently, each time step is standardized via layer normalization. For time step t, the overall mean of the feature dimension is subtracted from the input, which is then divided by the standard deviation. This is rescaled and shifted by the learnable parameters α and β as follows:

$$X_t^{norm} = \frac{X_t^{out} - \mu_t}{\sigma_t} \cdot \alpha + \beta, \tag{8}$$

where

$$\mu_t = \frac{1}{n} \sum_i X_{ti}^{out}, \tag{9}$$

$$\sigma_t = \sqrt{\frac{1}{n} (X_{ti}^{out} - \mu_t)^2}. \tag{10}$$

Next, a feedforward neural network is applied in a time-step-wise manner. This part typically consists of two fully connected layers parameterized by weight matrices $W_1 \in \mathbb{R}^{n \times \phi n}$, $W_2 \in \mathbb{R}^{\phi n \times n}$; bias vectors $b_1 \in \mathbb{R}^{\phi n}$, $b_2 \in \mathbb{R}^n$; and a residual connection as follows:

$$f(X_t^{nrom} W_1 + b_1) W_2 + b_2 + X_t^{nrom}, \tag{11}$$

where $f(\cdot)$ is an element-wise activation function, such as a ReLU or Gaussian error linear unit. Here, ϕ is a scaling factor for the inner dimensions of the feedforward module. Finally, another layer normalization is applied.

The encoder, decoder, and feedforward contexts were employed to accelerate translation and offer the most current translation findings, sentiment analysis, and other additional operations. The success of self-attention in these tasks motivated the first study on self-attention in speech recognition [52]. As a result, an attention-based encoder–decoder paradigm was devised. Although self-attention was first employed for machine translation, its versatility enabled it to be utilized for voice recognition as well [53–56]. Attention-based encoder–decoder models rapidly learn the mapping between the auditory frame and the letter sequence. These models generate a label at each output time step based on the input and target label histories. Despite not requiring an external language model, the attention model has a lower character error rate (CER) than CTC. However, the model performs poorly in real-world conditions for various voice recognition tasks, owing to the ease with which noise and other variables may impair the expected alignment in the attention mechanism. Additionally, learning the model from start is difficult, owing to the misalignment of extended input sequences [57,58].

This study used cascaded local self-attention CTC training criteria to improve performance and accelerate learning for the above-mentioned difficulties. When scaling to larger sequences, transformers scale quadratically in the input length. This problem is solved using a unique speech enhancement transformer model based on local attention [59,60]. Local attention is especially well suited for speech augmentation because the predictions do not require long-range correlations, as in natural language processing. Moreover, sufficient information is frequently stored within a few seconds of the target period. Local attention is naturally interwoven with this demand.

The above approach results in huge advances in speech augmentation, where typical sample lengths can involve up to hundreds of thousands of tokens or hours of speech. This small focus incurs only a fraction of the processing and memory overhead associated with attention throughout the entire feature. The windowed technique also allows a more compact packing of padded features in mini-batches, thereby saving costs. Consequently, this module acquires detailed local contextual information from the surrounding area. As the foundational model, we employed cascaded local self-attention with a context size of 12.

3. Experimental Evaluation

3.1. Dataset

In this study, the proposed architecture was evaluated on the OuluVS2 [27] dataset. This dataset comprises 52 speakers making three types of utterances (Digits, Phrases, and TIMIT), three times each (except TIMIT), simultaneously recorded from five distinct viewpoints (0°, 30°, 45°, 60°, and 90°) for a total of 780 samples per utterance. There are ten classes in total: "Please excuse me", "Goodbye", "Hello", "How are you", "Nice to meet you", "See you", "I am sorry", "Thank you", "Have a nice time", and "You are welcome". The impact of various mouth ROIs was evaluated by processing the lips from scratch rather than from existing data, and the 90° data were omitted from the experiment because the lips could not be recognized during the extraction process. For the recognition task, we used the Phrase dataset in this investigation. In particular, we utilized the same data split as in other previous studies [21,22,31], to provide a fair comparison. Twelve speakers were used for testing (s06, s08, s09, s15, s26, s30, s34, s43, s44, s49, s51, and s52; 10 men and 2 women) and 40 for training from the database (s06, s08, s09, s15, s26, s30, s34, s43, s44, s49, and s51). Note that s29 is not included in the list.

3.2. Data Preprocessing and Augmentation

A DLib face detector [61] was used in the data-preparation step to recognize the targeted face and mouth. A HoG feature-based linear classifier [33] was used in the detector. The diagonal edges' (x, y) coordinates were obtained and used to build a bounding box around the mouth. As a result, the iBug program was used to forecast facial landmarks [62], considering 68 landmarks and an online Kalman filter. This method is widely used to extract the lip points that match with those in the training dataset by reading lip motions. These algorithms were utilized to extract a mouth region from each frame, and to perform an affine transformation to equalize the RGB channels throughout the training set, resulting in a mean and variance of zero. Moreover, we employed a data augmentation approach for training data to avoid overfitting [18]. The training process considered both standard and horizontally mirrored picture sequences. The degradation rate for these occurrences was 0.925. Finally, to avoid variance, we identified the movement speed and repeated each frame with a probability of 0.05. All models were trained and evaluated on the OuluVS2 dataset, using identical preprocessing and augmentation methods.

3.3. Implementation

To evaluate the performance of the CTC decoder, all models used Keras, based on TensorFlow backend on Linux Ubuntu; the computer had an Intel® Core™ i7-7700K processor, along with 64GB RAM and an NVIDIA GeForce RTX 2080-Ti GPU. The hyperparameters specified in Table 1 are the values for each layer of the proposed model. The network parameters—other than the initialized GRU matrix and hyperparameters—were initialized for all models. To perform the optimization of models, adaptive moment estimation (Adam) [63], stochastic gradient descent (SGD) [64], RMSprop [65], AdaMax, and Nadam [64] optimizers were used in mini-batches of sizes 8 and 0.0001, trained at the learning rate. The proposed model was trained in a multi-scale 3D CNN with SAM; channel-wise dropped pixels and spatial dropout for the dropped channel were used, and the proposed model contained the baseline model, trained on the dataset until it was overfitted. The moving average strategy was used to smooth it down for better viewing.

Regarding the accuracy of the proposed model, the genuine value was represented by the shadow part of the image, while the curve represented the smoothed value. We selected a smaller batch size of 75 images owing to the computer's restricted capabilities, causing the real value fluctuation to be uneven. Smoothing was performed to alleviate this problem and to make the curves comprehensible.

Table 1. Hyperparameters of the proposed architecture.

Layer	Output Shape	Size/Stride/Pad		Dimension Order
Input Layer	40 × 100 × 50 × 3	-		
Convolution 3D Layer	40 × 50 × 25 × 64	[3 × 5 × 5]/(1, 2, 2)/(1, 2, 2)		
		[1 × 2 × 2] max pool/(1 × 2 × 2)		
3D Dense Block (1)	40 × 25 × 13 × 96	[3 × 1 × 1] 3D Conv	(×6)	
		[3 × 3 × 3] 3D Conv		
3D Transition Block (1)	40 × 12 × 6 × 6	[3 × 1 × 1] 3D Conv		
		[1 × 2 × 2] average pool/(1 × 2 × 2)		
3D Dense Block (2)	40 × 12 × 6 × 38	[3 × 1 × 1] 3D Conv	(×12)	
		[3 × 3 × 3] 3D Conv		
3D Transition Block (2)	40 × 6 × 3 × 3	[3 × 1 × 1] 3D Conv		T × C × H × W
		[1 × 2 × 2] average pool/(1 × 2 × 2)		
3D Dense Block (3)	40 × 6 × 3 × 35	[3 × 1 × 1] 3D Conv	(×24)	
		[3 × 3 × 3] 3D Conv		
3D Transition Block (3)	40 × 3 × 1 × 1	[3 × 1 × 1] 3D Conv		
		[1 × 2 × 2] average pool/(1 × 2 × 2)		
3D Dense Block (4)	40 × 3 × 1 × 33	[3 × 1 × 1] 3D Conv	(×16)	
		[3 × 3 × 3] 3D Conv		
Multi-scale 3D CNN (1)	40 × 3 × 1 × 32	[3 × 5 × 5]/(1, 2, 2)/(1, 2, 2)		
Multi-scale 3D CNN (2)	40 × 3 × 1 × 64	[3 × 5 × 5]/(1, 2, 2)/(1, 2, 2)		
Multi-scale 3D CNN (3)	40 × 3 × 1 × 192	[3 × 5 × 5]/(1, 2, 2)/(1, 2, 2)		
Spatial Attention (1)	40 × 3 × 1 × 32	[1 × 2 × 2] max pool/(1 × 2 × 2)		
		[1 × 2 × 2] average pool/(1 × 2 × 2)		
		[3 × 7 × 7]/(1, 2, 2)/(1, 2, 2)		
Spatial Attention (2)	40 × 3 × 1 × 64	[1 × 2 × 2] max pool/(1 × 2 × 2)		
		[1 × 2 × 2] average pool/(1 × 2 × 2)		
		[3 × 7 × 7]/(1, 2, 2)/(1, 2, 2)		
Spatial Attention (3)	40 × 3 × 1 × 96	[1 × 2 × 2] max pool/(1 × 2 × 2)		
		[1 × 2 × 2] average pool/(1 × 2 × 2)		
		[3 × 7 × 7]/(1, 2, 2)/(1, 2, 2)		
Bidirectional GRU Layer	40 × 512	256		T × F
Bidirectional GRU Layer	40 × 512	256		T × F
Local Self-Attention Layer	40 × 512	15		T × F
Dense Layer	40 × 28	27 + blank		T × F
SoftMax Layer	40 × 28			T × V

3.4. Performance Evaluation Metrics

We used standard automated speech-recognition assessment criteria as the evaluation metrics. The learning loss of each model was calculated to determine its learning status during the training operation. Furthermore, we compared each model's performance and computational efficiency by examining its parameters, epoch period, and CER.

For the misclassification analysis, it is necessary to compare the original text and the predicted text. The five variables used in the equation are the characters (C), the total number of ground truth characters (N), the false predicted characters (S), the non-selected characters (I), and the number of deleted characters (D). CTC beam search is performed for maximum probability prediction, and the CER equation is as follows:

$$\text{CER}\ (\%) = \left(\frac{C_S + C_D + C_I}{C_N}\right) \times 100, \quad (12)$$

We compared the CER for parameter count and computational efficiency during the study period. The results are presented using a confusion matrix.

4. Results

4.1. Learning Loss and Convergence Rate

Figures 5–7 compare the learning loss and convergence speed rates for the convolutional, recurrent, and transcription layers, respectively. Figure 5 shows the learning loss (training and validation) on the OuluVS2 dataset for the convergence rates of the three types of CNNs in the convolutional layer. The three models have different visual feature extraction modules at the front end, and the same recurrent and transcription layers at the back end. Model A consists of a densely connected 3D CNN, Model B combines the multi-scale 3D structure following Model A, and Model C is configured by combining a SAM with Model B. In addition, Figure 5 shows that the training and validation losses of all three models are similar from all four angles. However, the gap between the training and validation losses was the highest in Model A, and its degree of overfitting was higher than those of the other models. Furthermore, although Model C increased the number of parameters by 30 M compared to Model A, it exhibited lower overfitting results (the smallest among all models) (Figure 5). This is because Model A comprised a model with outstanding performance based on the DenseNet-121 [66] structure, thereby minimizing the number of model parameters, successfully suppressing overfitting, and saving computation. However, the combination of multi-scale 3D CNN (Model B) and SAM (Model C) yielded improved results because this combination identified better by focusing on the most distinguishable and beneficial areas of the input image. Therefore, the learning and convergence speeds of Model C were high, and the gap was small. These findings indicate that the proposed model had the smallest difference between the training and validation losses, preventing overfitting on the OuluVS2 dataset.

Figure 6 shows the learning loss (training and validation) on the OuluVS2 dataset for the convergence rates of the four types of RNN in the recurrent layer. The convolutional and transcription layers had the same structure, and only the configuration of the recurrent layers differed. The Bi-GRU exhibited the fastest learning convergence speed and best prediction accuracy, as shown in Figure 6 and Figure 9e–f. In particular, all four RNN unit types outperformed the RNN. The experimental results and prediction accuracy are similar to the findings reported in Section 5 of [44], where LSTM and GRU displayed improved validation accuracy and prediction accuracy compared to traditional RNNs (Table 2), owing to their resistance to the vanishing gradient problem. Compared with LSTM and Bi-LSTM, both GRU and Bi-GRU demonstrated faster convergence and lower losses. The bidirectional models outperformed the unidirectional models on the training set for both GRU and LSTM; they also outperformed their unidirectional counterparts on the validation dataset. Consequently, Bi-GRU exhibited the best overall performance.

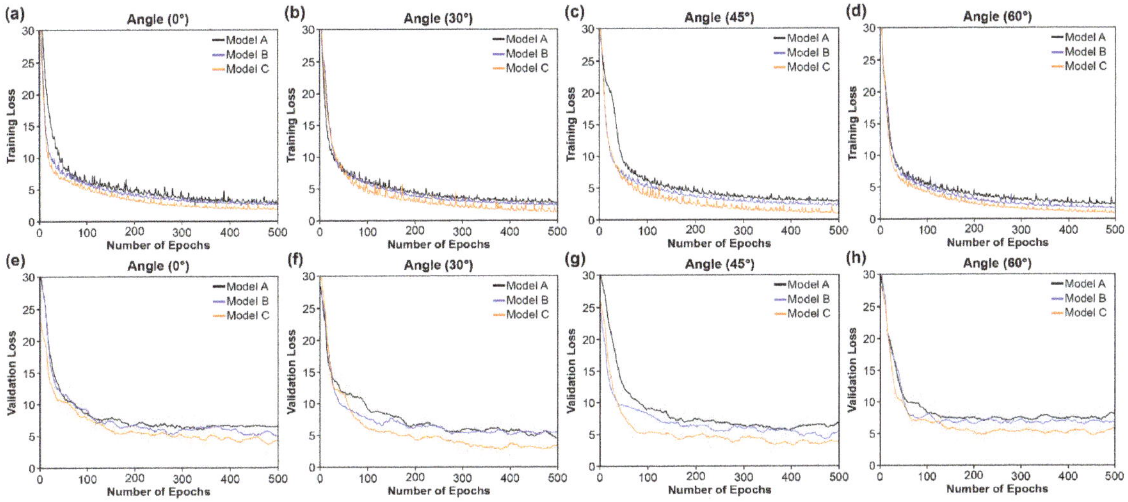

Figure 5. Training and validation loss comparing convergence speed of convolutional layers (Models A, B, and C): (**a**–**d**) Training loss at (**a**) 0°; (**b**) 30°; (**c**) 45°; and (**d**) 60°; (**e**–**h**) Validation loss at (**e**) 0°; (**f**) 30°; (**g**) 45°; and (**h**) 60°.

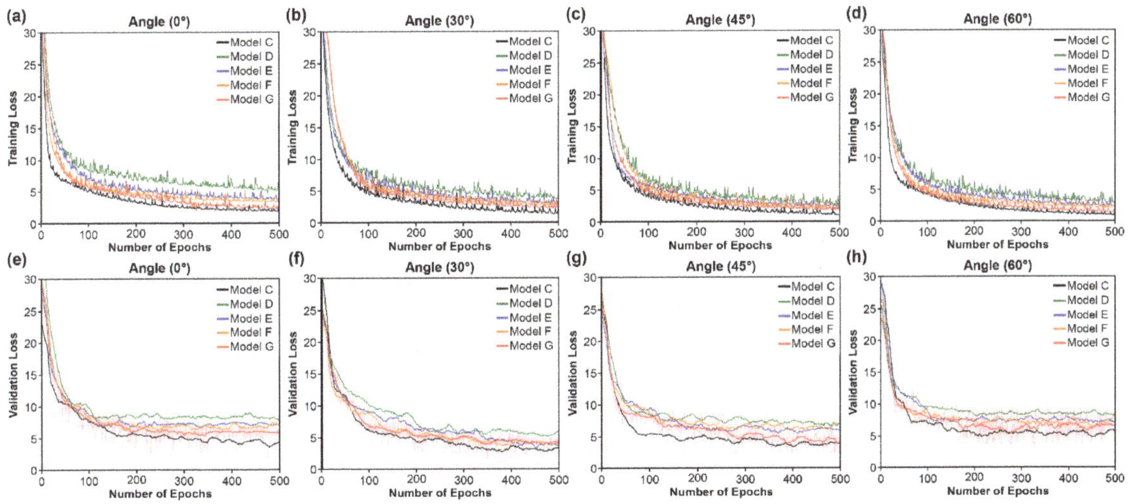

Figure 6. Training and validation loss comparing convergence speed of recurrent layers (Models C, D, E, F, and G). (**a**–**d**) Training loss at (**a**) 0°; (**b**) 30°; (**c**) 45°; (**d**) 60°. (**e**–**h**) Validation loss at (**e**) 0°; (**f**) 30°; (**g**) 45°; (**h**) 60°.

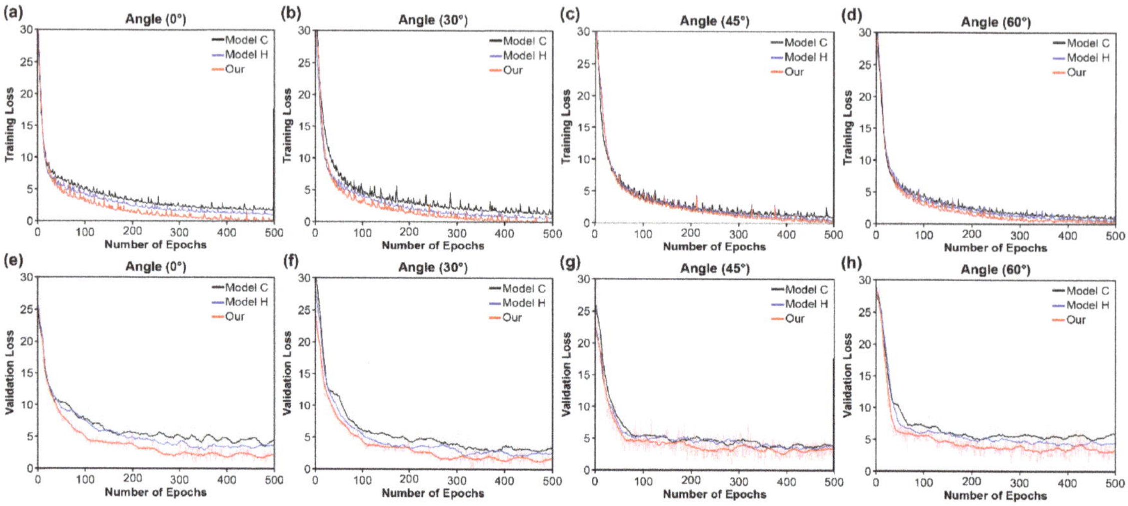

Figure 7. Training and validation loss comparing convergence speed of transcription layers (Model C, Model H, and the proposed model). (**a–d**) Training loss at (**a**) 0°; (**b**) 30°; (**c**) 45°; (**d**) 60°; (**e–h**) Validation loss at (**e**) 0°; (**f**) 30°; (**g**) 45°; (**h**) 60°.

Table 2. Performance of the proposed model compared to various models on the OuluVS2 dataset.

Model	Method	Top 10 Accuracy (%)				
		0°	30°	45°	60°	Mean
A *	3D dense connection CNN + Bi-GRU + CTC	90.44	88.73	86.93	87.72	88.45
B *	3D dense connection CNN + Multi-scale 3D CNN + Bi-GRU + CTC	92.72	91.02	88.02	88.09	89.62
C *	3D dense connection CNN + Multi-scale 3D CNN + SAM + Bi-GRU + CTC	94.14	92.86	91.34	89.97	92.08
D *	3D dense connection CNN + Multi-scale 3D CNN + SAM + RNN + CTC	88.51	85.74	83.93	83.04	85.31
E *	3D dense connection CNN + Multi-scale 3D CNN + SAM + LSTM + CTC	89.42	87.42	86.01	85.71	87.14
F *	3D dense connection CNN + Multi-scale 3D CNN + SAM + Bi-LSTM + CTC	89.78	88.84	87.26	86.18	88.02
G *	3D dense connection CNN + Multi-scale 3D CNN + SAM + GRU + CTC	92.85	91.23	90.91	89.67	91.14
H *	3D dense connection CNN + Multi-scale 3D CNN + SAM + Bi-GRU + Global self-attention + CTC	95.08	93.29	92.81	90.93	93.03
Our *	3D dense connection CNN + Multi-scale 3D CNN + SAM + Bi-GRU + Local self-attention + CTC	98.31	97.89	97.21	96.78	97.55

* Model trained with data augmentation.

The learning loss (training and validation) on the OuluVS2 dataset is shown in Figure 7 for the convergence rates of the proposed model's three types of CTC loss functions in the transcription layer. The convergence rate for learning was slower than that in the other two situations, when only the basic CTC loss function was used. In particular, as the angle of the detected lip changed, the convergence rate further decreased, while the two cases of cascaded self-attention exhibited similar convergence rate tendencies for all of the angles. The two self-attention modules learned with similar convergence rate tendencies. However, in all of the four results shown in Figure 7, the local self-attention module exhibited a faster convergence rate than the global self-attention modules. First, the principle of the CTC loss function assumes conditional independence for each label, and, since each output unit denotes the probability of seeing a single label at a given moment, it provides a high premium to the nearby local information [50]. Thus, ineffectiveness in predicting visemes is a possible reason for the difference in convergence rates.

The cascaded self-attention CTC module (which generates an output sequence with long-term temporal correlation) increases the speed of convergence, as compared to the CTC decoder (which assumes the input is conditionally independent). The attention approach is used in the CTC decoder's pre-alignment stage to remove unnecessary paths. The CTC decoder is then used to align the video frames and text labels, thereby allowing the attention mechanism to focus on the video–text pairs in the correct order. As a result, fewer irrelevant samples are created, resulting in the observed speedup. Second, the local self-attention module's windowed method results in more compact packaging of the padded features in mini-batches, and, hence, further cost reductions. Consequently, this local self-attention requires only a fraction of the computing and memory costs of attention over the entire feature, while providing rich local contextual information in the small region.

4.2. Optimization

The update rules of the optimization algorithms are usually defined by the hyperparameters that influence their behavior (e.g., the learning rate). The optimizer's responsibility is to update the weight parameters prior to reducing the error or loss function, which is the difference between the actual and predicted values. This requires several iterations with varying weights. However, choosing an optimizer for network training can be tricky. Deep learning employs iterative rules to modify or evaluate the data, utilizing numerous aspects and techniques. Therefore, training models as quickly as possible is vital to complete the iterative cycle and, as a result, enhance the prediction accuracy and speed. Consequently, in this part, we study the following optimizers used to train deep learning neural networks: SGD, RMSprop, Adam, Nesterov-accelerated Adam (Nadam), and AdaMax. After validating that AdaDelta and AdaGrad diverged without learning throughout the learning process, we omitted them from the experiments.

SGD realizes one update at a time to avoid duplication, making it significantly faster and easier to learn than other deep learning neural networks [67]. These frequent updates of the method with high variance introduce significant fluctuation in the objective function. This variation allows the parameters to move into new, possibly better, local minima. However, as SGD continues to overshoot, converging to the precise minimum is challenging. The parameters of AdaDelta have varying learning speeds, and the learning process comes to a halt after a certain point. This problem was addressed using the RMSprop method [65]. For each sample in each iteration, RMSprop uses a variable learning rate that is changed according to the results. RMSprop calculates the average of the first-order moments of the gradients and accelerates convergence by ignoring distant previous locations. Moreover, the squares of gradients and the average of the second-order moments are considered by AdaDelta and RMSprop. In the Adam optimizer, the adaptive optimization method is applied. Based on the parameters to be used, this optimizer dynamically modifies the learning rate for each sample in the dataset. Adam is a fast thinker with a limited memory span. Therefore, SGD, AdaDelta, and RMSprop [65] were used to create this algorithm.

Nadam combines Adam and Nesterov momentum. This method was developed similarly to Adam, with the exception that the flat momentum is replaced with the Nesterov momentum. The substitution causes a more considerable increase in performance than that in momentum. [63,68]. Alternatively, AdaMax, an extension of the Adam optimizer, was developed [63]. To update the weight parameters in AdaMax, the infinity norm of the moment is used, instead of the second-order moment estimate. Therefore, the size of the parameter update in AdaMax has a simpler constraint structure than in Adam, and the weight-updating rules are stable.

We used the Bi-GRU classifier to compare the training results and determine the most successful optimizer. Figure 8 depicts the loss curves of the optimizers. In particular, Adam performed better among the optimizers at all of the four angles. The Adam optimizer's loss converged at the quickest pace, implying that it trained the Bi-GRU classifier more successfully than the other algorithms. The results show that Adam was the best optimizer for training the Bi-GRU architecture's lip-based classification model. Therefore, this approach was employed in further trials in this study to train the Bi-GRU classifier.

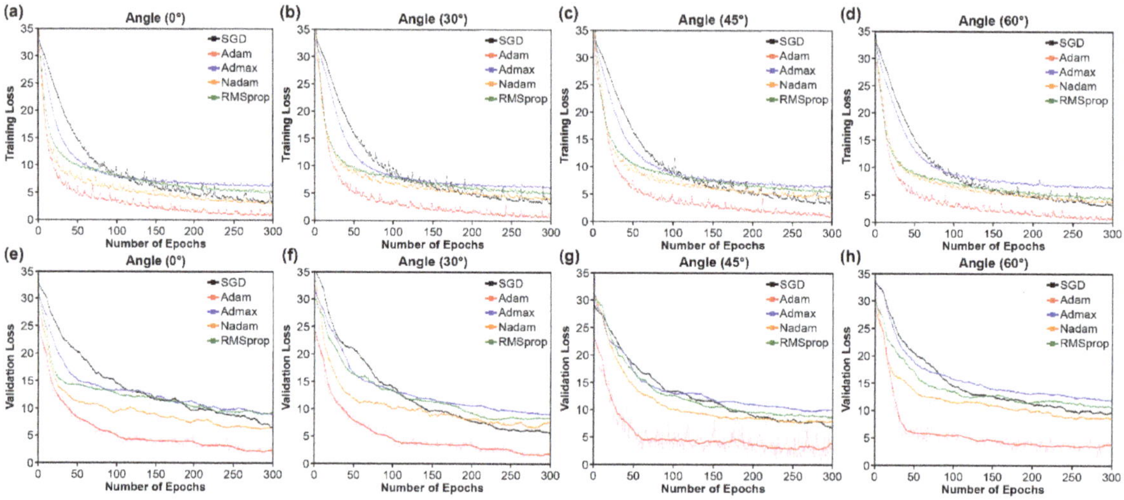

Figure 8. Loss curves comparing various optimizers. (**a–d**) Training loss at (**a**) 0°; (**b**) 30°; (**c**) 45°; and (**d**) 60°; (**e–h**) validation loss at (**e**) 0°; (**f**) 30°; (**g**) 45°; and (**h**) 60°.

4.3. *Performance and Accuracy*

The results presented in this section correspond to the OuluVS2 dataset phrases. Tables 2 and 3 show that the proposed model outperformed existing deep learning models by attaining state-of-the-art (SOTA) results: 3.31% (0°), 4.79% (30°), 5.51% (45°), 6.18% (60°), and 4.95% (mean). These results show an improvement over the previous SOTA results in all of the conditions. Figure 9 compares the accuracy results between the models by dividing them into three layers: convolutional layer (Figure 9a–d), recurrent layer (Figure 9e–h), and transcription layer (Figure 9i–l).

In the case of the convolutional layer (Figure 9a–d and Table 2), on average, the performance improved by 3.63% for all of the four angles when MLFF 3D CNN and SAM were combined than when only the DenseNet-121 structure was used. By combining the SAM with MLFF 3D CNN, a 2.46% improvement was observed owing to improved recognition among the inter-spatial relationships of features. This helped to better identify and focus on the most distinguishable and informative areas of the input image.

Table 3. Performance of existing models on the OuluVS2 dataset.

Year	Model	0° (%)	30° (%)	45° (%)	60° (%)	Mean (%)
2014	RAW-PLVM [69]	73.00	75.00	76.00	75.00	74.75
2016	CNN * [21]	85.60	82.50	82.50	83.30	83.48
	CNN + LSTM [31]	81.10	80.00	76.90	69.20	76.80
	CNN + LSTM, Cross-view Training [31]	82.80	81.10	85.00	83.60	83.13
	PCA Network + LSTM + GMM–HMM [22]	74.10	76.80	68.70	63.70	70.83
	CNN pretrained on BBC dataset * [52]	93.20	-	-	-	-
2017	CNN pretrained on BBC dataset + LSTM * [70]	94.10	-	-	-	-
	End-to-End Encoder + BLSTM [24]	94.70	89.70	90.60	87.50	90.63
	Multi-view SyncNet + LSTM * [71]	91.10	90.80	90.00	90.00	90.48
	End-to-End Encoder + BLSTM [13]	84.50	-	-	-	84.50
	End-to-End Encoder + BLSTM [72]	91.80	87.30	88.80	86.40	88.58
2018	CNN + Bi-LSTM [73]	90.30	84.70	90.60	88.60	88.55
	CNN + Bi-LSTM [73]	95.00	93.10	91.70	90.60	92.60
	Maxout-CNN-BLSTM * [74]	87.60	-	-	-	-
	CNN + LSTM with view classifier * [23]	-	86.11	83.33	81.94	-
	CNN + LSTM without view classifier * [23]	-	86.67	85.00	82.22	-
2019	VGG-M + LSTM * [75]	91.38	-	-	-	91.38
2020	CNN(2D + 3D) without view classifier [76]	91.02	90.56	91.20	90.00	90.70
	CNN with view classifier [76]	91.02	90.74	92.04	90.00	90.95
2021	CNN without view classifier [77]	91.02	90.56	91.20	90.00	90.70
	CNN with view classifier * [77]	91.02	91.38	92.21	90.09	91.18

* Model trained with data augmentation.

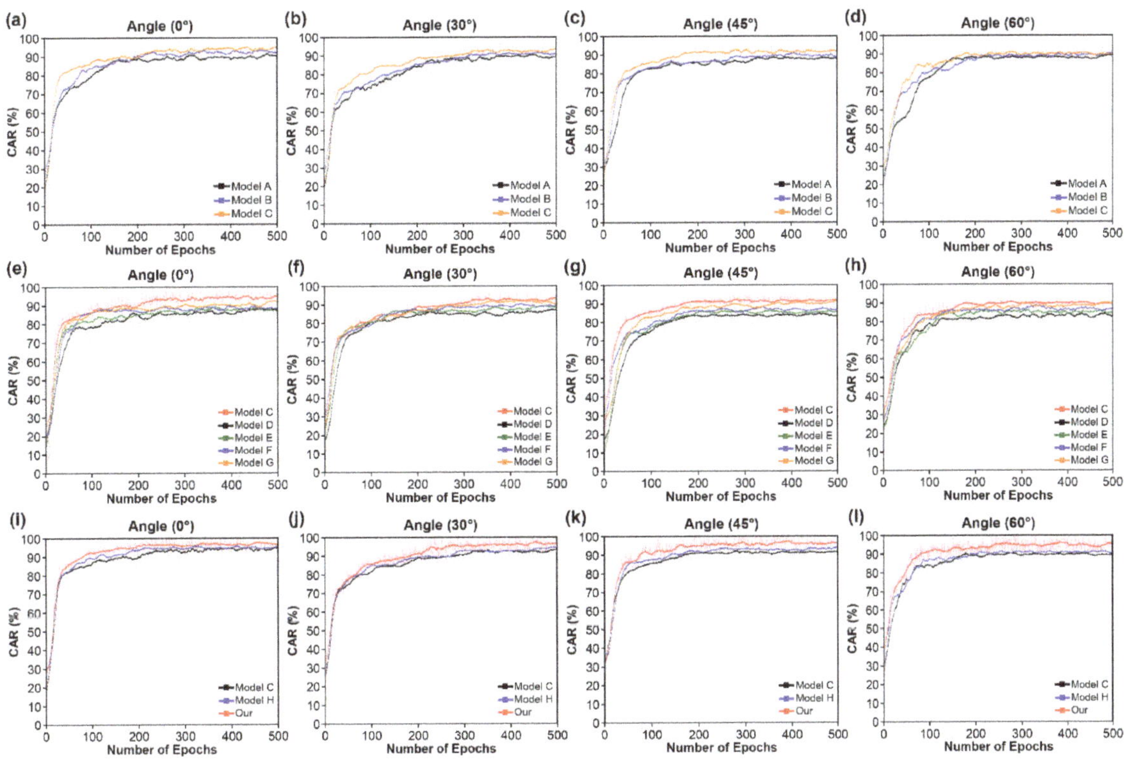

Figure 9. Training steps for character error rate (CER) comparing our proposed model to the baseline and other models: (a–d) Convolutional layer; (e–h) recurrent layer; (i–l) transcription layer.

In the case of the recurrent layer (Figure 9e–h and Table 2), five RNN units (RNN, LSTM, Bi-LSTM, GRU, and Bi-GRU) were compared. For all of the four angles, LSTM and GRU exhibited higher accuracy than the standard RNN. This is because of their robustness against gradient disappearance, which allows them to successfully learn long-range dependent input data. Therefore, the average accuracy of LSTM increased by 1.83% compared to when RNN was used. Similarly, the average accuracy of GRU increased by 4.17%. However, despite its similar performance, Bi-LSTM's accuracy increased by 2.71% compared to RNN, and Bi-GRU's accuracy improved by 6.77% when unidirectional models were used, compared to bidirectional models. The bidirectional models also achieved better results on the validation dataset than their unidirectional counterparts. Thus, the best overall performance was achieved using the Bi-GRU.

In the case of the transcription layer (Figure 9i–l and Table 2), we compared the performance by combining the global and local self-attention mechanisms with the basic CTC function in the cascade method. For all of the four angles, the two CTC loss functions exhibited higher performance than the basic CTC loss function. When using the global self-attention method, accuracy improved by 0.95%, while the local self-attention method improved by 5.47%. The performance of the two models is better than that of the CTC loss function because they overcome the disadvantage of assuming a conditionally independent input. Moreover, the performance difference between the two methods exists because the local self-attention module led to a more compact packing of the padded features in mini-batches, resulting in additional savings. Therefore, this local self-attention required a fraction of the compute and memory costs associated with attention over the entire feature and rich local contextual information in the local region. Thus, the proposed model surpasses current models, including the experimental model, in terms of accuracy, which can be attributed to the three layers. The training approach with three layers is illustrated in Figure 9, using the OuluVS2 dataset.

4.4. Statistical Analysis and Model Efficiency

We performed statistical analysis using the standard *t*-test to compare the significance of the combined modules. Models A and B of the convolutional layer were compared, based on Model C (Figure 10a–d), and Models C, D, E, F, and G were compared in the current layer (Figure 10a–d). In addition, in the transcription layer, Models C and H and the proposed model were compared (Figure 10e–h). For all four angles in Figure 10a–d, the proposed model showed that the modules in the convolutional layer have significant differences. That is, the performance increased by combining the MLFF 3D CNN and the SAM with the DenseNet-121 model. In addition, in the recurrent layer, the use of the Bi-GRU classifier (Model C) exhibited the highest performance and significant results compared to the four RNN-type units. However, in the case of Model G, because the unidirectional GRU model was used, there was no significant difference compared to Model C, which is a bidirectional model. Figure 10e–h shows the statistical analysis of the transcription layer. The performance of the two models using the self-attention mechanism in the cascade method was higher and significant than that for learning based on the basic CTC loss function. Consequently, the proposed model exhibited significant performance improvement.

In practical applications, the primary limitations of the VSR systems are their size and computing capacity. We explored the models' computational efficiency by examining their accuracy over various training settings and epochs. The system's performance as a function of the number of parameters is shown in Figure 11a–d. Furthermore, Figure 11e–h depict the results of the average epoch–time comparison of the nine models for 500 epochs. As demonstrated in Table 4, each model on the OuluVS2 dataset has a unique set of parameters and epoch time. Compared to Model D, which presented the lowest accuracy among the compared models, the proposed model had a parameter count difference of approximately 29 M. The average accuracy was improved by 12.24%. In comparison to Model F, which had the most parameters, the proposed approach decreased the number of parameters by roughly 11 M, while increasing accuracy by 9.53%. In addition, the difference in learning time compared to Model D, with the smallest number of parameters, differed by 5.54 s on average per epoch, which is not significant. Furthermore, the difference in learning time compared to that of Model F, which has the most parameters, was 13.05 s. Thus, the proposed model is capable of enhancing accuracy and decreasing learning time without considerably increasing the number of parameters.

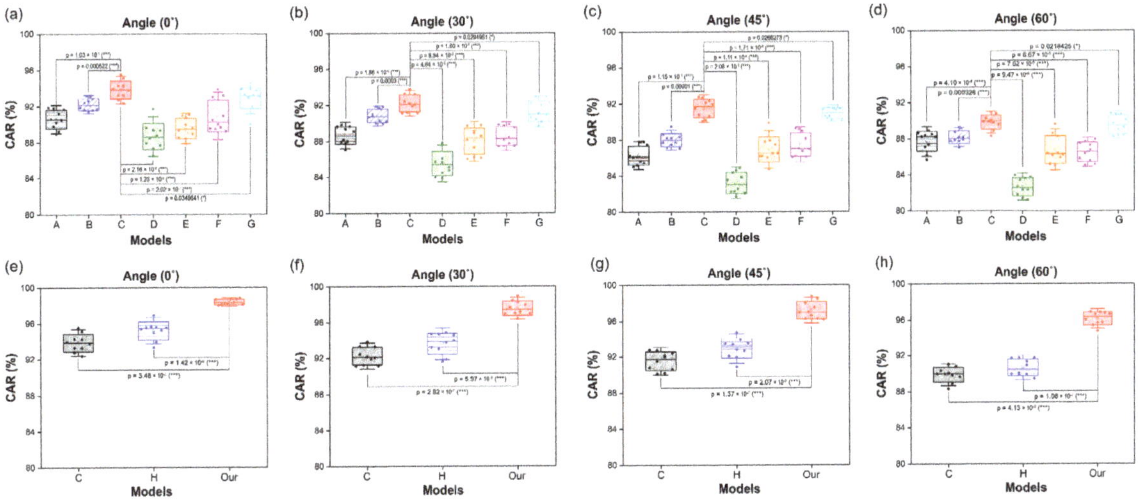

Figure 10. Comparison between different models and the proposed model based on mean accuracy of the last 10 epochs: (**a–d**) Convolutional layer and recurrent layer; (**e–h**) Transcription layer. Error bars represent standard deviation. Asterisks represent statistical significance-based t-tests between each group (* for $p < 0.05$, ** for $p < 0.01$, and *** for $p < 0.001$).

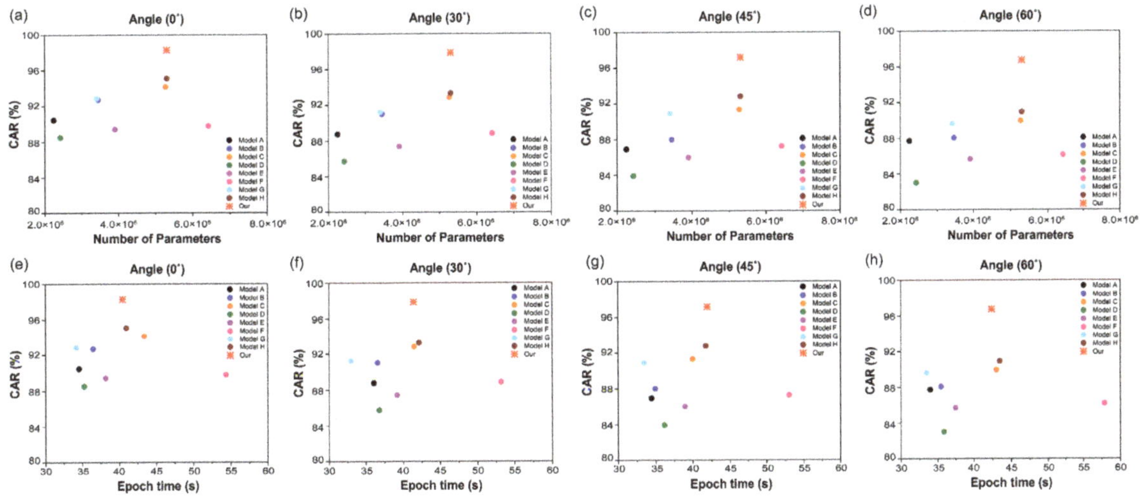

Figure 11. Comparison of character accuracy rate (CAR) between the proposed model and other models according to the (**a–d**) number of parameters and (**e–h**) average epoch time.

Table 4. Comparison between the number of parameters and epoch times of the proposed method and different methods.

Model	Method	Number of Parameters	Epoch Time (s)			
			0°	30°	45°	60°
A	3D dense connection CNN + Bi-GRU + CTC	2,247,537	34.57	36.07	34.43	33.97
B	3D dense connection CNN + Multi-scale 3D CNN + Bi-GRU + CTC	3,456,369	36.48	36.58	34.93	35.43
C	3D dense connection CNN + Multi-scale 3D CNN + SAM + Bi-GRU + CTC	5,273,457	43.37	41.44	40.01	43.03
D	3D dense connection CNN + Multi-scale 3D CNN + SAM + RNN + CTC	2,429,362	35.27	36.78	36.15	35.86
E	3D dense connection CNN + Multi-scale 3D CNN + SAM + LSTM + CTC	3,905,458	38.16	39.13	38.94	37.45
F	3D dense connection CNN + Multi-scale 3D CNN + SAM + Bi-LSTM + CTC	6,421,426	54.35	53.18	53.04	57.86
G	3D dense connection CNN + Multi-scale 3D CNN + SAM + GRU + CTC	3,413,426	34.18	32.98	33.48	33.48
H	3D dense connection CNN + Multi-scale 3D CNN + SAM + Bi-GRU + Global self-attention + CTC	5,306,290	40.95	42.13	41.78	43.48
Our	3D dense connection CNN + Multi-scale 3D CNN + SAM + Bi-GRU + Local self-attention + CTC	5,306,290	40.46	41.39	41.97	42.41

4.5. *Confusion Matrix*

We compared the confusion matrices of the two models that exhibited outstanding performance in the three layers with that of the proposed model for the four angles. Specifically, we evaluated Model C (Figure 12), which exhibited the highest accuracy in the convolutional and recurrent layers; Model H (Figure 13), which exhibited excellent performance in the transcription layer; and the proposed model (Figure 14). When comparing the results shown in Figure 12, the proposed model realizes fewer incorrect predictions. In addition, Model C had more erroneous predictions than the other two models for the four angles. The number was particularly high for "Hello", "Thank you", and "See you" because they are visually similar from the same viewpoint, furthermore, "Thank you" and "See you" have identical viseme sequences around the beginning and end of the utterance, which explains why these phase pairings have a higher number of false predictions. Because they are visually comparable from the same viewpoint, the three pairs of sentences with the highest error rate are the most demanding and confusing pairings with a high error rate, as indicated by the confusion matrix [13,24,31].

However, when the global self-attention mechanism was combined with the transcription layer, Model H exhibited better overall confusion pair results than Model C in 10 phases. Model H clearly demonstrated that confusion decreased compared to Model C. Despite the decrease in confusion, some pairs show particularly high confusion rates at each angle. As can be observed in Figure 13a, the predictions between "Nice to meet you" and "How are you" were the lowest, and, as shown in Figure 13b,c, were confused with "Nice to meet you" and "How are you" for "Thank you." In addition, unlike the other three angles, the 60° angle (Figure 13d) showed substantial confusion, wherein "Thank you" and "How are you" exhibited the lowest predictions. Therefore, Model H, similar to Model C, increased the number of confusions, due to the similarity of the visual view as the angle increased. The last pronunciation, such as "you", showed low predictions within a similar phase.

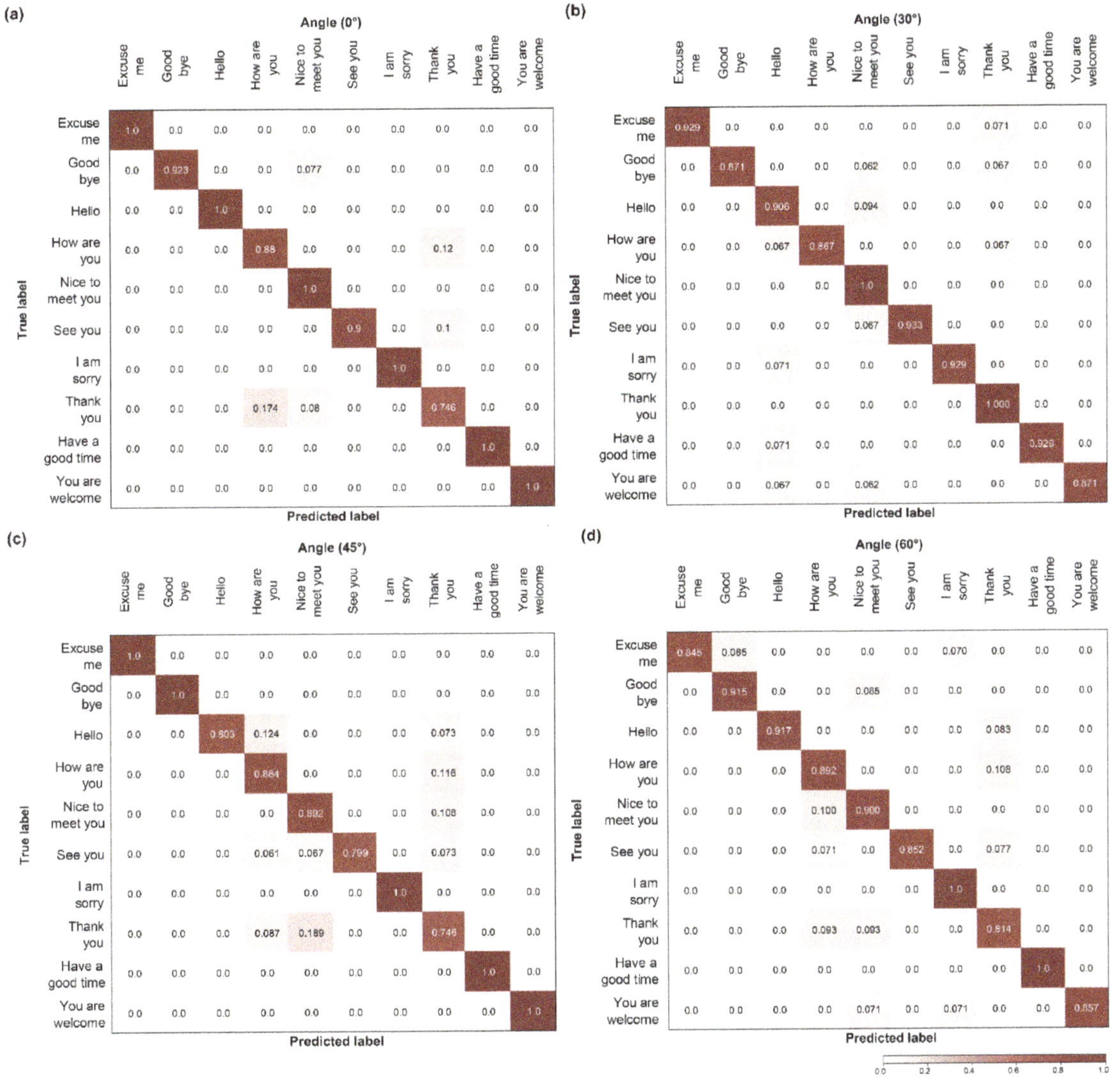

Figure 12. Comparison of confusion matrix models: (**a**–**d**) Model C.

Unlike the two models, the proposed model yields low confusion at all of the angles using the local self-attention mechanism. In particular, for the 60° angle, both Models C (Figure 12d) and H (Figure 13d) presented high confusion numbers. In contrast, the proposed model (Figure 14d) presented low confusion numbers, similar to other angles. In addition, the confusion between "Hello", "Thank you", and "See you" observed in the other two models was reduced, and the predicted value increased. By comparing the confusion matrices, we can easily define which of the models performs better. Thus, we can establish that the proposed model outperformed the others on the OuluVS2 dataset, distinguishing all comparable pronunciations in phase.

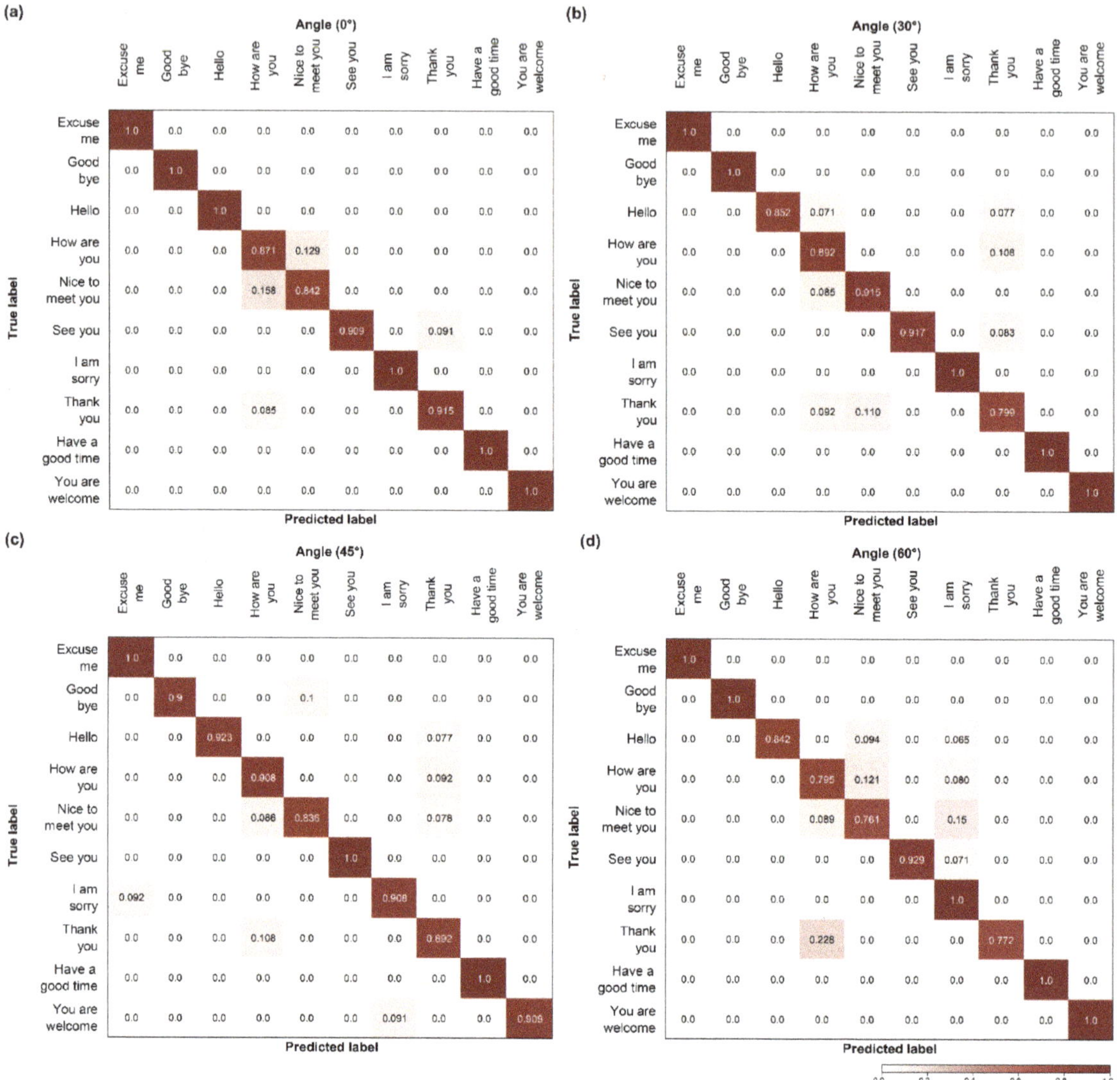

Figure 13. Comparison of confusion matrix models: (**a**–**d**) Model H.

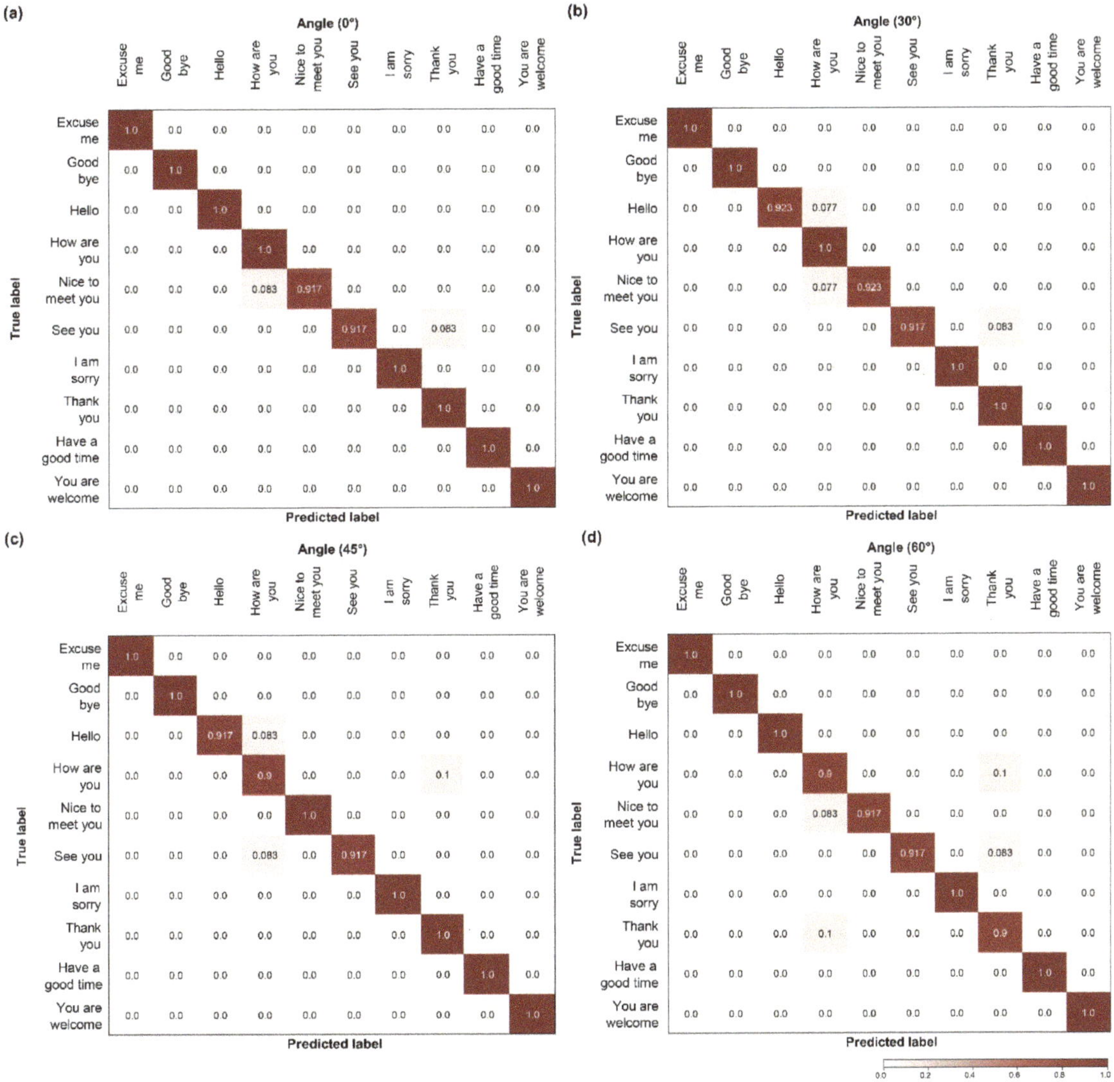

Figure 14. Comparison of confusion matrix models: (**a**–**d**) the proposed model.

5. Discussion and Conclusions

Lipreading is difficult to execute because it cannot be purely performed from the frontal perspective. Professional lip readers claim that a non-faceted approach, instead of a front-view, provides more information than a front-view with more pronounced lip protrusions and lip rounding. Consequently, the most significant limitation in using lipreading technology in real-world applications is its performance when reading lips from multiple angles. Therefore, we developed a multi-angle/multi-view VSR architecture that performs VSR by detecting both frontal and non-frontal lip images.

This study provides an end-to-end infrastructure for recording multi-view video surveillance. We obtained an accurate viseme prediction using SAM, multiple CNNs, and cascaded local self-attention-CTC. This is the first time that a 3D CNN, 3D dense connection CNN, and SAM have been combined with a multi-scale 3D CNN to extract lip motion characteristics as encoders. Following the decoder's Bi-GRU, a transcription layer based on cascaded local self-attention-CTC was used to extract exhaustive local contextual information from the surrounding environment.

The advantages of each level of the proposed architecture can be summarized as follows. The 3D dense connection CNN helps in reducing gradient vanishing and deepening the network (to use features) in an efficient manner. It also helps in reducing model parameters and preventing overfitting, thereby conserving computational resources. Finally, the multi-scale 3D CNN is applied to the two dropout layers, using features at different levels to effectively analyze the motion context in the temporal and spatial domains, with fine motion and high spatial correlation. SAM and multi-scale 3D CNNs are combined and concatenated to provide a single output. Consequently, SAM exploits the inter-spatial interaction of characteristics to better select and focus on the most identifiable and practical portions of an input picture. Moreover, cascaded local self-attention-CTC, following the decoder's Bi-GRU, requires only a fraction of the computation and memory costs of attention over the entire feature, leading to compact packaging of padded features in mini-batches and significant savings. Hence, this module can be used to acquire detailed local contextual information from the surrounding area.

We compared the outcomes of various deep learning models for predicting the sequence of phrases. The proposed architecture outperformed the others in terms of SOTA CER (Tables 2 and 3). We also compared the convergence rate, optimization, accuracy, statistical analysis, model efficiency, and confusion of the learning process for the three layers (convolution, recurrent, and transcription). The proposed model exhibited a faster convergence speed and higher accuracy compared to the other models, without a significant difference in the number of parameters and epoch time.

The proposed model attained SOTA performance on the OuluVS2 dataset without requiring external data or even data augmentation. The given mouth ROIs, on the other hand, were appropriately cropped, which may not be the case when employing automated mouth ROI identification techniques. Additionally, it would be interesting to investigate the effect of automated mouth ROI cropping on multi-view lipreading because the accuracy of automatic detectors is known to degrade with non-frontal views. Finally, because the model can be readily expanded to other streams, we expect to incorporate an audio stream to see how well it performs in audio-visual multi-view speech recognition.

Developing a multi-view VSR system that exclusively relies on visual data is crucial. Speech recognition in loud situations, hearing impairment, and biometric identification are some applications for which such a system will be practical. It could also be helpful for people with speech difficulties. However, because speech involves auditory and visual information, it is still challenging to perform ASR simply by using VSR. As a result, we plan to widen our approach in the future to include performance optimization and identification of potential uses for audio and visual data.

Author Contributions: Methodology, S.J.; software, S.J.; validation, S.J.; formal analysis, S.J.; investigation, S.J.; data curation, S.J.; writing—original draft preparation, S.J.; writing—review and editing, S.J. and M.S.K.; visualization, S.J.; supervision, M.S.K.; funding acquisition, M.S.K. All authors have read and agreed to the published version of the manuscript.

Funding: This research was funded by National Research Foundation of Korea (NRF) grant funded by the Korea government (MSIT), grant number NRF-2018X1A3A1069795.

Institutional Review Board Statement: Not applicable.

Informed Consent Statement: Not applicable.

Data Availability Statement: The database used in this article was OuluVS2. For details, please refer to [27].

Conflicts of Interest: The authors declare no conflict of interest. The funder had no role in the design of the study; in the collection, analyses, or interpretation of data; in the writing of the manuscript, or in the decision to publish the results.

References

1. Antonakos, E.; Roussos, A.; Zafeiriou, S. A survey on mouth modeling and analysis for sign language recognition. In Proceedings of the 2015 11th IEEE International Conference and Workshops on Automatic Face and Gesture Recognition (FG), Ljubljana, Slovenia, 4–8 May 2015; Volume 1, pp. 1–7.
2. Seymour, R.; Stewart, D.; Ming, J. Comparison of image transform-based features for visual speech recognition in clean and corrupted videos. *EURASIP J. Image Video Process.* **2007**, *2008*, 1–9. [CrossRef]
3. Potamianos, G. Audiovisual automatic speech recognition: Progress and challenges. *J. Acoust. Soc. Am.* **2008**, *123*, 3939. [CrossRef]
4. Zhou, Z.; Zhao, G.; Hong, X.; Pietikäinen, M. A review of recent advances in visual speech decoding. *Image Vis. Comput.* **2014**, *32*, 590–605. [CrossRef]
5. Akhtar, Z.; Micheloni, C.; Foresti, G.L. Biometric liveness detection: Challenges and research opportunities. *IEEE Secur. Priv.* **2015**, *13*, 63–72. [CrossRef]
6. Suwajanakorn, S.; Seitz, S.M.; Kemelmacher-Shlizerman, I. Synthesizing Obama: Learning lip sync from audio. *ACM Trans. Graphics (ToG)* **2017**, *36*, 1–13. [CrossRef]
7. Koller, O.; Camgoz, N.C.; Ney, H.; Bowden, R. Weakly supervised learning with multi-stream CNN-LSTM-HMMs to discover sequential parallelism in sign language videos. *IEEE Trans. Pattern Anal. Mach. Intell.* **2019**, *42*, 2306–2320. [CrossRef]
8. Zhou, H.; Zhou, W.; Zhou, Y.; Li, H. Spatial-temporal multi-cue network for continuous sign language recognition. In Proceedings of the AAAI Conference on Artificial Intelligence, New York City, NY, USA, 7–12 February 2020; Volume 34, pp. 13009–13016.
9. Fenghour, S.; Chen, D.; Guo, K.; Xiao, P. Lip reading sentences using deep learning with only visual cues. *IEEE Access* **2020**, *8*, 215516–215530. [CrossRef]
10. Yang, C.; Wang, S.; Zhang, X.; Zhu, Y. Speaker-independent lipreading with limited data. In Proceedings of the 2020 IEEE International Conference on Image Processing (ICIP), Abu Dhabi, United Arab Emirates, 25–28 October 2020; pp. 2181–2185.
11. Lu, Y.; Yan, J. Automatic lip reading using convolution neural network and bidirectional long short-term memory. *Int. J. Pattern. Recognit. Artif. Intell.* **2020**, *34*, 2054003. [CrossRef]
12. Chen, X.; Du, J.; Zhang, H. Lipreading with DenseNet and resBi-LSTM. *Signal Image Video Process.* **2020**, *14*, 981–989. [CrossRef]
13. Petridis, S.; Li, Z.; Pantic, M. End-to-end visual speech recognition with LSTMs. In Proceedings of the 2017 IEEE International Conference on Acoustics, Speech and Signal Processing (ICASSP), New Orleans, LA, USA, 5–9 March 2017; pp. 2592–2596.
14. Xu, K.; Li, D.; Cassimatis, N.; Wang, X. LCANet: End-to-end lipreading with cascaded attention-CTC. In Proceedings of the 2018 13th IEEE International Conference on Automatic Face & Gesture Recognition (FG 2018), Xi'an, China, 15–19 May 2018; pp. 548–555.
15. Margam, D.K.; Aralikatti, R.; Sharma, T.; Thanda, A.; Roy, S.; Venkatesan, S.M. LipReading with 3D-2D-CNN BLSTM-HMM and word-CTC models. *arXiv* **2019**, arXiv:1906.12170.
16. Bauman, S.L.; Hambrecht, G. Analysis of view angle used in speechreading training of sentences. *Am. J. Audiol.* **1995**, *4*, 67–70. [CrossRef]
17. Lan, Y.; Theobald, B.-J.; Harvey, R. View independent computer lip-reading. In Proceedings of the 2012 IEEE International Conference on Multimedia and Expo, Melbourne, VIC, Australia, 9–13 July 2012; pp. 432–437.
18. Assael, Y.M.; Shillingford, B.; Whiteson, S.; De Freitas, N. Lipnet: End-to-end sentence-level lipreading. *arXiv* **2016**, arXiv:1611.01599.
19. Santos, T.I.; Abel, A.; Wilson, N.; Xu, Y. Speaker-independent visual speech recognition with the Inception V3 model. In Proceedings of the 2021 IEEE Spoken Language Technology Workshop (SLT), Shenzhen, China, 19–22 January 2021; pp. 613–620.
20. Lucey, P.; Potamianos, G. Lipreading using profile versus frontal views. In Proceedings of the 2006 IEEE Workshop on Multimedia Signal Processing, Victoria, BC, Canada, 3–6 October 2006; pp. 24–28.
21. Saitoh, T.; Zhou, Z.; Zhao, G.; Pietikäinen, M. Concatenated frame image based CNN for visual speech recognition. In *Asian Conference on Computer Vision*; Springer: Berlin/Heidelberg, Germany, 2016; pp. 277–289.

22. Zimmermann, M.; Ghazi, M.M.; Ekenel, H.K.; Thiran, J.-P. Visual speech recognition using PCA networks and LSTMs in a tandem GMM-HMM system. In *Asian Conference on Computer Vision*; Springer: Berlin/Heidelberg, Germany, 2016; pp. 264–276.
23. Koumparoulis, A.; Potamianos, G. Deep view2view mapping for view-invariant lipreading. In Proceedings of the 2018 IEEE Spoken Language Technology Workshop (SLT), Athens, Greece, 18–21 December 2018; pp. 588–594.
24. Petridis, S.; Wang, Y.; Li, Z.; Pantic, M. End-to-end multi-view lipreading. *arXiv* **2017**, arXiv:1709.00443.
25. Zimmermann, M.; Ghazi, M.M.; Ekenel, H.K.; Thiran, J.-P. Combining multiple views for visual speech recognition. *arXiv* **2017**, arXiv:1710.07168.
26. Sahrawat, D.; Kumar, Y.; Aggarwal, S.; Yin, Y.; Shah, R.R.; Zimmermann, R. "Notic My Speech"—Blending speech patterns with multimedia. *arXiv* **2020**, arXiv:2006.08599.
27. Anina, I.; Zhou, Z.; Zhao, G.; Pietikäinen, M. OuluVS2: A multi-view audiovisual database for non-rigid mouth motion analysis. In Proceedings of the 2015 11th IEEE International Conference and Workshops on Automatic Face and Gesture Recognition (FG), Ljubljana, Slovenia, 4–8 May 2015; Volume 1, pp. 1–5.
28. Estellers, V.; Thiran, J.-P. Multipose audio-visual speech recognition. In Proceedings of the 2011 19th European Signal Processing Conference, Barcelona, Spain, 29 August–2 September 2011; pp. 1065–1069.
29. Isobe, S.; Tamura, S.; Hayamizu, S. Speech recognition using deep canonical correlation analysis in noisy environments. In Proceedings of the ICPRAM, Online. 4–6 February 2021; pp. 63–70.
30. Komai, Y.; Yang, N.; Takiguchi, T.; Ariki, Y. Robust AAM-based audio-visual speech recognition against face direction changes. In Proceedings of the 20th ACM international conference on Multimedia, Nara, Japan, 29 October–2 November 2012; pp. 1161–1164.
31. Lee, D.; Lee, J.; Kim, K.-E. Multi-view automatic lip-reading using neural network. In Proceedings of the Asian Conference on Computer Vision, Taipei, Taiwan, 20–24 November 2016; Springer: Berlin/Heidelberg, Germany; pp. 290–302.
32. Jeon, S.; Elsharkawy, A.; Kim, M.S. Lipreading architecture based on multiple convolutional neural networks for sentence-level visual speech recognition. *Sensors* **2022**, *22*, 72. [CrossRef]
33. Zhang, P.; Wang, D.; Lu, H.; Wang, H.; Ruan, X. Amulet: Aggregating multi-level convolutional features for salient object detection. In Proceedings of the IEEE International Conference on Computer Vision, Venice, Italy, 22–29 October 2017; pp. 202–211.
34. Larochelle, H.; Hinton, G.E. Learning to combine foveal glimpses with a third-order Boltzmann machine. *Adv. Neural Inf. Process. Syst.* **2010**, *23*, 1243–1251.
35. Mnih, V.; Heess, N.; Graves, A. Recurrent models of visual attention. In Proceedings of the Advances in Neural Information Processing Systems, Montreal, QC, Canada, 8–13 December 2014; pp. 2204–2212.
36. Vaswani, A.; Shazeer, N.; Parmar, N.; Uszkoreit, J.; Jones, L.; Gomez, A.N.; Kaiser, Ł.; Polosukhin, I. Attention is all you need. In Proceedings of the Advances in Neural Information Processing Systems, Long Beach, CA, USA, 4–9 December 2017; pp. 5998–6008.
37. Hu, J.; Shen, L.; Sun, G. Squeeze-and-excitation networks. In Proceedings of the IEEE Conference on Computer Vision and Pattern Recognition, Salt Lake City, UT, USA, 18–23 June 2018; pp. 7132–7141.
38. Woo, S.; Park, J.; Lee, J.-Y.; Kweon, I.S. CBAM: Convolutional block attention module. In Proceedings of the European Conference on Computer Vision (ECCV), Munich, Germany, 8–14 September 2018; pp. 3–19.
39. Zhang, T.; He, L.; Li, X.; Feng, G. Efficient end-to-end sentence-level lipreading with temporal convolutional networks. *Appl. Sci.* **2021**, *11*, 6975. [CrossRef]
40. Hlaváč, M.; Gruber, I.; Železný, M.; Karpov, A. Lipreading with LipsID. In Proceedings of the International Conference on Speech and Computer, St. Petersburg, Russia, 7–9 October 2020; Springer: Berlin/Heidelberg, Germany; pp. 176–183.
41. Luo, M.; Yang, S.; Shan, S.; Chen, X. Pseudo-convolutional policy gradient for sequence-to-sequence lip-reading. In Proceedings of the 2020 15th IEEE International Conference on Automatic Face and Gesture Recognition (FG 2020), Buenos Aires, Argentina, 16–20 November 2020; pp. 273–280.
42. Hinton, G.E.; Srivastava, N.; Krizhevsky, A.; Sutskever, I.; Salakhutdinov, R.R. Improving neural networks by preventing co-adaptation of feature detectors. *arXiv* **2012**, arXiv:1207.0580.
43. Tompson, J.; Goroshin, R.; Jain, A.; LeCun, Y.; Bregler, C. Efficient object localization using convolutional networks. In Proceedings of the IEEE Conference on Computer Vision and Pattern Recognition, Boston, MA, USA, 7–12 June 2015; pp. 648–656.
44. Hochreiter, S.; Schmidhuber, J. Long short-term memory. *Neural Comput.* **1997**, *9*, 1735–1780. [CrossRef]
45. Chung, J.; Gulcehre, C.; Cho, K.; Bengio, Y. Empirical evaluation of gated recurrent neural networks on sequence modeling. *arXiv* **2014**, arXiv:1412.3555.
46. Fischer, T.; Krauss, C. Deep learning with long short-term memory networks for financial market predictions. *Eur. J. Operat. Res.* **2018**, *270*, 654–669. [CrossRef]
47. Tran, Q.-K.; Song, S.-K. Water level forecasting based on deep learning: A use case of Trinity River-Texas-The United States. *J. KIISE* **2017**, *44*, 607–612. [CrossRef]
48. Chung, J.S.; Zisserman, A. Learning to lip read words by watching videos. *Comput. Vis. Image Understand.* **2018**, *173*, 76–85. [CrossRef]
49. Graves, A.; Fernández, S.; Gomez, F.; Schmidhuber, J. Connectionist temporal classification: Labelling unsegmented sequence data with recurrent neural networks. In Proceedings of the 23rd international Conference on Machine Learning, Pittsburgh, PA, USA, 25–29 June 2006; pp. 369–376.
50. Chung, J.S.; Senior, A.; Vinyals, O.; Zisserman, A. Lip reading sentences in the wild. In Proceedings of the 2017 IEEE Conference on Computer Vision and Pattern Recognition (CVPR), Honolulu, HI, USA, 26 July 2017; pp. 3444–3453.

51. Cheng, J.; Dong, L.; Lapata, M. Long short-term memory-networks for machine reading. *arXiv* **2016**, arXiv:1601.06733.
52. Shen, T.; Zhou, T.; Long, G.; Jiang, J.; Pan, S.; Zhang, C. Disan: Directional self-attention network for RNN/CNN-free language understanding. In Proceedings of the AAAI Conference on Artificial Intelligence, New Orleans, LA, USA, 2–7 February 2018; Volume 32.
53. Zhang, Y.; Chan, W.; Jaitly, N. Very deep convolutional networks for end-to-end speech recognition. In Proceedings of the 2017 IEEE International Conference on Acoustics, Speech and Signal Processing (ICASSP), New Orleans, LA, USA, 5–9 March 2017; pp. 4845–4849.
54. Kim, S.; Hori, T.; Watanabe, S. Joint CTC-attention based end-to-end speech recognition using multi-task learning. In Proceedings of the 2017 IEEE International Conference on Acoustics, Speech and Signal Processing (ICASSP), New Orleans, LA, USA, 5–9 March 2017; pp. 4835–4839.
55. Chiu, C.-C.; Sainath, T.N.; Wu, Y.; Prabhavalkar, R.; Nguyen, P.; Chen, Z.; Kannan, A.; Weiss, R.J.; Rao, K.; Gonina, E.; et al. State-of-the-art speech recognition with sequence-to-sequence models. In Proceedings of the 2018 IEEE International Conference on Acoustics, Speech and Signal Processing (ICASSP), Calgary, AB, Canada, 15–20 April 2018; pp. 4774–4778.
56. Zeyer, A.; Irie, K.; Schlüter, R.; Ney, H. Improved training of end-to-end attention models for speech recognition. *arXiv* **2018**, arXiv:1805.03294.
57. Chen, Z.; Droppo, J.; Li, J.; Xiong, W. Progressive joint modeling in unsupervised single-channel overlapped speech recognition. *IEEE/ACM Trans. Audio Speech Lang. Process.* **2017**, *26*, 184–196. [CrossRef]
58. Erdogan, H.; Hayashi, T.; Hershey, J.R.; Hori, T.; Hori, C.; Hsu, W.N.; Kim, S.; Le Roux, J.; Meng, Z.; Watanabe, S. Multi-channel speech recognition: Lstms all the way through. In Proceedings of the CHiME-4 Workshop, San Francisco, CA, USA, 13 September 2016; pp. 1–4.
59. Liu, P.J.; Saleh, M.; Pot, E.; Goodrich, B.; Sepassi, R.; Kaiser, L.; Shazeer, N. Generating Wikipedia by summarizing long sequences. *arXiv* **2018**, arXiv:1801.10198.
60. Huang, C.-Z.A.; Vaswani, A.; Uszkoreit, J.; Shazeer, N.; Simon, I.; Hawthorne, C.; Dai, A.M.; Hoffman, M.D.; Dinculescu, M.; Eck, D. Music transformer. *arXiv* **2018**, arXiv:1809.04281.
61. King, D.E. Dlib-ml: A machine learning toolkit. *J. Mach. Learn. Res.* **2009**, *10*, 1755–1758.
62. Sagonas, C.; Tzimiropoulos, G.; Zafeiriou, S.; Pantic, M. 300 faces in-the-wild challenge: The first facial landmark localization challenge. In Proceedings of the IEEE International Conference on Computer Vision Workshops, Sydney, NSW, Australia, 2–8 December 2013; pp. 397–403.
63. Kingma, D.P.; Ba, J. Adam: A method for stochastic optimization. *arXiv* **2014**, arXiv:1412.6980.
64. Bottou, L. Stochastic gradient descent tricks. In *Neural Networks: Tricks of the Trade*; Springer: Berlin/Heidelberg, Germany, 2012; pp. 421–436.
65. Tieleman, T.; Hinton, G. Lecture 6.5-rmsprop: Divide the gradient by a running average of its recent magnitude. *COURSERA Neural Netw. Mach. Learn.* **2012**, *4*, 26–31.
66. Huang, G.; Liu, Z.; Van Der Maaten, L.; Weinberger, K.Q. Densely connected convolutional networks. In Proceedings of the IEEE conference on computer vision and pattern recognition, Honolulu, HI, USA, 21–26 July 2017; pp. 4700–4708.
67. Schaul, T.; Antonoglou, I.; Silver, D. Unit tests for stochastic optimization. *arXiv* **2013**, arXiv:1312.6055.
68. Sutskever, I.; Martens, J.; Dahl, G.; Hinton, G. On the importance of initialization and momentum in deep learning. In Proceedings of the International Conference on Machine Learning, Atlanta, GA, USA, 16–21 June 2013; PMLR. pp. 1139–1147.
69. Zhou, Z.; Hong, X.; Zhao, G.; Pietikäinen, M. A compact representation of visual speech data using latent variables. *IEEE Trans. Pattern Anal. Mach. Intell.* **2013**, *36*, 1. [CrossRef] [PubMed]
70. Chung, J.S.; Zisserman, A. Out of time: Automated lip sync in the wild. In Proceedings of the Asian Conference on Computer Vision, Taipei, Taiwan, 20–24 November 2016; Springer: Berlin/Heidelberg, Germany; pp. 251–263.
71. Chung, J.S.; Zisserman, A. Lip reading in profile. In Proceedings of the British Machine Vision Conference (BMVC), Imperial College London, London, UK, 4–7 September 2017; pp. 1–11.
72. Petridis, S.; Wang, Y.; Li, Z.; Pantic, M. End-to-end audiovisual fusion with LSTMs. *arXiv* **2017**, arXiv:1709.04343.
73. Han, H.; Kang, S.; Yoo, C.D. Multi-view visual speech recognition based on multi task learning. In Proceedings of the 2017 IEEE International Conference on Image Processing (ICIP), Beijing, China, 17–20 September 2017; pp. 3983–3987.
74. Fung, I.; Mak, B. End-to-end low-resource lip-reading with maxout CNN and LSTM. In Proceedings of the 2018 IEEE International Conference on Acoustics, Speech and Signal Processing (ICASSP), Calgary, AB, Canada, 15–20 April 2018; pp. 2511–2515.
75. Fernandez-Lopez, A.; Sukno, F.M. Lip-reading with limited-data network. In Proceedings of the 2019 27th European Signal Processing Conference (EUSIPCO), A Coruña, Spain, 2–6 September 2019; pp. 1–5.
76. Isobe, S.; Tamura, S.; Hayamizu, S.; Gotoh, Y.; Nose, M. Multi-angle lipreading using angle classification and angle-specific feature integration. In Proceedings of the 2020 International Conference on Communications, Signal Processing, and Their Applications (ICCSPA), Sharjah, United Arab Emirates, 16–18 March 2021; pp. 1–5.
77. Isobe, S.; Tamura, S.; Hayamizu, S.; Gotoh, Y.; Nose, M. Multi-angle lipreading with angle classification-based feature extraction and its application to audio-visual speech recognition. *Future Internet* **2021**, *13*, 182. [CrossRef]

sensors

Article

Speaker Adaptation on Articulation and Acoustics for Articulation-to-Speech Synthesis

Beiming Cao [1,2], Alan Wisler [3] and Jun Wang [2,4,*]

1 Department of Electrical and Computer Engineering, University of Texas at Austin, Austin, TX 78712, USA
2 Department of Speech, Language, and Hearing Sciences, University of Texas at Austin, Austin, TX 78712, USA
3 Department of Mathematics and Statistics, Utah State University, Logan, UT 84322, USA
4 Department of Neurology, Dell Medical School, University of Texas at Austin, Austin, TX 78712, USA
* Correspondence: jun.wang@austin.utexas.edu

Abstract: Silent speech interfaces (SSIs) convert non-audio bio-signals, such as articulatory movement, to speech. This technology has the potential to recover the speech ability of individuals who have lost their voice but can still articulate (e.g., laryngectomees). Articulation-to-speech (ATS) synthesis is an algorithm design of SSI that has the advantages of easy-implementation and low-latency, and therefore is becoming more popular. Current ATS studies focus on speaker-dependent (SD) models to avoid large variations of articulatory patterns and acoustic features across speakers. However, these designs are limited by the small data size from individual speakers. Speaker adaptation designs that include multiple speakers' data have the potential to address the issue of limited data size from single speakers; however, few prior studies have investigated their performance in ATS. In this paper, we investigated speaker adaptation on both the input articulation and the output acoustic signals (with or without direct inclusion of data from test speakers) using the publicly available electromagnetic articulatory (EMA) dataset. We used Procrustes matching and voice conversion for articulation and voice adaptation, respectively. The performance of the ATS models was measured objectively by the mel-cepstral distortions (MCDs). The synthetic speech samples were generated and are provided in the supplementary material. The results demonstrated the improvement brought by both Procrustes matching and voice conversion on speaker-independent ATS. With the direct inclusion of target speaker data in the training process, the speaker-adaptive ATS achieved a comparable performance to speaker-dependent ATS. To our knowledge, this is the first study that has demonstrated that speaker-adaptive ATS can achieve a non-statistically different performance to speaker-dependent ATS.

Keywords: articulation-to-speech synthesis; silent speech interface; speaker adaption; voice conversion

Citation: Cao, B.; Wisler, A.; Wang, J. Speaker Adaptation on Articulation and Acoustics for Articulation-to-Speech Synthesis. *Sensors* **2022**, *22*, 6056. https://doi.org/10.3390/s22166056

Academic Editors: Bruce Denby, Tamás Gábor Csapó and Michael Wand

Received: 30 June 2022
Accepted: 8 August 2022
Published: 13 August 2022

1. Introduction

Laryngectomees are people who have their larynx partially or totally removed in surgeries (laryngectomy), due to the treatment of laryngeal cancer [1]. Especially, for people who have undergone total laryngectomy, their ability to produce normally voiced speech is lost. Currently, they have three main options for their daily communication: esophageal speech [2], tracheo-esophageal puncture (TEP) speech [3] and electro-larynx (EL) [4,5]. The major common disadvantage of these approaches is they generate unnatural or hoarse voices, which discourages their use and causes social isolation [6]. Silent speech interfaces (SSIs) are devices that enable speech communication when a human's phonatory abilities are impeded [7–9]. SSIs convert (silent) articulatory motion to speech, which have the potential of recovering speech ability for people who are unable to produce speech sounds but are still able to articulate(e.g., laryngectomees). There are currently two algorithmic designs in silent speech interfaces: the recognition-and-synthesis and the directly articulation-to-speech (ATS) synthesis. The recognition-and-synthesis design [7,10] recognizes textual information from non-audio articulatory signals with silent speech recognition [11], and then use a text-to-speech to convert recognized text to speech [12]. ATS is a procedure that directly maps a human's articulatory bio-signals to speech. As an end-to-end model and compared to the recognition-and-synthesis design, ATS has become a popular software design for silent speech interfaces, because of its advantages of low-latency and easier implementation [8].

Currently, most of the ATS studies focus on speaker-dependent (SD) design, in which only the data from testing speakers are used to train the ATS model [13]. SD-ATS usually suffer the restriction from insufficient training data since it is difficult to record a large amount of articulatory data from the same speakers. The main reason is the current articulatory information capture approaches normally require directly [14–16] or indirectly [17,18] attaching hardware such as sensors to subjects' articulators. Hours of data recording sessions will generally cause subjects to fatigue. Compared to speaker-dependent systems, speaker-independent (SI) systems require no training data from testing speakers [13] by using data collected from other speakers for training. Speaker-independent systems could be a solution for insufficient training data from individual subjects. However, due to the inter-speaker variability, they usually suffer lower performance than well-trained SD systems. Therefore, speaker adaption approaches may be an alternative solution for ATS. Speaker adaptation approaches adapt speaker-independent systems to the target speakers (users) [13], which take the advantages of both speaker-independent (large training dataset) and speaker-dependent (target speaker information involved) systems. Speaker adaptation approaches have been actively studied and demonstrated to be effective in automatic speech recognition (ASR) and text-to-speech (TTS) applications [13,19], but have been relatively less studied in ATS [20,21]. To highlight these concepts and assist the description in this paper, we list the major difference between these terminologies below.

- Speaker-dependent ATS (SD-ATS) is where training and testing data are from the same speakers;
- Speaker-independent ATS (SI-ATS) is where training and testing data are from different speakers;
- Speaker-adaptive ATS (SA-ATS) is where training data are from other speakers and the target speaker.

Speaker adaptation for ATS is challenging because the inter-speaker variations take place in both the input articulation and the output acoustics. In addition, to maintain the identity of the output speech from SSI, the output side of ATS (speech voice) has to be as similar as possible to the target speaker's original voice. This characteristic restricted the usage of some averaging-based [19] and warping-based speaker adaptation approaches on the output audio, such as cepstral mean and variance normalization (CMVN) [22] and vocal tract length normalization (VTLN) [23]. A valid approach to perform adaptation on acoustic output is setting adapting other speakers' acoustic data to the target speaker [24]. Therefore, in this study, we proposed a voice conversion-based audio adaptation approach for ATS.

In this study, we performed an investigation on speaker adaptation of ATS with voice conversion [24] and Procrustes matching [11]. The dataset used was a publicly available, electromagnetic articulograph (EMA) and audio data set (Haskins Production Rate Comparison database) [25]. The experiments were conducted in three sessions. The first session is the speaker-independent ATS (SI-ATS) as the baseline performance, and the speaker-dependent ATS as the target performance. Then we applied speaker adaptation on acoustics and articulation to the SI-ATS. In this session, Procrustes matching [11,26,27] was applied for the adaptation of the articulation, and voice conversion [24] models were adopted to convert the acoustic features of the training speakers to that are similar to the target speakers. Finally, we directly added the both articulatory and acoustic data of the target speakers to the training set to train a kind of speaker-adaptive (SA) model, then applied voice conversion and Procrustes matching on that to see if it could further improve the performance.

The ATS and voice conversion models used are long short-term memory (LSTM)-recurrent neural network (RNN). The Waveglow vocoder [28] was employed as the vocoder to convert the predicted acoustic features to speech waveforms. Due to the real-time decoding preference of ATS, advanced sequence-to-sequence models were not used in this study. Audible speech samples were generated and presented from the best ATS models in each experiment stage. Detailed discussions were made based on the experimental results.

The contributions of this paper include: (1) proposed and verified applying voice conversion for acoustics adaptation for speaker-independent (SI) ATS; (2) validated the Procrustes matching in SI-ATS application, which has only been shown effective in speaker-independent silent speech recognition [11,29]; (3) applied Waveglow vocoder [28] in EMA-based ATS application for the first time; (4) presented audible synthetic speech samples that were generated from multi-speaker (speaker-independent and speaker-adaptive) ATS.

2. Related Works

In the silent speech interface area, multiple techniques have been used for capturing articulatory motion data for the SSI purpose including: electromagnetic articulograph (EMA) [10,11,14,15], permanent magnet articulograph (PMA) [16,30–32], ultrasound image (UI) [18,20,33,34], surface electromyography (sEMG) [17,35], non-audible murmur (NAM) [36]. Doppler signals have been explored in the SSI application as well [37,38]. Kapur et al. used neuromuscular signals captured with electrodes as the input of SSI [39]. Recently, frequency-modulated continuous-wave radar has been investigated for SSI application as well [40]. Sebkhi et al. [41] have proposed an inertial measurement unit (IMU)-based PMA device that is suitable for SSI usage. As mentioned previously, most of the studies above used a speaker-dependent design, one of which is speaker-dependent and session-independent [35].

Only a few recent works studied speaker-independent and speaker-adaptive ATS systems. Shandiz et al. [20] have conducted studies on embedding speaker information into the ultrasound-based ATS to improve the performance on multiple speakers, in which the data from the testing speakers were involved in the training set. Similarly, Ribeiro et al. [21] also conducted multi-speaker ATS with ultrasound image data for a validation of their newly proposed dataset. The authors of [42] presented a study on speaker-independent mel-cepstrum estimator, in which the speaker-independent acoustic feature estimator was improved by embedding d-vectors and using pre-averaged acoustic. This study focused on speaker-independent systems [42], but the model predicted mel-cepstrum coefficients only, without generating speech samples. Although these ATS performances have been improved, no one has achieved a comparable performance by speaker-dependent ATS. In addition, no previous study was able to generate audible speech samples in their SI- or SA-ATS models.

This present study explored speaker-independent ATS and generated speech samples from that. The speaker adaptation was performed in a strategy of adapting voice from training speakers to that from the targeting speakers, which requires training one specific ATS model for one target speaker. This strategy is different to that in [20,42], which embedded speaker information to train one ATS model that aims to work for all testing speakers.

3. Dataset

The dataset used in this study is a dataset collected by the Haskins Lab, Yale University [25], which is an open access dataset, in which the electromagnetic articulography (EMA) data [14] and audio data were synchronously recorded from eight native American English speakers (four males, four females). The stimuli are the 720 phonetically balanced Harvard sentences from [43]. Each speaker read the 720 sentences at least two times, one in a normal speaking rate, one in a fast speaking rate. After that, they read a varying number of sentences in the normal speaking rate. In total, 1553 to 1738 sentences were recorded from each speaker, the duration of recorded data from each speaker is about 1 h. Additional details on the amount of data available for each speaker are provided in Table 1.

Table 1. Number of sentences and duration recorded from each speaker.

Speaker	Phrase Num.	Duration (min)
F01	1738	61.75
F02	1560	60.58
F03	1617	58.63
F04	1618	59.71
M01	1553	55.67
M02	1554	57.47
M03	1610	60.31
M04	1620	59.65
Sum.	12,870	472.82
Ave.	1609	59.22

The EMA data were recorded with the NDI Wave system, 8 sensors were attached to the tongue tip (TT), tongue blade (TB), tongue rear (TR), upper lip (UL), lower lip (LL), mouth left (corner) (ML), jaw, and jaw left (canine) [25]. Three-dimensional (x: posterior –> anterior, y: right –> left, z: inferior –> superior) articulatory movement of the sensors were recorded in a sampling rate of 100 Hz. The trajectories of sensors have been filtered with a 20 Hz Butterworth lowpass filter after recording. The audio data were recorded at a sampling rate of 44,100 Hz. In this study, we used 6 of 8 sensors for the experiments: tongue tip (TT), tongue blade (TB), tongue rear (TR), upper lip (UL), lower lip (LL), and jaw (JAW), which is consist with the setup in the mngu0 EMA dataset [44]. The audio data were downsampled from 44,100 Hz to 22,050 Hz, to make it consistent with the trained Waveglow vocoder [28] used in this study.

Other than the dataset used in this study, EMA-MAE corpus [45] is another EMA dataset that was collected from multiple speakers. EMA-MAE corpus is the EMA dataset that was collected from a relatively large number of speakers (40 speakers in total). About 30 to 45 min data were collected from each speaker, and part of that are isolated words. Therefore, the EMA-MAE dataset was not used in this study, due to the smaller amount of data from single speakers.

4. Methods

4.1. Articulation-to-Speech Synthesis

Figure 1 provides an overview of the implementation of articulation-to-speech synthesis models in this study. Articulatory movement of articulators (tongue, lips and jaw) was captured with sensors and sampled into frames, then fed to the ATS to predict the acoustic feature for speech synthesis. To maintain the real-time implementation of ATS, the advanced sequence-to-sequence models were excluded in this study. The ATS model used in this study is the long short-term memory-recurrent neural networks (LSTM-RNN), which has been shown to outperform typical deep neural networks (DNN) [15,42]. The bidirectional-LSTM (BLSTM) model has high performance in preliminary experiments, but the BLSTM-based ATS models do not support real-time SSI implementation.

The vocoder used in this study is the Waveglow vocoder, which is a flow-based network capable of generating high-quality speech from mel-spectrograms [28]. WaveGlow combines insights from the invertible implementation Glow [46] and the high performance neural vocoder WaveNet [47]. It has been demonstrated that WaveNet could generate higher-quality speech samples than the conventional source-filter vocoders [12,47–49] but in relatively high latency. WaveGlow showed a similar performance to WaveNet, but in a very low latency [28]. In addition, [34] demonstrated that Waveglow vocoder outperformed conventional vocoders [50–52] in ultrasound image-based ATS. Therefore, WaveGlow vocoder was chosen as the vocoder in this study, and the trained Waveglow model for English (WaveGlow-EN) provided by NVIDIA was directly adopted without additional training.

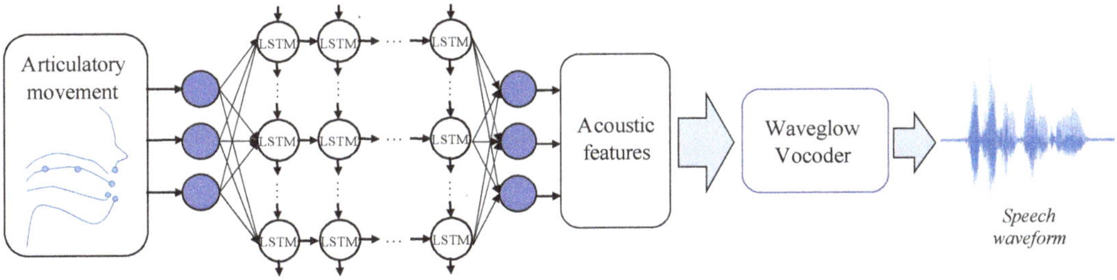

Figure 1. The overview illustration of a generic articulation-to-speech synthesis model.

The acoustic features are same as the default setup of Waveglow which were 80-dimensional mel-spectrograms, the fast Fourier transform (FFT) size was 1024, hop size (step size) was 256. The articulatory data were consisted of the 3-dimensional (3D) spatial location of six sensors at a sampling rate of 100 Hz, as mentioned the sensors were attached to six articulators: tongue tip (TT), tongue blade (TB), tongue rear (TR), upper lip (UL), lower lip (LL), and jaw (JAW). The first- and second-order derivatives were concatenated to the movement frames as the input frames, therefore the dimension of the ATS input is 54 (3-dim. $\times$ 6 sensors $\times$ 3). Although the left–right dimension is not as significant as the other two dimensions (front-back, and up-down) in speech production, 3D EMA data have demonstrated higher performance the 2D in preliminary experiments. Finally, the articulatory data of each phrase were scaled to the same length to the extracted acoustic features accordingly by interpolation.

The experimental results were measured with the mel-cepstral distortions (MCDs) [53]. For the MCD computation, the mel-spectrogram features were converted to the mel-frequency cepstral coefficients (MFCC) by applying discrete cosine transform (DCT). With the first 13 MFCCs, the MCDs were computed with the Equation (1) [54], the first MFCC was not included in the computation since it represents system energy gain rather than speech quality information (Equation (1)). In Equation (1), $C_{m,d}$ indicates the d-th ($1 \leq d \leq D$) MFCC dimension at time step m ($0 \leq m \leq T$). D is equal to 13, which is the total dimensional of MFCC included. T is the total number of MFCC frames generated.

$$MCD = \frac{10}{ln10} \sum_{m=0}{}^{T} \sqrt{2 \sum_{d=1}^{D} (C_{m,d} - C_{m,d}^{gen})^2} \quad (1)$$

4.2. Acoustic Adaptation Using Voice Conversion

Voice conversion (VC) is a type of voice transformation which aims to convert speech utterances of a source speaker to sound as if it was uttered by a target speaker [55]. Therefore, VC could be a suitable technology for adapting the voice of training speakers to the target speakers' voice [24]. Figure 2 shows the schema of the VC-based speaker adaptation for a single target speaker. The eight speakers take turns to be the target speaker in the cross-validation loop. Then train voice conversion models with the phrases in the training set of target and training speakers. The acoustic features of parallel phrases were aligned to the same length by the dynamic time warping (DTW) [56]. With the aligned acoustic features, VC models were trained for each of the target-training speaker pairs. After that, the acoustic features of training speakers were converted to target speakers' acoustic features by the VC models, and used for the speaker-independent ATS model training.

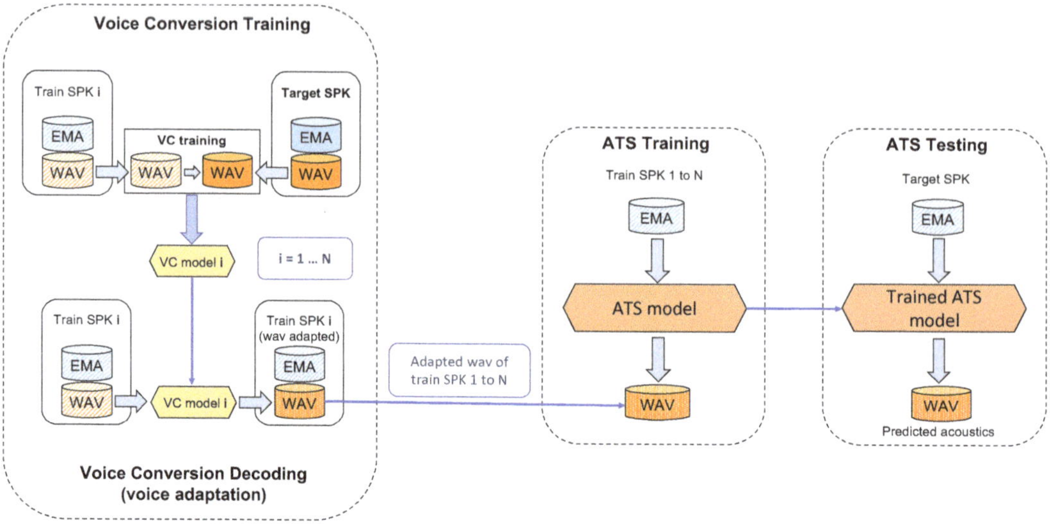

Figure 2. The pipeline of ATS Speaker adaptation using voice conversion. For each target speaker, the other N (seven) speakers were training speakers.

4.3. Articulation Adaptation Using Procrustes Matching

Procrustes matching [27] is a robust statistical two-dimensional shape analysis technique [29,57]. In Procrustes analysis, shapes are composed of ordered series of landmarks on articulators (Figure 3a,b). Shapes from different participants have different sizes, relative locations, and different angles of tongue and lips, which leads to inter-speaker variations. In this study, Procrustes matching was conducted in y (vertical) and z (anterior-posterior) dimensions, which reduced the inter-speaker physiological difference. Procrustes matching has shown improvement in the silent speech recognition studies [11,29,57]. In this study, we applied Procrustes matching to all the EMA data as a normalization method for ATS. Specifically, for instance, let (y_i, z_i) represents the i-th data point (spatial coordinates) of a sensor, then for each sentence the speaker spoke, the data points will construct a set of landmarks S (sensors). S can be represented as below:

$$S = \{(y_i, z_i)\}, \ i = 1, \ldots, n \tag{2}$$

n is the total number of data points. As mentioned, y is the vertical direction and z is the front-back direction. A full procedure of Procrustes matching includes: (1) translating all articulatory data of each speaker to the average position of all data points in the shape (averaged across speaker); (2) rotating all shapes of each speaker to the angle that the centroids of lower and upper lips movements defined the vertical axis [57]; (3) scaling all shapes to unit size. Previous tests indicated that scaling will cause a slight increase in the error rate in silent speech recognition, therefore scaling was eliminated from the Procrustes matching approach in this experiment. The translation and rotation operations in Procrustes matching are described with the equation below:

$$\begin{bmatrix} \bar{y}_i \\ \bar{z}_i \end{bmatrix} = \begin{bmatrix} \cos\theta & -\sin\theta \\ \sin\theta & \cos\theta \end{bmatrix} \begin{bmatrix} \beta_y \\ \beta_z \end{bmatrix} \begin{bmatrix} y_i - c_y \\ z_i - c_z \end{bmatrix} \tag{3}$$

(c_y, c_z) are centroids of the two shapes which were used as translation factors; (β_y, β_z) are the square roots of the sum of the squares of all data points along the y and z directions; θ is the angle to rotate [27]. An example of Procrustes matching is provided in Figure 3. Figure 3a illustrates the original motion trajectories of a sample speaker when producing the phrase "the birch canoe slid on the smooth planks". Figure 3b illustrates those same trajectories after Procrustes matching has been used to align them to those of a separate speaker.

Procrustes matching could be applied at two levels: sentence-level and speaker-level. Sentence-level is to obtain the parameters in Equation (2) from the same sentences produced by different speakers respectively. The speaker-level obtains the parameters from all sentences produced by one speaker. During testing for both levels, individual (test) shapes were translated and rotated according to the obtained parameters. Preliminary results have shown that sentence-level Procrustes matching outperforms the speaker-level matching. Therefore, only sentence-level Procrustes matching was reported in this paper.

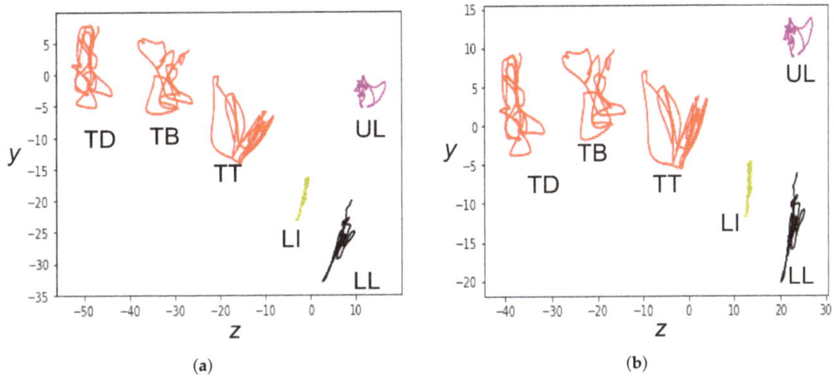

Figure 3. Example of shapes (motion path of the articulators) before and after Procrustes matching for producing "the birch canoe slid on the smooth planks". In this coordinate system, y is vertical and z is anterior-posterior. (**a**) Before Procrustes matching. (**b**) After Procrustes matching.

5. Experimental Setup

In the ATS experiments of this study, 50 sentences from each speaker's data were used as the testing set, another 50 sentences as the validation set, and the rest for training. The eight speakers took turns being chosen as the target speaker, and the other seven speakers were used as training speakers (leave-one-subject-out cross-validation). As introduced, the experiments in this study were conducted in three sessions: (1) speaker-independent (SI) and speaker-dependent (SD) ATS; (2) speaker adaptation for speaker-independent ATS on the acoustic (output) and articulation (input); (3) speaker-adaptive (SA) experiments by adding target speakers' data to the training set with and without further applying the speaker adaptations in session (2). In the speaker-dependent experiment, the model was trained, validated and tested with the same speakers. The speaker-independent experiments trained and validated models with seven training speakers, then tested with the left eighth speaker. The speaker-adaptive experiments directly adding the data from testing speakers to the training set of SI, validated and tested with data from testing speakers. The validation here indicates hyper-parameter exploration with the validation sets.

The detailed experimental setup of the deep learning models in this study were presented in Table 2. As mentioned, we use LSTM-RNN for the ATS model to maintain the real-time function of SSI, and BLSTM-RNN for the VC models for speaker adaptation. The training of all models was conducted in a batch size of single whole sentences. ATS models take 54-dim. EMA data as input and predict 80-dim. mel-spectrograms for Waveglow vocoder. To achieve the best baseline performance of both SD- and SI-ATS models before our improvement approaches (VC and Procrustes matching), we used distinct hyper-parameters for them, including learning rates and max epochs. The hyperparameters were chosen in a preliminary experiment, where a grid search of two to six layers LSTM and 128 to 512 nodes was performed. The hyper-parameter setups with the best performance were selected. Both input and output of VC were 80-dim. mel-spectrogram. All deep learning models were implemented with the Pytorch toolkit [58].

Table 2. Topologies of the neural networks in this study.

Acoustic Feature	
Mel-spectrogram	80-dim. vectors
Sampling rate	22,050 Hz
Windows length	1024
Step size	256
Articulatory Feature	54-dim. vectors
Articulatory movement (6 sensors)	(18-dim. vectors) + Δ + $\Delta\Delta$ (54-dim.)
SD-ATS LSTM Topology	
Input	54-dim. articulatory
Output.	80-dim. acoustic feature
No. of LSTM nodes each hidden layer	256
Depth	3-depth layers
Batch size	1 sentence (one whole sentence per batch)
Max Epochs	50
Learning rate	0.0003
Optimizer	Adam
SI-ATS LSTM Topology	
Input	54-dim. articulatory
Output.	80-dim. acoustic feature
No. of LSTM nodes each hidden layer	256
Depth	3-depth layers
Batch size	1 sentence (one whole sentence per batch)
Max Epochs	30
Learning rate	0.00001
Optimizer	Adam
VC BLSTM Topology	
Input	80-dim. acoustic feature
Output.	80-dim. acoustic feature
No. of LSTM nodes each hidden layer	128
Depth	3-depth layers
Batch size	1 sentence (one whole sentence per batch)
Max Epochs	30
Learning rate	0.00005
Optimizer	Adam
Toolkit	Pytorch

5.1. Speaker-Dependent (Target) and Speaker-Independent (Baseline) ATS

We firstly conducted speaker-dependent (SD) and speaker-independent (SI) ATS experiments for all speakers as the target (ceiling) and baseline performances, respectively. The speaker-dependent ATS uses training, validation, and testing data from the same speakers. Although no inter-speaker variation in SD experiments, normalization on the input articulatory data could help accelerate training and improve performance. Therefore, SD-ATS with and without z-score normalization were performed. The z-score normalization on the input EMA data was conducted by firstly computing the dimension-wise mean and standard deviation (STD) from the training set, then applying the mean and STD to the training, validation, and testing set ($X_{\text{norm}} = (X-\text{ mean})/\text{STD}$). Preliminary results have indicated that z-score normalization provides consistent improvement on the SD-ATS performance.

In speaker-independent ATS experiments, the training data are the mixture of training sets from seven training speakers. To maintain the concept of speaker-independent, the validation data are the 50-sentence validation set of training speakers (7 speakers $\times$ 50 = 350 sentences). Same as SD-ATS, z-score normalization improved SI-ATS as well.

As mentioned, we also applied Procrustes matching on the input EMA data. One thing that is worth noting is that when applying both of them together (z-score and Procrustes matching), the translation operation in Procrustes matching was eliminated by the z-score normalization, thus only the rotation operation affected the performance. In addition, z-score normalization will be applied in all following experiments by default since it has been demonstrated effective.

5.2. Acoustic Adaptation for SI-ATS Using Voice Conversion

Starting from the baselines speaker-independent ATS (with and without Procrustes matching), we adopted voice conversion models for acoustic adaptation (Figure 2). For the purpose of developing high performance and easy implementation, we used parallel voice conversion models in which the data from the source and target speaker shared the same stimulus. Across all eight speakers, 1428 parallel phrases were found in the dataset. These 1428 phrases were used for the VC model development, in which we use 14 for validating VC model training, 14 for testing, and the rest 1400 for training.

The eight speakers took turns to be the target speaker in the cross-validation loop. Figure 2 shows the pipeline of the VC-based speaker adaptation for a single target speaker. Firstly we trained voice conversion models with the phrases in the training set of target and training speakers. The acoustic features of parallel phrases were aligned to the same length by the dynamic time warping (DTW) [56]. With the aligned acoustic features, VC models were trained for each of the target-training speaker pairs. The VC models were bi-directional LSTM (Table 2) since no real-time implementation was required at this stage (voice conversion), and the BLSTM outperformed LSTM in the preliminary experiment. After that, the acoustic features of training speakers were converted to target speakers' acoustic features by the VC models, and used for the later multi-speaker ATS model training.

The speaker-independent ATS experiments with this VC speaker adaptation were essentially not speaker-independent, since the audio data from the target speakers were used during the adaptation (VC). However, for the convenience for describing the different setups and for distinguishing the ATS experiments in which both articulation and audio data from the target speaker were used for training, we still call these experiments "speaker-independent with voice conversion" in the rest of this paper (SI-VC in Results section).

5.3. Speaker Adaptive ATS Including Training Data from Target Speakers

In this session, we directly added the training set from target speakers to the dataset that trained speaker-independent with and without voice conversion, for a further speaker adaptation to see if that could outperform speaker-dependent ATS (target performance). As mentioned we named this a type of speaker-adaptive model in this study (SA). The Procrustes matching (after z-score normalization) was used by default in this stage. In this session, we maintained the method with target speakers, in which one ATS model was trained for one target speaker (rather than one ATS model that works for all speakers). The main difference was the validation sets were from the current target speaker, rather than all of them. After that, we applied the voice conversion approach on this SA-ATS.

6. Results

Figure 4 shows the average mel-cepstral distortions (MCDs) across all speaker and Table 3 details the MCD values of each speaker. Note that lower MCD values generally indicate that the speech output of the ATS model is more similar to the participant's actual speech, and thus indicates a higher performance. As can be observed, on average, the speaker-independent ATS with Procrustes matching (SI-P) outperforms that without Procrustes matching (SI), across all speakers except M01 and M03. Speaker-independent ATS with voice conversion adaptation (SI-VC) showed consistent improvement in the speaker-independent experiments (Figure 4). When both of the Procrustes matching and voice conversion were applied, we saw additional improvements in MCD (SI-VC and SI-VC-

P). After adding the testing speakers' data to the ATS training set (SA-P), the average MCD decreased significantly and slightly outperformed speaker-dependent ATS (on average). Voice conversion brought further improvement (SA-VC-P), but much less dramatic than in speaker-independent experiments. Procrustes matching was used here by default since it was verified effective for speaker normalization in the previous session. A Mann-Whitney U test indicated the significant difference of the proposal SA approaches (SA-P and SA-VC-P) outperformed the baseline approach (SI) ($p < 0.001$ for both SA-P and SA-VC-P) and there were no significant differences with the target performance (SD).

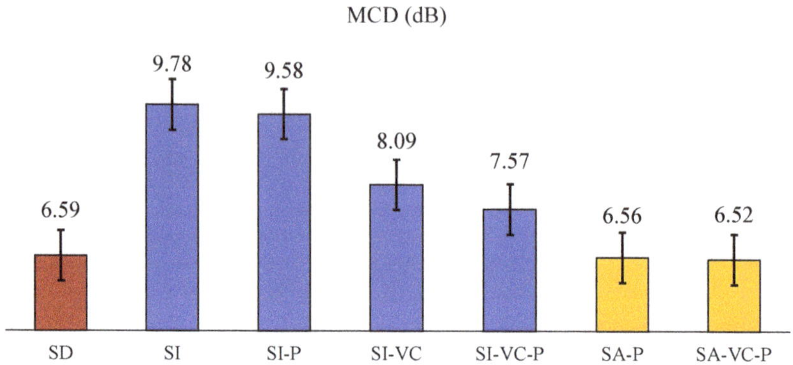

Figure 4. Average MCDs of the experiments in this study. **SD**: speaker-dependent. **SI**: speaker-independent. **SI-P** speaker-independent with Procrustes matching. **SI-VC**: speaker-independent ATS with voice conversion. **SA**: data from targets speakers were directly added to the ATS training set.

Table 3. MCD of ATS experiments on each speaker.

	SD	SI	SI-P	SI-VC	SI-VC-P	SA-P	SA-VC-P
Train:	**Tar SPK**	**Src SPK**	**Src SPK (P)**	**VC-Src SPK**	**VC-Src SPK (P)**	**Src + Tar SPK (P)**	**Tar + VC-Src SPK (P)**
Test:	**Tar SPK**	**Tar SPK**	**Tar SPK (P)**	**Tar SPK**	**Tar SPK (P)**	**Tar SPK (P)**	**Tar SPK (P)**
F01	4.98	7.80	7.48	6.63	5.79	5.26	5.08
F02	5.47	8.41	8.21	6.82	6.45	5.51	5.23
F03	6.02	9.04	8.66	8.03	6.99	6.11	6.20
F04	5.99	8.37	8.35	7.87	7.19	6.14	6.33
M01	8.96	10.41	10.44	9.45	9.33	8.22	8.23
M02	7.54	10.66	10.05	9.25	8.85	7.29	7.21
M03	6.59	8.18	8.37	7.95	7.55	6.87	6.85
M04	7.14	8.83	8.69	8.71	8.38	7.11	7.03
Mean	6.59	8.96	8.78	8.09	7.57	6.56	6.52
STD	1.27	1.04	0.98	1.03	1.21	0.99	1.05

The audio speech samples were generated from the experiments [59]. Figure 5 provides illustrations of the predicted (or original) mel-spectrogram and the synthetic speech waveforms from different ATS models including speaker-independent ATS (with and without Procrustes matching), ATS with speaker adaptation of voice conversion (with and without directly adding the training set from target speakers), and the speaker-dependent ATS. Visually, it appears that the most significant improvements in the frequency resolution were brought by the voice conversion and directly adding the testing speakers' data to the training set since there is less visible stratification across the harmonics in the SI-ATS and SI-P-ATS spectrograms. By contrast, the difference in frequency resolution and synthetic waveform between SI and SI-P, SA-VC-P, and SD are not equivalently significant. Selected synthetic speech samples are available at [59]. Speech samples from a speaker-independent

(SI) ATS have rarely been presented. Perceptually, the SI-ATS speech samples in this study sound like audible but less intelligible speech.

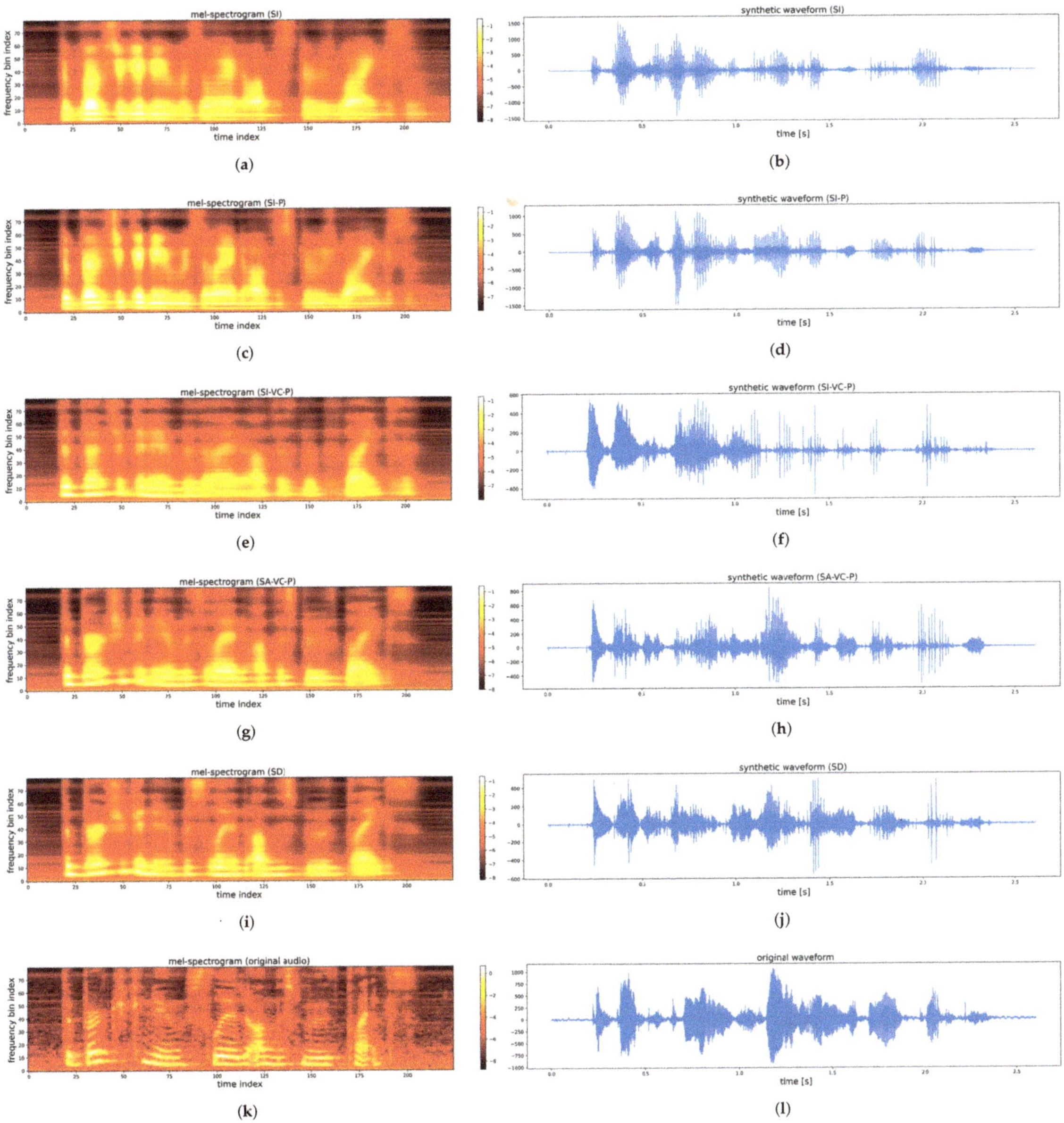

Figure 5. Examples of original and speaker-dependent ATS predicted mel-spectrograms and the synthetic waveforms. (**a**) Mel-spectrogram from SI ATS. (**b**) Speech waveform from SI ATS. (**c**) Mel-spectrogram from SI-P ATS. (**d**) Speech waveform from SI-P ATS. (**e**) Mel-spectrogram from SI-VC-P ATS. (**f**) Speech waveform from SI-VC-P ATS. (**g**) Mel-spectrogram from SA-VC-P ATS. (**h**) Speech waveform from SA-VC-P ATS. (**i**) Mel-spectrogram from SD ATS. (**j**) Speech waveform from SD ATS. (**k**) Original mel-spectrogram. (**l**) Original speech waveform.

The MCDs of voice conversion across all source-target speaker pairs were presented in Table 4. The values in this table tell us the similarities between the source speakers and the target speakers following the voice conversion process (lower MCD ≈ more similar). Speakers with more similar speech characteristics will likely exhibit lower MCD values.

Table 4. MCD of voice conversion (dB) during the speaker adaptation on acoustics. The diagonal cells are empty, because voice conversion is not applicable to the same speaker.

Source \ Target	F01	F02	F03	F04	M01	M02	M03	M04
F01		6.32	7.23	6.71	7.08	6.86	7.91	8.69
F02	9.26		6.75	7.60	7.66	7.55	8.46	9.15
F03	7.43	7.01		7.02	6.85	7.04	7.83	8.77
F04	7.15	6.24	7.45		7.24	7.66	8.02	9.25
M01	6.64	6.40	6.32	7.38		7.03	7.68	8.52
M02	6.47	6.64	6.50	7.56	6.78		7.77	8.70
M03	6.97	6.63	6.65	7.51	6.70	7.05		8.50
M04	7.01	6.32	7.05	7.42	7.96	7.17	7.71	
Average	7.30	6.51	6.85	7.31	7.18	7.19	7.91	8.80

7. Discussion

7.1. Acoustic and Articulation Adaptation Performances

Voice conversion has brought more significant improvement than the Procrustes matching (Table 3). The Procrustes matching has brought additional and consistent improvement when combined with voice conversion. Speaker-independent ATS has speaker variation in both the input articulation and output acoustics. Therefore, it is natural that adapting both articulation and acoustics outperform adapting only one of them. The Procrustes matching is an average-based normalization approach, while the voice conversion in this study is a "personalized" adaptation that converts all training speakers' voice to that of the target speakers. Therefore, it is expected that voice conversion improved speaker-independent (SI) ATS more than the Procrustes matching. In practice, the voice conversion approach proposed in this study is expected to reduce the effort of articulatory data collection, since it only adopts audio data from the target speakers. Collecting acoustic data only is less challenging than collecting synchronized acoustic and articulatory data. Audio data could also be collected remotely, which is normally impractical for current SSI articulatory data collection approaches. In addition, voice conversion requires less training data than ATS, audible speech could be generated by a VC model trained with only 10–20 sentences [24].

As shown in the results, the inclusion of the data (both acoustic and articulatory) from target speakers is a dominating advantage in training ATS (SA approaches), which has significantly outperformed the speaker-independent experiments. The VC adaptation has also shown less improvement here (6.56 –> 6.52 dB). Although not statistically significant, both SA-P and SA-VC-P outperformed speaker-dependent ATS on average. It is worth noting that the performance of each approach still varied significantly across speakers, as seen in Table 3. Although the performance differences were somewhat marginal, they illustrate the potential efficacy of speaker adaptation methods. These results demonstrated that the data (both acoustic and articulatory) from target speakers is still a strong advantage in training ATS, which also indicated the challenge in outperforming SD-ATS with speaker adaptation approaches. Further improvements in the effectiveness of both SI and SA methods are likely to come as datasets with larger groups of speakers. Such as in speaker-independent ASR systems that generally use tens or hundreds of speakers in their training data.

It notes that, although our speaker-adaptive ATS obtained comparable performances with SD-ATS, it does not mean speaker-adaptive ATS could outperform SD-ATS with an increased number of subjects and data size from single speakers. As the number of subjects increases, the inter-speaker variability in both articulation and acoustics increases. Although SD-ATS may be still the first choice when developing new ATS algorithms, our findings suggest SA-ATS may be a promising alternative solution.

7.2. Performance Variation across Speakers

Given the similar data amount, the eight speakers in the dataset have shown different performances in both SD- and SI-ATS (Table 3). Due to the inter- and intra-speaker variation, the MCDs have shown obvious differences across speakers in all experiments. Speakers with lower intra-speaker variation may show higher performance in speaker-dependent ATS. Speakers that have higher similarity to other speakers seem to have higher performance in speaker-independent ATS (e.g., F01). Speakers' data with higher intra- and inter-speaker variation may demonstrate lower performance in SD and SI experiments, respectively (e.g., M01 and M02).

7.3. Observations from the Synthetic Speech Samples

Interestingly, in the synthetic speech samples presented in [59], it was observed that the speaker-independent ATS generated "gender-confused" speech samples. We expect this because the training set includes data from both genders. While the speech samples from speaker-independent ATS with VC adaptation show obvious gender characteristics since the training data from the opposite gender were converted. Therefore, the voice conversion adaptation may also improve the gender characteristic in the synthetic speech. Gender-dependency in speaker-independent ATS might be a topic that is worth further investigation in the future.

7.4. Relationship between VC and ATS Performances

Table 4 has shown the MCDs of voice conversion model development. The rows and columns are the source and target speakers of voice conversion. The Pearson correlation coefficients between the mean VC performance (Table 4) and the SI-VC performance across all speakers (Table 3) was 0.84 and was decreased to 0.78 for the correlation between VC and SI-VC-P (Table 3). The improvement from Procrustes matching reduced the correlation between VC and the SI-VC-P. It is possible that an SI-ATS model for a target speaker got strong acoustics adaptation by the voice conversion, while the Procrustes matching for that speaker was not strong enough accordingly to form a good mapping between the adapted articulation and the acoustics. Therefore, a matched adaptation of acoustics and articulation might be more effective in the task of speaker adaptation for ATS.

7.5. Feasibility of Articulation Conversion

Other than the Procrustes matching, an alternative approach for articulation adaptation is the articulation conversion, which has a similar procedure to the voice conversion in this study. In the articulation conversion, the articulatory data from the training speakers were converted to that of target speakers with the trained articulation conversion models. However, the articulation conversion generated articulatory movement with a spatial RMSE larger than 3 mm, which led to a performance decrease in the following ATS experiments. The EMA data are low-frequency time domain signals, which might be more challenging to precisely predict than the high-frequency frequency domain acoustic features. As an end-to-end model, ATS might be very sensitive to the variation in the input articulation. Therefore, we did not use articulation conversion in the current study. More studies are needed to confirm the feasibility of this articulation conversion approach.

8. Conclusions

In this study, we investigated speaker adaptation approaches for articulation-to-speech synthesis using voice conversion and Procrustes matching. Procrustes matching was first applied to reduce the speaker variations in the articulation. Then a framework of using voice conversion for ATS voice adaptation was proposed and validated, in which voice conversion (VC) models were trained for reducing the acoustic variations between training and testing speakers. The experimental results have shown the effectiveness of both Procrustes matching and voice conversion; the performance was further improved when both were used in conjunction. Additionally, we performed speaker-adaptive (SA) ATS experiments in which the data from the target speakers (both acoustic and articulatory) were included in the training set (both with and without VC adaptation) and achieved a similar performance to the speaker-dependent ATS. To our knowledge, this is the first study that demonstrated the potential of speaker-adaptive ATS by showing a comparable performance to that of speaker-dependent ATS. This study is also the first to demonstrate audible speech output from speaker-independent and speaker-adaptive ATS systems.

Author Contributions: Conceptualization, B.C., A.W. and J.W.; methodology, B.C., A.W. and J.W.; software, B.C.; validation, B.C., A.W. and J.W.; formal analysis, B.C.; investiation, B.C.; writing—original draft preparation, B.C.; funding acquisition, J.W.; project administration, J.W.; resources, J.W.; supervision, J.W.; Writing—review and editing, B.C., A.W. and J.W. All authors have read and agreed to the published version of the manuscript.

Funding: This research was funded by the National Institute on Deafness and Other Communication Disorders (NIDCD) of the National Institutes of Health (NIH) under award number R01DC016621 (PI: Wang).

Institutional Review Board Statement: Ethical review and approval were waived for this study because only publicly available data were used.

Informed Consent Statement: Not applicable to this study as only publicly available data were used.

Data Availability Statement: Publicly available datasets were analyzed in this study. This data can be found here: https://yale.app.box.com/s/cfn8hj2puveo65fq54rp1ml2mk7moj3h/folder/30415804819 (accessed on 30 June 2022).

Conflicts of Interest: The authors declare no conflict of interest.

References

1. Braz, D.S.A.; Ribas, M.M.; Dedivitis, R.A.; Nishimoto, I.N.; Barros, A.P.B. Quality of life and depression in patients undergoing total and partial laryngectomy. *Clinics* **2005**, *60*, 135–142. [CrossRef] [PubMed]
2. Nijdam, H.; Annyas, A.; Schutte, H.; Leever, H. A New Prosthesis for Voice Rehabilitation after Laryngectomy. *Arch. Oto-Rhino-Laryngol.* **1982**, *237*, 27–33. [CrossRef]
3. Singer, M.I.; Blom, E.D. An Endoscopic Technique for Restoration of Voice after Laryngectomy. *Ann. Otol. Rhinol. Laryngol.* **1980**, *89*, 529–533. [CrossRef] [PubMed]
4. Liu, H.; Ng, M.L. Electrolarynx in Voice Rehabilitation. *Auris Nasus Larynx* **2007**, *34*, 327–332. [CrossRef] [PubMed]
5. Kaye, R.; Tang, C.G.; Sinclair, C.F. The Electrolarynx: Voice Restoration after Total Laryngectomy. *Med. Devices* **2017**, *10*, 133–140. [CrossRef]
6. Eadie, T.L.; Otero, D.; Cox, S.; Johnson, J.; Baylor, C.R.; Yorkston, K.M.; Doyle, P.C. The Relationship between Communicative Participation and Postlaryngectomy Speech Outcomes. *Head Neck* **2016**, *38*, E1955–E1961. [CrossRef]
7. Denby, B.; Schultz, T.; Honda, K.; Hueber, T.; Gilbert, J.M.; Brumberg, J.S. Silent Speech Interfaces. *Speech Commun.* **2010**, *52*, 270–287. [CrossRef]
8. Schultz, T.; Wand, M.; Hueber, T.; Krusienski, D.J.; Herff, C.; Brumberg, J.S. Biosignal-based Spoken Communication: A Survey. *IEEE/ACM Trans. Audio Speech Lang. Process.* **2017**, *25*, 2257–2271. [CrossRef]
9. Gonzalez-Lopez, J.A.; Gomez-Alanis, A.; Martín-Doñas, J.M.; Pérez-Córdoba, J.L.; Gomez, A.M. Silent Speech Interfaces for Speech Restoration: A Review. *IEEE Access* **2020**, *8*, 177995–178021. [CrossRef]
10. Cao, B.; Sebkhi, N.; Bhavsar, A.; Inan, O.T.; Samlan, R.; Mau, T.; Wang, J. Investigating Speech Reconstruction for Laryngectomees for Silent Speech Interfaces. In Proceedings of the Interspeech 2021, 22nd Annual Conference of the International Speech Communication Association, Brno, Czechia, 30 August–3 September 2021; pp. 651–655.
11. Kim, M.; Cao, B.; Mau, T.; Wang, J. Speaker-Independent Silent Speech Recognition from Flesh-Point Articulatory Movements Using an LSTM Neural Network. *IEEE/ACM Trans. Audio Speech Lang. Process. (TASLP)* **2017**, *25*, 2323–2336. [CrossRef] [PubMed]

12. Zen, H.; Senior, A.; Schuster, M. Statistical parametric speech synthesis using deep neural networks. In Proceedings of the 2013 IEEE International Conference on Acoustics, Speech and Signal Processing, Vancouver, BC, Canada, 26–31 May 2013; pp. 7962–7966.
13. Huang, X.; Lee, K.F. On speaker-independent, speaker-dependent, and speaker-adaptive speech recognition. *IEEE Trans. Speech Audio Process.* **1993**, *1*, 150–157. [CrossRef]
14. Schönle, P.W.; Gräbe, K.; Wenig, P.; Höhne, J.; Schrader, J.; Conrad, B. Electromagnetic articulography: Use of alternating magnetic fields for tracking movements of multiple points inside and outside the vocal tract. *Brain Lang.* **1987**, *31*, 26–35. [CrossRef]
15. Cao, B.; Kim, M.; Wang, J.R.; Van Santen, J.; Mau, T.; Wang, J. Articulation-to-Speech Synthesis Using Articulatory Flesh Point Sensors' Orientation Information. In Proceedings of the Interspeech 2018, 19th Annual Conference of the International Speech Communication Association, Hyderabad, India, 2–6 September 2018; pp. 3152–3156.
16. Gonzalez, J.A.; Cheah, L.A.; Bai, J.; Ell, S.R.; Gilbert, J.M.; Moore, R.K.; Green, P.D. Analysis of Phonetic Similarity in a Silent Speech Interface Based on Permanent Magnetic Articulography. In Proceedings of the Fifteenth Annual Conference of the International Speech Communication Association, Singapore, 14–18 September 2014.
17. Diener, L.; Bredehoeft, S.; Schultz, T. A Comparison of EMG-to-Speech Conversion for Isolated and Continuous Speech. In Proceedings of the 13th ITG Symposium on Speech Communication, Oldenburg, Germany, 10–12 October 2018; pp. 66–70.
18. Csapó, T.G.; Grósz, T.; Gosztolya, G.; Tóth, L.; Markó, A. DNN-Based Ultrasound-to-Speech Conversion for a Silent Speech Interface. In Proceedings of the Interspeech 2017, 18th Annual Conference of the International Speech Communication Association, Stockholm, Sweden, 20–24 August 2017; pp. 3672–3676.
19. Yamagishi, J.; Nose, T.; Zen, H.; Ling, Z.H.; Toda, T.; Tokuda, K.; King, S.; Renals, S. Robust speaker-adaptive HMM-based text-to-speech synthesis. *IEEE Trans. Audio Speech Lang. Process.* **2009**, *17*, 1208–1230. [CrossRef]
20. Shandiz, A.H.; Tóth, L.; Gosztolya, G.; Markó, A.; Csapó, T.G. Neural Speaker Embeddings for Ultrasound-Based Silent Speech Interfaces. In Proceedings of the Interspeech 2021, 22nd Annual Conference of the International Speech Communication Association, Brno, Czechia, 30 August–3 September 2021; pp. 1932–1936. [CrossRef]
21. Ribeiro, M.S.; Sanger, J.; Zhang, J.X.; Eshky, A.; Wrench, A.; Richmond, K.; Renals, S. TaL: A synchronised multi-speaker corpus of ultrasound tongue imaging, audio, and lip videos. In Proceedings of the 2021 IEEE Spoken Language Technology Workshop (SLT), Shenzhen, China, 19–22 January 2021; pp. 1109–1116.
22. Liu, F.H.; Stern, R.M.; Huang, X.; Acero, A. Efficient cepstral normalization for robust speech recognition. In Proceedings of the workshop on Human Language Technology, Plainsboro, NJ, USA, 21–24 March 1993; pp. 69–74.
23. Eide, E.; Gish, H. A parametric approach to vocal tract length normalization. In Proceedings of the 1996 IEEE International Conference on Acoustics, Speech, and Signal Processing Conference Proceedings, Atlanta, GA, USA, 7–10 May 1996; Volume 1, pp. 346–348.
24. Toda, T.; Black, A.W.; Tokuda, K. Voice Conversion Based on Maximum-Likelihood Estimation of Spectral Parameter Trajectory. *IEEE Trans. Audio Speech Lang. Process.* **2007**, *15*, 2222–2235. [CrossRef]
25. Tiede, M.; Espy-Wilson, C.Y.; Goldenberg, D.; Mitra, V.; Nam, H.; Sivaraman, G. Quantifying kinematic aspects of reduction in a contrasting rate production task. *J. Acoust. Soc. Am.* **2017**, *141*, 3580. [CrossRef]
26. Gower, J.C. Generalized Procrustes Analysis. *Psychometrika* **1975**, *40*, 33–51. [CrossRef]
27. Dryden, I.L.; Mardia, K.V. *Statistical Shape Analysis*; Wiley: Chichester, UK, 1998.
28. Prenger, R.; Valle, R.; Catanzaro, B. Waveglow: A flow-based generative network for speech synthesis. In Proceedings of the ICASSP 2019—2019 IEEE International Conference on Acoustics, Speech and Signal Processing (ICASSP), Brighton, UK, 12–17 May 2019; pp. 3617–3621.
29. Wang, J.; Hahm, S. Speaker-Independent Silent Speech Recognition with Across-speaker Articulatory Normalization and Speaker Adaptive Training. In Proceedings of the Sixteenth Annual Conference of the International Speech Communication Association, Dresden, Germany, 6–10 September 2015.
30. Gonzalez, J.A.; Cheah, L.A.; Gomez, A.M.; Green, P.D.; Gilbert, J.M.; Ell, S.R.; Moore, R.K.; Holdsworth, E. Direct Speech Reconstruction from Articulatory Sensor Data by Machine Learning. *IEEE/ACM Trans. Audio Speech Lang. Process.* **2017**, *25*, 2362–2374. [CrossRef]
31. Kim, M.; Sebkhi, N.; Cao, B.; Ghovanloo, M.; Wang, J. Preliminary Test of a Wireless Magnetic Tongue Tracking System for Silent Speech Interface. In Proceedings of the 2018 IEEE Biomedical Circuits and Systems Conference (BioCAS), Cleveland, OH, USA, 17–19 October 2018; pp. 1–4.
32. Sebkhi, N.; Desai, D.; Islam, M.; Lu, J.; Wilson, K.; Ghovanloo, M. Multimodal Speech Capture System for Speech Rehabilitation and Learning. *IEEE Trans. Biomed. Eng.* **2017**, *64*, 2639–2649.
33. Hueber, T.; Benaroya, E.L.; Chollet, G.; Denby, B.; Dreyfus, G.; Stone, M. Development of a silent speech interface driven by ultrasound and optical images of the tongue and lips. *Speech Commun.* **2010**, *52*, 288–300. [CrossRef]
34. Csapó, T.G.; Zainkó, C.; Tóth, L.; Gosztolya, G.; Markó, A. Ultrasound-Based Articulatory-to-Acoustic Mapping with WaveGlow Speech Synthesis. In Proceedings of the Interspeech 2020, 21st Annual Conference of the International Speech Communication Association, Virtual Event, Shanghai, China, 25–29 October 2020; pp. 2727–2731.
35. Diener, L.; Felsch, G.; Angrick, M.; Schultz, T. Session-Independent Array-based EMG-to-Speech Conversion Using Convolutional Neural Networks. In Proceedings of the 13th ITG Symposium on Speech Communication, Oldenburg, Germany, 10–12 October 2018; pp. 276–280.

36. Nakajima, Y.; Kashioka, H.; Shikano, K.; Campbell, N. Non-Audible Murmur Recognition Input Interface Using Stethoscopic Microphone Attached to the Skin. In Proceedings of the 2003 IEEE International Conference on Acoustics, Speech, and Signal Processing, ICASSP'03, Hong Kong, 6–10 April 2003; Volume 5, p. V-708.
37. Toth, A.R.; Kalgaonkar, K.; Raj, B.; Ezzat, T. Synthesizing speech from Doppler signals. In Proceedings of the 2010 IEEE International Conference on Acoustics, Speech and Signal Processing, Dallas, TX, USA, 14–19 March 2010; pp. 4638–4641.
38. Lee, K.S. Silent speech interface using ultrasonic Doppler sonar. *IEICE Trans. Inf. Syst.* **2020**, *103*, 1875–1887. [CrossRef]
39. Kapur, A.; Kapur, S.; Maes, P. Alterego: A personalized wearable silent speech interface. In Proceedings of the 23rd International Conference on Intelligent User Interfaces, Tokyo, Japan, 7–11 March 2018; pp. 43–53.
40. Ferreira, D.; Silva, S.; Curado, F.; Teixeira, A. Exploring Silent Speech Interfaces Based on Frequency-Modulated Continuous-Wave Radar. *Sensors* **2022**, *22*, 649. [CrossRef] [PubMed]
41. Sebkhi, N.; Bhavsar, A.; Anderson, D.V.; Wang, J.; Inan, O.T. Inertial Measurements for Tongue Motion Tracking Based on Magnetic Localization With Orientation Compensation. *IEEE Sens. J.* **2020**, *21*, 7964–7971. [CrossRef] [PubMed]
42. Katsurada, K.; Richmond, K. Speaker-Independent Mel-cepstrum Estimation from Articulator Movements Using D-vector Input. In Proceedings of the Interspeech 2020, 21st Annual Conference of the International Speech Communication Association, Virtual Event, Shanghai, China, 25–29 October 2020; pp. 3176–3180.
43. Electrical, I.; Engineers, E. IEEE recommended practice for speech quality measurements. *IEEE Trans. Audio Electroacoust.* **1969**, *17*, 225–246.
44. Richmond, K.; Hoole, P.; King, S. Announcing the electromagnetic articulography (day 1) subset of the mngu0 articulatory corpus. In Proceedings of the Twelfth Annual Conference of the International Speech Communication Association, Florence, Italy, 27–31 August 2011; pp. 1505–1508.
45. Ji, A.; Berry, J.J.; Johnson, M.T. The Electromagnetic Articulography Mandarin Accented English (EMA-MAE) corpus of acoustic and 3D articulatory kinematic data. In Proceedings of the 2014 IEEE International Conference on Acoustics, Speech and Signal Processing (ICASSP), Florence, Italy, 4–9 May 2014; pp. 7719–7723.
46. Kingma, D.P.; Dhariwal, P. Glow: Generative flow with invertible 1x1 convolutions. In Proceedings of the Advances in Neural Information Processing Systems 31: Annual Conference on Neural Information Processing Systems 2018, NeurIPS 2018, Montreal, QC, Canada, 3–8 December 2018; pp. 10215–10224.
47. Oord, A.v.d.; Dieleman, S.; Zen, H.; Simonyan, K.; Vinyals, O.; Graves, A.; Kalchbrenner, N.; Senior, A.; Kavukcuoglu, K. Wavenet: A generative model for raw audio. *arXiv* **2016**, arXiv:1609.03499.
48. Arfib, D.; Keiler, F.; Zölzer, U.; Verfaille, V. Source-filter processing. *DAFX–Digital Audio Eff.* **2002**, *9*, 299–372.
49. Black, A.W.; Zen, H.; Tokuda, K. Statistical parametric speech synthesis. In Proceedings of the 2007 IEEE International Conference on Acoustics, Speech and Signal Processing-ICASSP'07, Honolulu, HI, USA, 15–20 April 2007; Volume 4, pp. IV-1229–IV-1232.
50. Imai, S.; Sumita, K.; Furuichi, C. Mel log spectrum approximation (MLSA) filter for speech synthesis. *Electron. Commun. Jpn. (Part I Commun.)* **1983**, *66*, 10–18. [CrossRef]
51. Kawahara, H. STRAIGHT, exploitation of the other aspect of VOCODER: Perceptually isomorphic decomposition of speech sounds. *Acoust. Sci. Technol.* **2006**, *27*, 349–353. [CrossRef]
52. Morise, M.; Yokomori, F.; Ozawa, K. WORLD: A vocoder-based high-quality speech synthesis system for real-time applications. *IEICE Trans. Inf. Syst.* **2016**, *99*, 1877–1884. [CrossRef]
53. Kubichek, R. Mel-Cepstral Distance Measure for Objective Speech Quality Assessment. In Proceedings of the IEEE Pacific Rim Conference on Communications Computers and Signal Processing, Victoria, BC, Canada, 19–21 May 1993; Volume 1, pp. 125–128.
54. Battenberg, E.; Mariooryad, S.; Stanton, D.; Skerry-Ryan, R.; Shannon, M.; Kao, D.; Bagby, T. Effective use of variational embedding capacity in expressive end-to-end speech synthesis. *arXiv* **2019**, arXiv:1906.03402.
55. Mohammadi, S.H.; Kain, A. An Overview of Voice Conversion Systems. *Speech Commun.* **2017**, *88*, 65–82. [CrossRef]
56. Müller, M. Dynamic time warping. In *Information Retrieval for Music and Motion*; Springer: Berlin/Heidelberg, Germany, 2007; pp. 69–84.
57. Hahm, S.; Daragh, H.; Wang, J. Recognizing Dysarthric Speech due to Amyotrophic Lateral Sclerosis with Across-Speaker Articulatory Normalization. In Proceedings of the ACL/ISCA Workshop on Speech and Language Processing for Assistive Technologies, Dresden, Germany, 11 September 2015; pp. 47–54.
58. Paszke, A.; Gross, S.; Massa, F.; Lerer, A.; Bradbury, J.; Chanan, G.; Killeen, T.; Lin, Z.; Gimelshein, N.; Antiga, L.; et al. Pytorch: An imperative style, high-performance deep learning library. In Proceedings of the Advances in Neural Information Processing Systems 32: Annual Conference on Neural Information Processing Systems 2019, NeurIPS **2019**, Vancouver, BC, Canada, 8–14 December 2019; pp. 8026–8037.
59. Cao, B. Demo of Speaker Adaptation of Articulation-to-Speech Synthesis. 2022. Available online: https://beimingcao.github.io/SI_ATS_demo/ (accessed on 30 June 2022).

sensors

MDPI

Article

Optimizing the Ultrasound Tongue Image Representation for Residual Network-Based Articulatory-to-Acoustic Mapping

Tamás Gábor Csapó [1,*], Gábor Gosztolya [2], László Tóth [3], Amin Honarmandi Shandiz [3] and Alexandra Markó [4]

[1] Department of Telecommunications and Media Informatics, Budapest University of Technology and Economics, H-1117 Budapest, Hungary
[2] ELRN-SZTE Research Group on Artificial Intelligence, H-6720 Szeged, Hungary
[3] Institute of Informatics, University of Szeged, H-6720 Szeged, Hungary
[4] MTA-ELTE Lendület Lingual Articulation Research Group, H-1088 Budapest, Hungary
* Correspondence: csapot@tmit.bme.hu

Abstract: Within speech processing, articulatory-to-acoustic mapping (AAM) methods can apply ultrasound tongue imaging (UTI) as an input. (Micro)convex transducers are mostly used, which provide a wedge-shape visual image. However, this process is optimized for the visual inspection of the human eye, and the signal is often post-processed by the equipment. With newer ultrasound equipment, now it is possible to gain access to the raw scanline data (i.e., ultrasound echo return) without any internal post-processing. In this study, we compared the raw scanline representation with the wedge-shaped processed UTI as the input for the residual network applied for AAM, and we also investigated the optimal size of the input image. We found no significant differences between the performance attained using the raw data and the wedge-shaped image extrapolated from it. We found the optimal pixel size to be 64 × 43 in the case of the raw scanline input, and 64 × 64 when transformed to a wedge. Therefore, it is not necessary to use the full original 64 × 842 pixels raw scanline, but a smaller image is enough. This allows for the building of smaller networks, and will be beneficial for the development of session and speaker-independent methods for practical applications. AAM systems have the target application of a "silent speech interface", which could be helpful for the communication of the speaking-impaired, in military applications, or in extremely noisy conditions.

Keywords: speech processing; ultrasound imaging; deep learning

Citation: Csapó, T.G.; Gosztolya, G.; Tóth, L.; Shandiz, A.H.; Markó, A. Optimizing the Ultrasound Tongue Image Representation for Residual Network-Based Articulatory-to-Acoustic Mapping. *Sensors* **2022**, *22*, 8601. https://doi.org/10.3390/s22228601

Academic Editor: Manuchehr Soleimani

Received: 9 September 2022
Accepted: 31 October 2022
Published: 8 November 2022

1. Introduction

Speech is used in our everyday human–computer interfaces when interacting with mobile or fixed electronic devices. Future speech interfaces will go beyond current human–machine communication systems because speech has several drawbacks: (1) it can be easily captured by a third party; (2) speech communication is problematic for the speaking-impaired (e.g., patients after laryngectomy); (3) speech understanding degrades rapidly in noisy environments.

There has been an increased interest in the analysis, processing, prediction, and synthesis of biosignals in the speech processing community. Such biosignals include: the speech waveform, information about the articulators (larynx, tongue, lips, teeth, etc.), neural pathways, or the brain itself. These biosignals can be used in scenarios such as articulatory-to-acoustic mapping (AAM) or acoustic-to-articulatory inversion (AAI). Such biosignals can typically be recorded with some external sensor or specific device, and processing this data causes various challenges. In the AAM field, articulatory data (i.e., information about the movement of the articulatory organs) are recorded while the subject is speaking, and machine learning methods (nowadays, typically deep neural networks (DNNs)) are applied for predicting the speech signal, while the network is conditioned on the articulatory input. Systems that can perform the automatic articulatory-to-acoustic mapping are often referred to as "silent speech interfaces" (SSIs) [1–3], with the final aim of a target application where silent (mouthed) articulation can be converted to audible speech. Such an SSI could be helpful for the communication of the speaking-impaired, in military applications, or in extremely noisy conditions.

In the area of AAM, several different types of articulatory acquisition equipments have been used, including ultrasound tongue imaging (UTI) [4–22], electromagnetic articulography (EMA) [23–27], permanent magnetic articulography (PMA) [28,29], surface electromyography (sEMG) [30–32], electro-optical stomatography (EOS) [33], lip video [5,6,34–36], continuous-wave radar [37], or multimodal combination [38]. There are basically two distinct methods of SSI solutions, namely "direct synthesis" and "recognition-and-synthesis" [2]. In the first case, the speech signal is generated without an intermediate step, directly from the articulatory data, typically using vocoders [4,7,9,11,12,15–17,25,26,29–31]. In the second case, silent speech recognition (SSR) is applied on the biosignal, which extracts the content spoken by the person (i.e., the result is text). This step is then followed by text-to-speech (TTS) synthesis [5,6,10,23,24,28,32,33]. The drawback of the SSR+TTS approach might be that the errors made by the SSR component inevitably appear as errors in the final TTS output [2], and also that it causes a significant end-to-end delay. Furthermore, any information related to speech prosody is totally lost, while several studies have showed that certain prosodic components may be estimated reasonably well from the articulatory recordings (e.g., energy [11] and pitch [12]). Depending on the use-case scenario, the two approaches may have various advantages; for example, the smaller delay with the direct synthesis approach might enable conversational use and potential research on human-in-the-loop scenarios.

In this study, we focus on ultrasound tongue images as the articulatory input, with the direct synthesis approach used for AAM.

1.1. Representations of Ultrasound Tongue Images

For investigating the tongue movement using ultrasound, a B-mode scan is typically used with a (micro)convex transducer [39]. In a real-time B-scan ultrasound transducer, a row of identical piezoelectric crystals emit sound waves and receive their reflected echoes (for an illustration, see the left-hand side of Figure 1). The received echoes are converted to an electrical signal, and are then sent to the internal computer of the ultrasound machine. The internal computer reconstructs the returning echoes into a 2D grayscale image usually shaped like a 90–120 degree wedge (see the right-hand side of Figure 1). Typically, during recordings, a midsagittal orientation is maintained with the shadows of the jaw and the hyoid bones visible at opposite sides of the scan wedge [39]. For linguistic studies, manual tracing or the automatic tracking of the tongue is frequently performed [40,41], but, for articulatory-to-acoustic mapping purposes, such a contour extraction is not typically used.

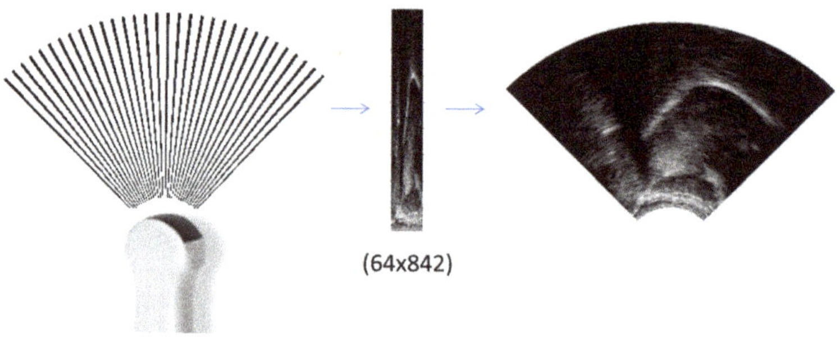

Figure 1. Ultrasond tongue image representations: raw scanlines during recording (**left**), array of raw scanline data (**middle**), and a wedge-formatted image (**right**).

In the first AAM studies that had ultrasound images for recording the articulatory movement, it was not possible to gain access to the raw echo data due to the restrictions of the equipment. Instead, the ultrasound scanlines were interpolated and organized as a "fan-shaped"/"wedge" representation, as described above. In the earliest UTI-based direct synthesis study by Denby et al. [4], the ultrasound images (recorded at 30 fps) were first reduced to a 14 by 40 grid and automatic contour tracking was carried out on the fan-shaped data to reduce dimensionality. A few years later, Hueber et al. [6] used fan-shaped images (with an Aloka SSD-1000 machine), but post-processing algorithms, such as image averaging and speckle reduction, were disabled. After this, with an analog system, an NTSC video was created, limiting the time resolution to 29.97 Hz fps. In their next experimental setup [5,7,9], a Terason T3000 ultrasound system was used with a dedicated software to record the wedge-shaped articulatory data at 320 × 240 pixels and 60 fps, doubling the time resolution. The fan-shaped ultrasound images were resized to 64 × 64 pixels and the EigenTongues decomposition technique [42] was applied for dimension reduction, keeping the first 30 coefficients. In the latest relevant study from the same research group [10], the ultrasound images were resized to 32 × 32 pixels, and these images were used with CNNs (without EigenTongues compression). Similarly, a 320 × 240 pixels ultrasound video was recorded for the Silent Speech Challenge (SSC) dataset [14]. Wei et al. [8], with an unspecified system, used a fan-shaped 64 × 48 pixels UTI input (compressed with PCA and autoencoders) for AAM and AAI. Kimura et al. [18] used a CONTEC CMS600P2 system and a display-digitizing unit for converting the signal sent to the display to a 30 fps MPEG-4 movie file, and resized the fan-shaped images to 128 × 128 pixels for the AAM input. In their next study [43], interpolated ultrasound videos were recorded with a resolution of 640 × 445 pixels. In most of the above studies, classical image processing of the ultrasound input is not performed, and the feature extraction is left to the DNN. This is similar to how other modalities are processed in related tasks such as lip images [34], MRI [44], or EMA [27].

In our earlier studies on ultrasound-based articulatory-to-acoustic mapping, we used raw scanline data as the input of the DNNs, recorded using a "Micro" system (developed by Telemed Ltd., Vilnius, Lithuania, and distributed by Articulate Instruments Ltd., Musselburgh, UK), a 2–4 MHz/64 element 20 mm radius convex ultrasound transducer at 80–85 fps [11–13,15–17,20,21]. In [11–13], data from a single female speaker were used, and the raw echo-returns of 64 × 946 were resized to 64 × 119 pixels using a bicubic interpolation. In [17], four speakers were used, and the raw images of 64 × 842 pixels were resized to 64 × 128. Instead of using the full raw scanline data, in [11], we investigated correlation-based feature selection, and, in [16], we tested the applicability of autoencoders for dimensionality reduction. Besides the above works by our research group, there were only a few studies that used raw scanlines. Ribeiro et al. [45] applied a raw ultrasound for the classification of phonetic segments. Here, 63 × 412 echo-return data (recorded using Ultrasonix SonixRP) were utilized as the input of DNNs and CNNs, and the raw data input was compared with PCA and 2D-DCT-based compression. A subsequent study [46] applied the raw scanlines of the "Micro" system, resized to 63 × 128 pixels.

The advantage of fan-shaped data is that they correspond to the physical/spatial orientation of the speaking organs of the subject; therefore, comparisons across sessions and speakers are relatively easy. Another benefit can be that CNNs might process the wedge-shaped data easier as they do not contain nonlinear distortions. On the other hand, the advantage of raw scanline data is that they can be acquired directly from the ultrasound equipment, without any postprocessing. Therefore, feature extraction can be left up to the machine learning algorithms. However, the disadvantage is that, because of the convex transducer, the raw data do not correspond to the original mid-sagittal slice, and non-linear interpolation is necessary to transform into real-world orientation. Therefore, a comparison across sessions and speakers using the raw scanline data is a challenge.

1.2. Contributions of This Paper

In our previous studies, we hypothesized that the use of a raw scanline ultrasound always results in lower errors during the prediction of spectral or excitation parameters [11–13,15–17]. However, this hypothesis was never tested explicitly (neither by us, nor by other research groups). In the current paper, we compared raw scanline data with the wedge-formatted ultrasound tongue image input for articulatory-to-acoustic mapping, applying deep neural networks. Furthermore, we investigated the effect of reducing the input image size.

2. Materials and Methods

2.1. Data Acquisition

The same dataset was used as in our previous studies [17,20]. Several Hungarian male and female subjects with normal speaking abilities were recorded while reading sentences aloud (altogether, 209 sentences each), of which, a female speaker (#048) was chosen for the current study. The tongue movement was recorded in midsagittal orientation using the "Micro" ultrasound system (Articulate Instruments Ltd.) with a 2–4 MHz/64 element 20 mm radius convex ultrasound transducer at 81.67 fps. The speech signal was recorded with a Beyerdynamic TG H56c tan omnidirectional condenser microphone. At the time of capturing an ultrasound frame, the "Micro" equipment generates a pulse at the "frame sync" output, which was digitized together with the speech signal with an M-Audio—MTRACK PLUS external sound card at 22 050 Hz (see Figure 2). The ultrasound data and the audio signals were synchronized using a custom tool that is looking at the rising edge of the peaks in the "frame sync" signal. More details about the recording set-up can be found in [11,17]. The overall duration of the recordings was approximately 15 min, which was partitioned into training, validation, and test sets in an 85:10:5 ratio.

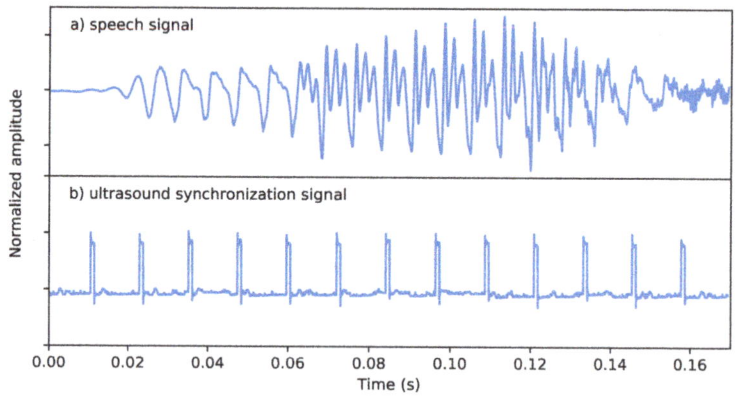

Figure 2. Ultrasound synchronization signal: the rising edge of the pulses indicates the capture time of ultrasound images.

2.2. Input 1: Ultrasound as Raw Scanlines (UTIraw)

In the first case, the raw scanline data (64 × 842 pixels, Figure 3/1) of the ultrasound were used. To check the optimal image resolution, they were further resized to 64 × 421, 64 × 210, 64 × 105, 64 × 53, 64 × 26, and 64 × 13 pixels using bicubic interpolation (with the `skimage.transform` function). The resized raw images served as the input of the deep neural networks, which can be seen in Figure 3 and will be introduced in Section 2.6.

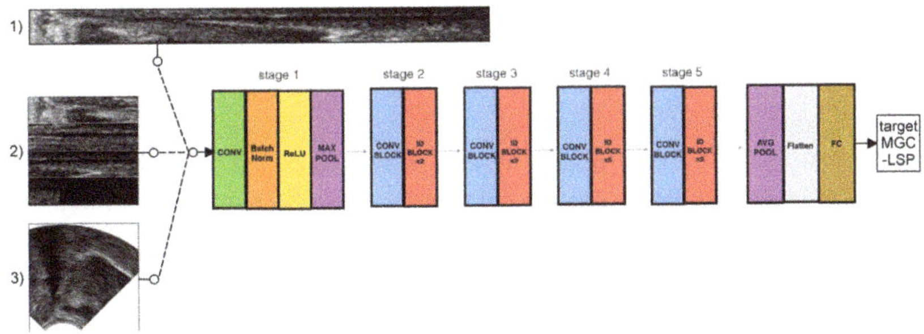

Figure 3. ResNet-50 architecture for articulatory-to-acoustic mapping using ultrasound tongue image (raw scanline vs. wedge) input and MGC-LSP target. ResNet image adopted from https://towardsdatascience.com/understanding-and-coding-a-resnet-in-keras-446d7ff84d33, accessed date: 11 May 2020.

2.3. *Input 2: Ultrasound as Raw Scanlines, Reshaped to Square (UTIraw-Padding)*

In the second case, the scanline data (64 × 842 pixels) of the ultrasound were used, after being transposed to a 512 × 512 square for ResNet input (see Figure 3/2). To check the optimal image resolution, they were further resized to 256 × 256, 128 × 128, 64 × 64, 32 × 32, 16 × 16, and 8 × 8 pixels using bicubic interpolation (with the `skimage.transform` function).

2.4. *Input 3: Ultrasound as a Wedge-Shape (UTIwedge)*

In the third case, the raw scanline data (left-hand side of Figure 1) were interpolated to achieve a wedge-shape. For this, we used the `pcolormesh` function of `matplotlib` to smooth and interpolate the data for a continuous wedge-shape, including aliasing (right-hand side of Figure 1). The necessary details for the interpolation (e.g., angle between scanlines, zero offset) were extracted from the AAA software (V219.08, Articulate Instruments Ltd.) that was used for the recordings. The generated image sequences (840 × 510 pixels) were saved to MP4 video using `ffmpeg`, keeping the original scaling of the pixel values. After this, the middle of the images was cropped to a 512 × 512 square box (region of interest), and this was used as the input of the ResNet (see Figure 3/3). The further image resizing steps were the same as those for the raw scanline data in Section 2.3, i.e., resized to 256 × 256, 128 × 128, 64 × 64, 32 × 32, 16 × 16, and 8 × 8 pixels using bicubic interpolation (with the `skimage.transform` function).

2.5. *Target: Spectral Features of the Vocoder*

To create the speech synthesis targets, the speech recordings were analyzed using mel-generalized log spectral approximation (MGLSA) [47] at a frame shift of 22,050 Hz/81.67 fps = 270 samples in order to be synchronous with the ultrasound data. As shown in Figure 2, this was achieved using the hardware sync output of the "Micro" equipment. This resulted in 25-dimensional spectral features (mel-generalized cepstrum–line spectral pair representation (MGC-LSP)) [48]. The vocoder spectral parameters served as the training targets of the DNNs, similarly to our earlier experiments in articulatory-to-acoustic mapping [11,17].

2.6. *Training of Deep Neural Networks*

We applied the ResNet-50 network [49] for the deep learning experiments. In our earlier studies, we either used fully connected deep neural networks [11,12], convolutional networks [15,17,20], LSTMs [15], 3D-CNNs [21], or GANs [22], but here, we opted for a more advanced network. The advantage of ResNet is that, by using skip connections, deeper convolutional networks can be trained than with simple DNNs or CNNs. By using ResNet-50, the network is spatially deep enough to capture most information from the ultrasound-based articulatory data. As ResNet was originally developed for image classification, the original output layer is "softmax", which was replaced here by a "linear" activation for the current regression task.

For all cases, we trained a speaker-specific ResNet model using the training data (180 sentences). Altogether, 21 networks were trained (3 data representations × 7 image sizes × 1 speaker). The cost function applied for the MGC-LSP regression task was the normalized mean-squared error (NMSE), and the optimizer was ADAM. We trained the network using backpropagation, and applied early stopping to avoid over-fitting. The network was trained at most for 100 epochs, but the training was stopped when the validation loss did not decrease within 10 epochs.

3. Results

After training the above ResNet models, we evaluated them by comparing the input image representations and the output spectral features.

3.1. Demonstration Samples

A sample Hungarian sentence (not being present in the training data) was chosen for demonstrating how the systems deal with the prediction of MGC-LSP spectral parameters. Figure 4 shows the output spectral features with the three input representations and seven image sizes.

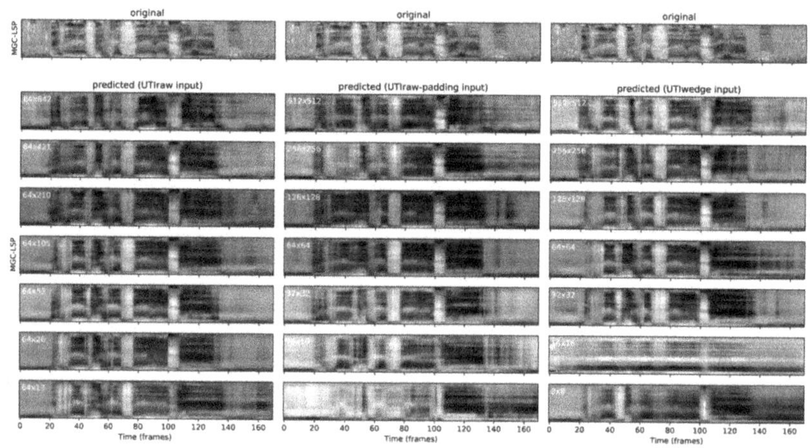

Figure 4. Demonstration samples: predicted MGC-LSP spectral features as a function of input image representation and size. Sentence: "Az Északi szél és a Nap".

In the first column, we can compare the results when using ultrasound as a raw scanlines input between 64 × 842–64 × 13 pixels. The predicted spectrograms follow the original sentence for the most part, but we can observe some artifacts: in the case of large input sizes (64 × 842, 64 × 421 and 64 × 210), the spectrogram is oversmoothed (i.e., formants are only weakly visible); and with a very small input size (64 × 13), unwanted frequency components appear at the end of the sentence, after frame 130. The remaining three figures in the middle (64 × 105, 64 × 53, and 64 × 26) seem to be the closest to the original spectrogram.

The second column shows the results when using the ultrasound of raw scanlines input, reshaped to a square, between 512 × 512–8 × 8 pixels. The tendencies are similar to the first column: the largest (512 × 512) and smallest (8 × 8) images cause oversmoothing, whereas those in between follow the spectral features or the original sentence with finer details. Interestingly, the 128 × 128 image size resulted in some distortion at the end of the sentence, between frames 140–160.

In the third column of Figure 4, we can see the effect ofthe ultrasound as a wedge-shape when used as an input of the ResNet, again between 512 × 512–8 × 8 pixels. The middle images sizes (128 × 128, 64 × 64, and 32 × 32) resulted in a relatively well-predicted spectrogram between frames 20–140; but after frame 140, distortion is visible in the case of 64 × 64. In the case of this demonstration sentence, the spectral prediction with 16 × 16 is extremely weak and almost constant, whereas in the case of the 8 × 8 image size, the formant movements of the original spectrogram are at least roughly visible.

Overall, the best MGC-LSP spectrogram predictions could be achieved with input image sizes of 64 × 53, 64 × 64, and 32 × 32 pixels on this single demonstration sentence. To obtain more general evaluations, we measured errors on the whole validation set, which will be introduced in the next section.

3.2. Comparison of Raw Scanline Data and Wedge Format

Figure 5 presents the validation loss results that we obtained after training the ResNet-50 network separately for the three data representations as a function of the input image size. When comparing (1) raw data (*UTIraw*), (2) raw data in square form (*UTIraw-padding*), and (3) wedge-shaped ultrasound data (*UTIwedge*), we can see similar tendencies in the validation error (which is NMSE measured on the validation data). All of the errors with the raw scanlines and the wedge-formatted images are in the range of 0.44–0.55. The best results (lowest errors) were achieved with the (1) raw scanline representation. This is followed by the (2) raw data in square form, while the (3) wedge-shaped ultrasound data have the weakest results—but the values do not seem to be significantly different.

Therefore, we can conclude that the wedge representation of ultrasound tongue images (when extrapolated directly from the original raw scanlines) can result in roughly the same errors during articulatory-to-acoustic mapping.

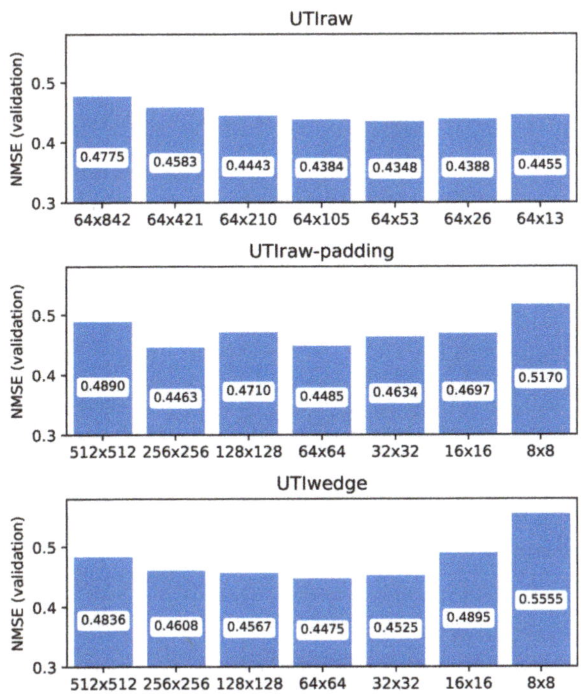

Figure 5. Final validation loss after ResNet-50 training as a function of input image representation and size. UTIraw: ultrasound as raw scanlines; UTIraw-padding: ultrasound as raw scanlines, reshaped to square; UTIwedge: ultrasound as a wedge shape.

3.3. Relation of Input Image Size and NMSE

We can investigate the three subfigures in Figure 5 as a function of image size. The tendencies are the same for all three data representations: the original image sizes (either 64 × 842 or 512 × 512 pixels) achieved a validation NMSE of around 0.48–0.49. When the image size is decreased (64 × 421 or 256 × 256 pixels), the validation error of the network will be lower. The optimal image size is around 64 × 64, resulting in a validation NMSE of around 0.44–0.45. Here, we can find some differences with the three data representations: (1) in the case of the raw scanline input (top subfigure), the image size causing the lowest error is 64 × 53 pixels; (2) if the scanlines are in square representation, then the lowest error is achieved with 256 × 256 pixels, but 64 × 64 results in almost the same values; (3) in the case of the wedge input, then, again, 64 × 64 pixels is the optimal size. If we further decrease the image size (64 × 26/64 × 13/32 × 32/etc.), then the error gets higher, until we reach the weakest results: NMSE is 0.45 with 64 × 13, and 0.52/0.56 with 8 × 8 pixels input images.

Based on the above comparison, we can conclude that the optimal image sizes are 64 × 53 and 64 × 64 when taking into account the validation error.

3.4. Training Time

Figure 6 shows the (wall clock) DNN training times expressed in seconds. For all three input representations, this was measured on an Intel i7-2600 3.4 GHz PC with 16 GB RAM and an NVidia Titan X video card. Note that the largest images (512 × 512, 64 × 842, and 64 × 421) were trained with a batch size of 2 in order to fit into GPU memory; whereas, for the other image sizes, a batch size of 64 was used. The other parameters of DNN training were the same for all networks.

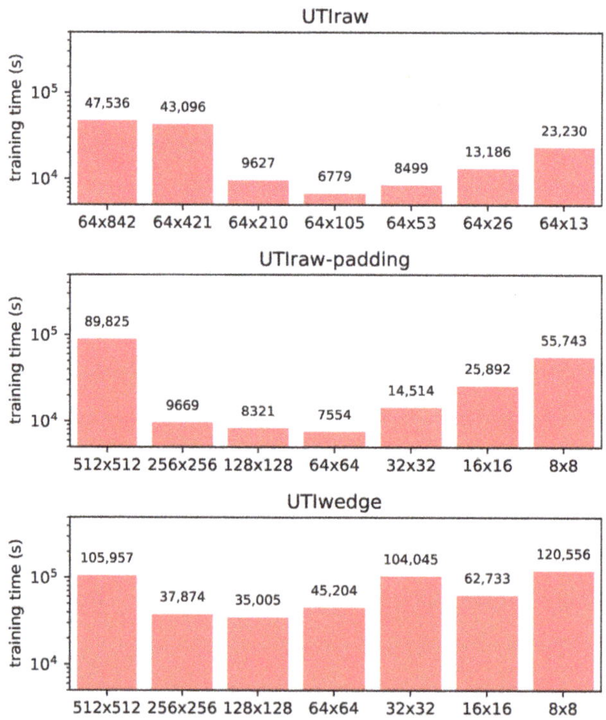

Figure 6. ResNet-50 wall-clock training time (in seconds) as a function of image size.

We can observe the tendency that networks with a middle-sized input image are faster to train. In particular, the original images (512 × 512 and 64 × 842) are highly disadvantageous when trained with ResNet-50 because of memory limitations (i.e., a smaller batch size). Based on the training time, the optimal image size is around 64 × 64 and 64 × 105 pixels (except for *UTIwedge*, where the training with the 128 × 128 input image size was the fastest). With *UTIwedge* representation, with all image sizes, the training time was significantly larger than with *UTIraw* or *UTIraw-padding*.

4. Discussion

For articulatory-to-acoustic mapping, ultrasound tongue imaging is often applied as an input, as shown in Section 1. Mostly, (micro)convex transducers are used, which provide a wedge-shape visual image. However, this is optimized for the visual inspection of the human eye (which is perfect for linguistic or medical studies), and the signal is often post-processed by the equipment (which might be a problem for engineering studies). Examples for such early systems are: Acoustic Imaging Performa 30 Hz ultrasound machine [4], Aloka SSD-1000 machine [6], Terason T3000 ultrasound [5,7,9], and the CONTEC CMS600P2 system [18].

With newer ultrasound equipment, it is now possible to gain full access to the raw scanline data (i.e., ultrasound echo return). A good example for this is the "Micro" system (developed by Telemed Ltd., Vilnius, Lithuania, and distributed by Articulate Instruments Ltd., Musselburgh, UK),which is available since 2016, and was also used for our recordings in the MTA-ELTE Lendület Lingual Articulation Research Group [11,17,50]. In addition, it was used for large-scale databases, such as UltraSuite [51] and UltraSuite-TaL [52]. The advantage of the "Micro" ultrasound equipment in this context is that we can use the data without any internal post-processing of the device, and the feature extraction can be left up to the machine learning algorithms. For other scenarios, e.g., automatic tongue contour tracking from ultrasound images, preprocessing the features has been shown to be useful [53], but, for contour tracking in the above study, deep learning approaches have not been used, which could help the feature learning.

The raw scanline data used in this study refer to the digitized, envelope-detected beam vectors of the "Micro" ultrasound system. When the ultrasound is recorded internally in the device, the envelopes of raw beamformed RF signals are generated from the delay and sum of channel signals. After further demodulation, low-pass filtering, and amplitude operation, the scanline data can be obtained, and the final B-mode images can also be generated by image processing and coordinate transformation. Therefore, the significant information differences should exist between the raw beamformed RF signals and raw scanline data or final B-mode images, rather than raw scanline data and final B-mode images. However, there is no control of beamforming in "Micro" and we cannot have access to the above RF signal (p.c., Articulate Instruments Ltd.). With other ultrasound equipment (e.g., "Art" system of Articulate Instruments Ltd.), one can record and process the RF output, but, in this case, the hardware synchronization with the speech signal has to be solved.

Although a large number of studies have already applied ultrasound tongue imaging for articulatory-to-acoustic mapping, the optimal data representations and input image sizes have not been deeply investigated before. In the current study, we compared the raw scanline representation (digitized, envelope-detected beam vectors) with the wedge-shaped processed UTI as the input for the residual network applied for AAM, and showed that all input representations can result in a similar validation error while training DNNs. We expect that, with a higher resolution ultrasound (e.g., higher fps, larger spatial resolution, or 3D/4D ultrasound [54]), the synthesized speech would be more natural, i.e., result in a lower MSE during DNN training.

However, a comparison across sessions and speakers (or designing speaker-independent AAM systems) using the raw scanline data is a challenge. Because of the convex transducer, the raw data do not correspond to the original mid-sagittal slice, and non-linear interpolation is necessary to transform into real-world orientation. Therefore, for comparisons across sessions and speakers, the wedge-shape ultrasound images might be more useful than the raw scanline data. By using tracing methods on wedge-shaped ultrasound images, it is also possible to obtain a raw-like data representation [55], but this conversion cannot revert the postprocessing methods of the equipment, and the back-and-forth conversion obviously leads to some data loss.

In spite of the significant achievements of the last decade, potential SSI applications seem to still be far away from a practically working scenario. Part of the reason is the lack of fully developed cross-session and cross-speaker methodologies. With some articulatory tracking devices, there have already been such experiments, e.g., signal normalization and model adaptation for sEMG [56,57], domain-adversarial DNN training [32], inter-speaker analysis for EOS [58], region of interest detection and cropping for lip video [43], and articulation adaptation using Procrustes matching with EMA [27]. Ultrasound-based SSI systems, however, might be less robust, as slight changes in probe positioning causes shifts and rotations in the resulting image [59,60]. Therefore, the results of the current study can help future cross-session and cross-speaker experiments.

5. Conclusions

In this study, we compared the raw scanline input with the wedge-shaped ultrasound tongue image representation. In addition, we investigated the optimal input image size of a residual network applied for articulatory-to-acoustic mapping. We found that there is no significant difference between using the raw data (either in original form or transposed to a square) and the wedge shape that is directly extrapolated from the raw data. We also found that the optimal pixel size is 64 × 64 when taking into account the validation loss and network training time. Therefore, it is not necessary to use the full original 64 × 842 pixels raw scanline, but a smaller image is enough, which allows for the building of smaller networks using less training data. In addition, the smaller image size enables the use of multiple consecutive input images [11] or a recurrent neural network [15], as already applied in our earlier work.

The advantage of fan/wedge-shaped data is that they correspond to the physical/spatial orientation of the speaking organs of the subject; therefore, comparisons across sessions and speakers are relatively easy. In the future, we plan to apply the raw-to-wedge conversion methods for experimenting with speaker-independent articulatory-to-acoustic systems in order to develop practically working silent speech interface applications.

The Keras implementations are accessible at https://github.com/BME-SmartLab/UTI-optimization, last accessed on 30 October 2022.

Author Contributions: Conceptualization, T.G.C.; methodology, T.G.C., G.G. and L.T.; software, T.G.C.; investigation, T.G.C., G.G., L.T., A.H.S. and A.M.; data curation, T.G.C. and A.M.; writing—original draft preparation, T.G.C.; writing—review and editing, T.G.C., G.G., L.T., A.H.S. and A.M.; visualization, T.G.C.; project administration, T.G.C.; funding acquisition, T.G.C. All authors have read and agreed to the published version of the manuscript.

Funding: T.G. Csapó's research was partly supported by the APH-ALARM project (contract 2019-2.1.2-NEMZ-2020-00012) funded by the European Commission and the National Research, Development and Innovation Office of Hungary (FK 142163 grant), by the Bolyai János Research Fellowship of the Hungarian Academy of Sciences and the ÚNKP-22-5-BME-316 New National Excellence Program of the Ministry for Culture and Innovation from the source of the National Research, Development and Innovation Fund. The work of G. Gosztolya and L. Tóth were also supported by the Hungarian Ministry of Innovation and Technology NRDI Office (grant TKP2021-NVA-09) and by the Artificial Intelligence National Laboratory (RRF-2.3.1-21-2022-00004).

Institutional Review Board Statement: Not applicable.

Informed Consent Statement: Not applicable.

Data Availability Statement: Trained models can be downloaded at https://github.com/BME-SmartLab/UTI-optimization, last accessed on 30 October 2022.

Acknowledgments: The Titan X GPU used was donated by NVIDIA Corporation.

Conflicts of Interest: The authors declare no conflict of interest.

Abbreviations

The following abbreviations are used in this manuscript:

AAI	Acoustic-to-Articulatory Inversion
AAM	Articulatory-to-Acoustic Mapping
CNN	Convolutional Neural Network
DNN	Deep Neural Network
EMA	Electromagnetic Articulography
EOS	Electro-Optical Stomatography
LSTM	Long-Short Term Memory
MGC-LSP	Mel-Generalized Cepstrum–Line Spectral Pair
MGLSA	Mel-Generalized Log Spectral Approximation
PMA	Permanent Magnetic Articulography
sEMG	surface Electromyography
SSI	Silent Speech Interface
SSR	Silent Speech Recognition
TTS	Text-To-Speech
UTI	Ultrasound Tongue Imaging

References

1. Denby, B.; Schultz, T.; Honda, K.; Hueber, T.; Gilbert, J.M.; Brumberg, J.S. Silent speech interfaces. *Speech Commun.* **2010**, *52*, 270–287. [CrossRef]
2. Schultz, T.; Wand, M.; Hueber, T.; Krusienski, D.J.; Herff, C.; Brumberg, J.S. Biosignal-Based Spoken Communication: A Survey. *IEEE/ACM Trans. Audio, Speech Lang. Process.* **2017**, *25*, 2257–2271. [CrossRef]
3. Gonzalez-Lopez, J.A.; Gomez-Alanis, A.; Martin Donas, J.M.; Perez-Cordoba, J.L.; Gomez, A.M. Silent Speech Interfaces for Speech Restoration: A Review. *IEEE Access* **2020**, *8*, 177995–178021. [CrossRef]
4. Denby, B.; Stone, M. Speech synthesis from real time ultrasound images of the tongue. In Proceedings of the ICASSP, Montreal, QC, Canada, 17–21 May 2004; pp. 685–688. [CrossRef]
5. Denby, B.; Cai, J.; Hueber, T.; Roussel, P.; Dreyfus, G.; Crevier-Buchman, L.; Pillot-Loiseau, C.; Chollet, G.; Manitsaris, S.; Stone, M. Towards a Practical Silent Speech Interface Based on Vocal Tract Imaging. In Proceedings of the 9th International Seminar on Speech Production (ISSP 2011), Montreal, QC, Canada, 20–23 June 2011; pp. 89–94.
6. Hueber, T.; Benaroya, E.L.; Chollet, G.; Dreyfus, G.; Stone, M. Development of a silent speech interface driven by ultrasound and optical images of the tongue and lips. *Speech Commun.* **2010**, *52*, 288–300. [CrossRef]
7. Hueber, T.; Benaroya, E.l.; Denby, B.; Chollet, G. Statistical Mapping Between Articulatory and Acoustic Data for an Ultrasound-Based Silent Speech Interface. In Proceedings of the Interspeech, Florence, Italy, 27–31 August 2011; pp. 593–596.
8. Wei, J.; Fang, Q.; Zheng, X.; Lu, W.; He, Y.; Dang, J. Mapping ultrasound-based articulatory images and vowel sounds with a deep neural network framework. *Multimed. Tools Appl.* **2016**, *75*, 5223–5245. [CrossRef]
9. Jaumard-Hakoun, A.; Xu, K.; Leboullenger, C.; Roussel-Ragot, P.; Denby, B. An Articulatory-Based Singing Voice Synthesis Using Tongue and Lips Imaging. In Proceedings of the Interspeech, San Francisco, CA, USA, 8–12 September 2016; pp. 1467–1471. [CrossRef]
10. Tatulli, E.; Hueber, T. Feature extraction using multimodal convolutional neural networks for visual speech recognition. In Proceedings of the ICASSP, New Orleans, LA, USA, 5–9 March 2017; pp. 2971–2975. [CrossRef]
11. Csapó, T.G.; Grósz, T.; Gosztolya, G.; Tóth, L.; Markó, A. DNN-Based Ultrasound-to-Speech Conversion for a Silent Speech Interface. In Proceedings of the Interspeech, Stockholm, Sweden, 20–24 August 2017; pp. 3672–3676. doi: 10.21437/ Interspeech.2017-939. [CrossRef]
12. Grósz, T.; Gosztolya, G.; Tóth, L.; Csapó, T.G.; Markó, A. F0 Estimation for DNN-Based Ultrasound Silent Speech Interfaces. In Proceedings of the ICASSP, Calgary, AB, Canada, 15–20 April 2018; pp. 291–295.
13. Tóth, L.; Gosztolya, G.; Grósz, T.; Markó, A.; Csapó, T.G. Multi-Task Learning of Phonetic Labels and Speech Synthesis Parameters for Ultrasound-Based Silent Speech Interfaces. In Proceedings of the Interspeech, Hyderabad, India, 2–6 September 2018; pp. 3172–3176. [CrossRef]
14. Ji, Y.; Liu, L.; Wang, H.; Liu, Z.; Niu, Z.; Denby, B. Updating the Silent Speech Challenge benchmark with deep learning. *Speech Commun.* **2018**, *98*, 42–50. [CrossRef]
15. Moliner, E.; Csapó, T.G. Ultrasound-based silent speech interface using convolutional and recurrent neural networks. *Acta Acust. United Acust.* **2019**, *105*, 587–590. [CrossRef]

16. Gosztolya, G.; Pintér, Á.; Tóth, L.; Grósz, T.; Markó, A.; Csapó, T.G. Autoencoder-Based Articulatory-to-Acoustic Mapping for Ultrasound Silent Speech Interfaces. In Proceedings of the International Joint Conference on Neural Networks, Budapest, Hungary, 14–19 July 2019.
17. Csapó, T.G.; Al-Radhi, M.S.; Németh, G.; Gosztolya, G.; Grósz, T.; Tóth, L.; Markó, A. Ultrasound-based Silent Speech Interface Built on a Continuous Vocoder. In Proceedings of the Interspeech, Graz, Austria, 15–19 September 2019; pp. 894–898. [CrossRef]
18. Kimura, N.; Kono, M.C.; Rekimoto, J. Sottovoce: An ultrasound imaging-based silent speech interaction using deep neural networks. In Proceedings of the CHI'19: 2019 CHI Conference on Human Factors in Computing Systems, Glasgow, UK, 4–9 May 2019; pp. 1–11. [CrossRef]
19. Zhang, J.; Roussel, P.; Denby, B. Creating Song from Lip and Tongue Videos with a Convolutional Vocoder. *IEEE Access* **2021**, *9*, 13076–13082. [CrossRef]
20. Csapó, T.G.; Zainkó, C.; Tóth, L.; Gosztolya, G.; Markó, A. Ultrasound-based Articulatory-to-Acoustic Mapping with WaveGlow Speech Synthesis. In Proceedings of the Interspeech, Shanghai, China, 25–29 October 2020; pp. 2727–2731. [CrossRef]
21. Shandiz, A.H.; Tóth, L.; Gosztolya, G.; Markó, A.; Csapó, T.G. Neural speaker embeddings for ultrasound-based silent speech interfaces. In Proceedings of the Interspeech, Brno, Czech Republic, 30 August–3 September 2021; pp. 151–155. [CrossRef]
22. Shandiz, A.H.; Tóth, L.; Gosztolya, G.; Markó, A.; Csapó, T.G. Improving Neural Silent Speech Interface Models by Adversarial Training. In Proceedings of the 2nd International Conference on Artificial Intelligence and Computer Vision (AICV2021), Settat, Morocco, 28–30 June 2021.
23. Wang, J.; Samal, A.; Green, J.R.; Rudzicz, F. Sentence Recognition from Articulatory Movements for Silent Speech Interfaces. In Proceedings of the ICASSP, Kyoto, Japan, 25–30 March 2012; pp. 4985–4988.
24. Kim, M.; Cao, B.; Mau, T.; Wang, J. Speaker-Independent Silent Speech Recognition from Flesh-Point Articulatory Movements Using an LSTM Neural Network. *IEEE/ACM Trans. Audio Speech Lang. Process.* **2017**, *25*, 2323–2336. [CrossRef] [PubMed]
25. Cao, B.; Kim, M.; Wang, J.R.; Van Santen, J.; Mau, T.; Wang, J. Articulation-to-Speech Synthesis Using Articulatory Flesh Point Sensors' Orientation Information. In Proceedings of the Interspeech, Hyderabad, India, 2–6 September 2018; pp. 3152–3156. [CrossRef]
26. Taguchi, F.; Kaburagi, T. Articulatory-to-speech conversion using bi-directional long short-term memory. In Proceedings of the Interspeech, Hyderabad, India, 2–6 September 2018; pp. 2499–2503.
27. Cao, B.; Wisler, A.; Wang, J. Speaker Adaptation on Articulation and Acoustics for Articulation-to-Speech Synthesis. *Sensors* **2022**, *22*, 6056. [CrossRef] [PubMed]
28. Fagan, M.J.; Ell, S.R.; Gilbert, J.M.; Sarrazin, E.; Chapman, P.M. Development of a (silent) speech recognition system for patients following laryngectomy. *Med. Eng. Phys.* **2008**, *30*, 419–425. [CrossRef] [PubMed]
29. Gonzalez, J.A.; Cheah, L.A.; Gomez, A.M.; Green, P.D.; Gilbert, J.M.; Ell, S.R.; Moore, R.K.; Holdsworth, E. Direct Speech Reconstruction from Articulatory Sensor Data by Machine Learning. *IEEE/ACM Trans. Audio Speech Lang. Process.* **2017**, *25*, 2362–2374. [CrossRef]
30. Diener, L.; Janke, M.; Schultz, T. Direct conversion from facial myoelectric signals to speech using Deep Neural Networks. In Proceedings of the 2015 International Joint Conference on Neural Networks (IJCNN), Killarney, Ireland, 12–17 July 2015; IEEE: Piscataway, NJ, USA, 2015; pp. 1–7. [CrossRef]
31. Janke, M.; Diener, L. EMG-to-Speech: Direct Generation of Speech from Facial Electromyographic Signals. *IEEE/ACM Trans. Audio Speech Lang. Process.* **2017**, *25*, 2375–2385. [CrossRef]
32. Wand, M.; Schultz, T.; Schmidhuber, J. Domain-Adversarial Training for Session Independent EMG-based Speech Recognition. In Proceedings of the Interspeech, Hyderabad, India, 2–6 September 2018; pp. 3167–3171. [CrossRef]
33. Stone, S.; Birkholz, P. Silent-speech command word recognition using electro-optical stomatography. In Proceedings of the Interspeech, San Francisco, CA, USA, 8–12 September 2016; pp. 2350–2351.
34. Wand, M.; Koutník, J.; Schmidhuber, J. Lipreading with long short-term memory. In Proceedings of the ICASSP, Shanghai, China, 20–25 March 2016; pp. 6115–6119.
35. Ephrat, A.; Peleg, S. Vid2speech: Speech Reconstruction from Silent Video. In Proceedings of the ICASSP, New Orleans, LA, USA, 5–9 March 2017; pp. 5095–5099.
36. Sun, K.; Yu, C.; Shi, W.; Liu, L.; Shi, Y. Lip-Interact: Improving Mobile Device Interaction with Silent Speech Commands. In Proceedings of the UIST 2018—31st Annual ACM Symposium on User Interface Software and Technology, Berlin, Germany, 14–17 October 2018; pp. 581–593. [CrossRef]
37. Ferreira, D.; Silva, S.; Curado, F.; Teixeira, A. Exploring Silent Speech Interfaces Based on Frequency-Modulated Continuous-Wave Radar. *Sensors* **2022**, *22*, 649. [CrossRef] [PubMed]
38. Freitas, J.; Ferreira, A.J.; Figueiredo, M.A.T.; Teixeira, A.J.S.; Dias, M.S. Enhancing multimodal silent speech interfaces with feature selection. In Proceedings of the Interspeech, Singapore, 14–18 September 2014; pp. 1169–1173.
39. Stone, M. A guide to analysing tongue motion from ultrasound images. *Clin. Linguist. Phon.* **2005**, *19*, 455–501. [CrossRef] [PubMed]
40. Csapó, T.G.; Lulich, S.M. Error analysis of extracted tongue contours from 2D ultrasound images. In Proceedings of the Interspeech, Dresden, Germany, 6–10 September 2015; pp. 2157–2161.
41. Wrench, A.; Balch-Tomes, J. Beyond the Edge: Markerless Pose Estimation of Speech Articulators from Ultrasound and Camera Images Using DeepLabCut. *Sensors* **2022**, *22*, 1133. [CrossRef] [PubMed]

42. Hueber, T.; Aversano, G.; Chollet, G.; Denby, B.; Dreyfus, G.; Oussar, Y.; Roussel, P.; Stone, M. Eigentongue feature extraction for an ultrasound-based silent speech interface. In Proceedings of the ICASSP, Honolulu, HI, USA, 15–20 April 2007; pp. 1245–1248.
43. Kimura, N.; Su, Z.; Saeki, T.; Rekimoto, J. SSR7000: A Synchronized Corpus of Ultrasound Tongue Imaging for End-to-End Silent Speech Recognition. In Proceedings of the Language Resources and Evaluation Conference, Marseille, France, 20–25 June 2022; pp. 6866–6873.
44. Yu, Y.; Honarmandi Shandiz, A.; Tóth, L. Reconstructing Speech from Real-Time Articulatory MRI Using Neural Vocoders. In Proceedings of the EUSIPCO, Dublin, Ireland, 23–27 August 2021; pp. 945–949.
45. Ribeiro, M.S.; Eshky, A.; Richmond, K.; Renals, S. Speaker-independent Classification of Phonetic Segments from Raw Ultrasound in Child Speech. In Proceedings of the ICASSP, Brighton, UK, 12–17 May 2019; pp. 1328–1332. [CrossRef]
46. Ribeiro, M.S.; Eshky, A.; Richmond, K.; Renals, S. Silent versus modal multi-speaker speech recognition from ultrasound and video. In Proceedings of the Interspeech, Brno, Czech Republic, 30 August–3 September 2021; pp. 641–645.
47. Imai, S.; Sumita, K.; Furuichi, C. Mel Log Spectrum Approximation (MLSA) filter for speech synthesis. *Electron. Commun. Jpn. Part I Commun.* **1983**, *66*, 10–18. [CrossRef]
48. Tokuda, K.; Kobayashi, T.; Masuko, T.; Imai, S. Mel-generalized cepstral analysis—A unified approach to speech spectral estimation. In Proceedings of the ICSLP, Yokohama, Japan, 18–22 September 1994; pp. 1043–1046.
49. He, K.; Zhang, X.; Ren, S.; Sun, J. Deep Residual Learning for Image Recognition. In Proceedings of the 2016 IEEE Conference on Computer Vision and Pattern Recognition (CVPR), Las Vegas, NV, USA, 27–30 June 2016; pp. 770–778. [CrossRef]
50. Csapó, T.G.; Deme, A.; Gráczi, T.E.; Markó, A.; Varjasi, G. Synchronized speech, tongue ultrasound and lip movement video recordings with the "Micro" system. In Proceedings of the Challenges in Analysis and Processing of Spontaneous Speech, Budapest, Hungary, 14–17 May 2017.
51. Eshky, A.; Ribeiro, M.S.; Cleland, J.; Richmond, K.; Roxburgh, Z.; Scobbie, J.M.; Wrench, A. UltraSuite: A Repository of Ultrasound and Acoustic Data from Child Speech Therapy Sessions. In Proceedings of the Interspeech, Hyderabad, India, 2–6 September 2018; ISCA: Hyderabad, India, 2018; pp. 1888–1892. [CrossRef]
52. Ribeiro, M.S.; Sanger, J.; Zhang, J.X.X.; Eshky, A.; Wrench, A.; Richmond, K.; Renals, S. TaL: A synchronised multi-speaker corpus of ultrasound tongue imaging, audio, and lip videos. In Proceedings of the 2021 IEEE Spoken Language Technology Workshop (SLT), Online, 19–22 January 2021; pp. 1109–1116. [CrossRef]
53. Czap, L. Impact of preprocessing features on the performance of ultrasound tongue contour tracking, via dynamic programming. *Acta Polytech. Hung.* **2021**, *18*, 159–176. [CrossRef]
54. Lulich, S.M.; Berkson, K.H.; de Jong, K. Acquiring and visualizing 3D/4D ultrasound recordings of tongue motion. *J. Phon.* **2018**, *71*, 410–424. [CrossRef]
55. Czap, L. A Nyelvkontúr Automatikus Követése és Elemzése Ultrahang Felvételeken [Automatic Tracking and Analysis of the Tongue Contour on Ultrasound Recordings]. Habilitation Thesis, University of Miskolc, Miskolc, Hungary, 2020.
56. Maier-Hein, L.; Metze, F.; Schultz, T.; Waibel, A. Session independent non-audible speech recognition using surface electromyography. In Proceedings of the ASRU, San Juan, Puerto Rico, 27 November–1 December 2005; IEEE: San Juan, Puerto Rico, 2005; pp. 331–336. [CrossRef]
57. Janke, M.; Wand, M.; Nakamura, K.; Schultz, T. Further investigations on EMG-to-speech conversion. In Proceedings of the ICASSP, Kyoto, Japan, 25–30 March 2012; IEEE: Kyoto, Japan, 2012; pp. 365–368. [CrossRef]
58. Stone, S.; Birkholz, P. Cross-speaker silent-speech command word recognition using electro-optical stomatography. In Proceedings of the ICASSP, Barcelona, Spain, 4–8 May 2020; pp. 7849–7853.
59. Csapó, T.G.; Xu, K. Quantification of Transducer Misalignment in Ultrasound Tongue Imaging. In Proceedings of the Interspeech, Shanghai, China, 25–29 October 2020; pp. 3735–3739. [CrossRef]
60. Csapó, T.G.; Xu, K.; Deme, A.; Gráczi, T.E.; Markó, A. Transducer Misalignment in Ultrasound Tongue Imaging. In Proceedings of the 12th International Seminar on Speech Production, New Haven, CT, USA, 14–18 December 2020.

sensors

MDPI

Article

Exploring Silent Speech Interfaces Based on Frequency-Modulated Continuous-Wave Radar

David Ferreira [1,2,*], Samuel Silva [1,2,*], Francisco Curado [1,2] and António Teixeira [1,2]

1 Department of Electronics, Telecommunications & Informatics, University of Aveiro, 3810-193 Aveiro, Portugal; fcurado@ua.pt (F.C.); ajst@ua.pt (A.T.)
2 Institute of Electronics and Informatics Engineering of Aveiro (IEETA), 3810-193 Aveiro, Portugal
* Correspondence: davidcruzferreira@ua.pt (D.F.); sss@ua.pt (S.S.)

Abstract: Speech is our most natural and efficient form of communication and offers a strong potential to improve how we interact with machines. However, speech communication can sometimes be limited by environmental (e.g., ambient noise), contextual (e.g., need for privacy), or health conditions (e.g., laryngectomy), preventing the consideration of audible speech. In this regard, silent speech interfaces (SSI) have been proposed as an alternative, considering technologies that do not require the production of acoustic signals (e.g., electromyography and video). Unfortunately, despite their plentitude, many still face limitations regarding their everyday use, e.g., being intrusive, non-portable, or raising technical (e.g., lighting conditions for video) or privacy concerns. In line with this necessity, this article explores the consideration of contactless continuous-wave radar to assess its potential for SSI development. A corpus of 13 European Portuguese words was acquired for four speakers and three of them enrolled in a second acquisition session, three months later. Regarding the speaker-dependent models, trained and tested with data from each speaker while using 5-fold cross-validation, average accuracies of 84.50% and 88.00% were respectively obtained from Bagging (BAG) and Linear Regression (LR) classifiers, respectively. Additionally, recognition accuracies of 81.79% and 81.80% were also, respectively, achieved for the session and speaker-independent experiments, establishing promising grounds for further exploring this technology towards silent speech recognition.

Keywords: silent speech; continuous-wave radar; European Portuguese; machine learning

Citation: Ferreira, D.; Silva, S.; Curado, F.; Teixeira, A. Exploring Silent Speech Interfaces Based on Frequency-Modulated Continuous-Wave Radar. *Sensors* **2022**, *22*, 649. https://doi.org/10.3390/s22020649

Academic Editors: Bruce Denby, Tamás Gábor Csapó and Michael Wand

Received: 1 December 2021
Accepted: 10 January 2022
Published: 14 January 2022

1. Introduction

Speech is a natural and efficient form of human communication, and, as such, the research on speech technologies that can foster its use in domains such as Human–Computer Interaction (HCI) is highly relevant. While Automatic Speech Recognition (ASR) is commonly used in HCI environments, as in the case of Amazon's Alexa and Apple's Siri [1], there are still some scenarios that cannot take the most out of speech interaction, including situations where privacy is needed, environmental noise is present, silence is required, or in the most extreme cases, when health conditions incapacitate speakers to produce acoustic signals.

To tackle such scenarios, Silent Speech Interfaces (SSI) emerged as a possible alternative to consider, consisting of the process of speech communication in the absence of an audible/intelligible acoustic signal [2]. As speech production is a complex motor process, which starts in the brain and ends with respiratory, laryngeal, and articulatory motion, each step of its production process can be explored and physiologically measured through specialized sensors and methods to potentially infer what the speaker is trying to say without relying on the acoustic signal [3,4].

Although there are already a large set of proposed sensors and technologies for silent speech recognition, most of them have characteristics that limit their use in everyday life, e.g., by being intrusive, non-portable, affected by noise, user-dependent, or just not affordable [5–8].

In light of these challenges, it is important to explore novel and improved technologies that might bring SSI to a wider variety of scenarios and users. To this end, frequency-modulated continuous wave (FMCW) radar technology, already used in a wide variety of scenarios including the automotive industry [9,10] and service robots [11,12], emerges as a possible candidate to tackle some of these issues given its non-invasiveness, non-intrusiveness, portability, and independence of ambient lighting. Furthermore, recent evolutions have made it less costly and easily available commercially, making radar technology appear in many of our daily environments. The recent market launch of the first mobile phone with radar (Google pixel 4; 2019) dedicated to proximity detection and identification of manual commands from the user without direct contact with the device, points towards the vulgarization of this technology and opens up prospects of its future exploitation for novel applications. In this regard, the exploration of a technology that might already be present to bring SSI capabilities to the environment is a promising path to follow. However, while a strong potential can be anticipated, given these perceived advantages, radar technology has yet to prove its mettle for silent speech applications.

In previous work, the authors performed a preliminary study exploring the capabilities of contactless radar-based technology for silent speech interfaces [13]. The achieved results, obtained considering a corpus of 13 European Portuguese words and three speakers, demonstrated good overall performance establishing the feasibility of the proposed approach and yielding promising grounds for additional research. In this context, the main goal of the work presented here is to expand on previous work regarding radar-based SSI and contribute to the body of work in the field by (a) expanding the number of considered speakers, in regards to our previous work, from three to four; (b) assessing session independence by considering data obtained for the same speakers in two independent acquisition sessions; and (c) exploring speaker-independent performance.

The remainder of this document is structured as follows. Section 2 presents a brief overview on related work regarding non-invasive SSI, also covering previous research on SSIs for European Portuguese. Section 3 describes the adopted methods, from environment and acquisition settings to data exploration, feature extraction, and classification approaches. Section 4 reports the results for all the performed research experiments (i.e., per-speaker, intra-speaker, and inter-speaker). In Section 5, these are further analyzed and discussed. Finally, Section 6 presents some concluding remarks and ideas for further advancing this work.

2. Related Work on SSI

A distinguishing element of SSIs is speech recognition beyond the acoustic signal, exploring other biosignals associated with the different stages of the speech production process [2,4]. From brain waves to the visual aspects of speech, several approaches have been, and continue to be, proposed towards silent speech recognition (SSR). While relevant work also exists for invasive technologies, the overview provided in what follows focuses on non-invasive methods, as they are in line with our goals. Along with the presented overview, Table 1 is also presented, encompassing attained results from different considered technologies in the existing literature in the most recent years. Apart from researches that mainly focused on classification purposes, several others that studied the possibility of achieving session and speaker independence were also included, as they are topics incident on this work that will be subject to further exploration. Regarding the literature review process, Google Scholar was the search engine resorted to given its vast scope of scholarly literature.

In SSI development research, surface EMG (sEMG) is the most commonly used technology as it is easy to apply and is less prone to raise ethical concerns for volunteers [14,15]. Recent work has privileged the evolution and consideration of increasingly imperceptible and highly flexible sEMG electrodes (see, e.g., in [16–20]), and notable results include those of Liu et al. [16] and Dong et al. [15], where accuracies greater than 80% (for a vocabulary of six words) and 70.00% (for a vocabulary of three words) were, respectively, achieved. Nevertheless, there is still a high data variability between sessions due to the participants' skin impedance [21].

Non-audible murmur (NAM) microphones are another technology widely used in SSI development to capture and record murmured speech and other smooth vocal productions resultant from the acoustic output. Recognition rates of nearly 70.00% were achieved in a total of 21 tested utterances [22]. However, NAM is highly user-dependent due to participants' physiological differences, and its acquired waveforms typically lack in quality and intelligibility, consisting of open challenges that still need addressing [22].

Electroencephalography (EEG) enables measuring electrical brain signals in a non-invasive way. Recent studies (see, e.g., in [23,24]) attained accuracy rates around the 70% mark for corpora of five syllables and six imagined sounds (i.e., non-articulated, just imagined by the subjects, corresponding to the vocalization of the five vowels, a/i/u/e/o, and mute). Although allowing visualization of the activation of the different brain areas associated with speech production, this technology is highly sensitive to noise, and its recognition rates are user-dependent.

Table 1. Notable recent studies tackling silent speech recognition and the outcomes regarding speaker and session independence. Apart from stating the considered technology, for each study, its publication year and acquisition corpus (when applicable) are also presented.

	Tech.	Year	Corpus	Accuracy	Inter-Speaker	Intra-Speaker
Ma et al. [25]	sEMG	2019	10 words	72.00%	-	-
Rameau et al. [26]	sEMG	2020	2 isolated words	86.40%	-	-
Prorokovic et al. [27]	sEMG	2019	-	-	-	Lower WER than conventional methods
Meltzner et al. [14]	sEMG	2018	65 words 1200 word sequences	91.40% MFCC 94.20% with grammar models	Thousands of recorded hours are required from diverse population	-
Wand et al. [28]	sEMG	2018	-	-	-	Lower WER than conventional methods
Fernandes et al. [29]	sEMG	2019	2 isolated words	84.00% for 2 words	-	-
Liu et al. [16]	sEMG	2020	1 set of 5 words 1 set of 6 words	89.60% for set 1 92.70% for set 2	-	-
Kapur et al. [20]	sEMG	2018	15 words	92.01%	Future Work	-
Dong et al. [15]	sEMG	2019	3 words	71.70%	-	-
Shah et al. [22]	NAM	2018	21 utterances	64.33%	-	-
Sarmiento et al. [23]	EEG	2019	5 syllables	69.73% to 72.67%	-	-
Morooka et al. [24]	EEG	2018	6 sounds	79.70%	-	-
Chen et al. [30]	US	2018	-	-	Future Work	-
Zhao et al. [31]	US	2019	-	-	Future Work	-
Gosztolya et al. [32]	US	2019	-	-	Future Work	Future Work
Kimura et al. [33]	US	2019	4 commands	65.00%	-	-
Csapó et al. [34]	US	2019	9 sentences	78.84%	-	-
Sun et al. [35]	VID	2018	20 commands limited to usage context with lip exaggeration	98.90%	95.40%	-
Vougioukas et al. [36]	VID	2019	GRID database	73.40%	59.50%	-
Uttam et al. [37]	VID	2019	Oulu VS2 database	-	PESQ scores similar to speaker-dependent models	
Petridis et al. [38]	VID	2018	10 digits 10 phrases	-	70.50% 70.80%	-
Birkholz et al. [39]	UWB	2018	25 phonemes	89.00%	-	-
Dash et al. [40]	MEG	2019	5 phrases	79.93% imagination	22.10% without adapt. 55.08% with adapt.	-

sEMG = Surface Electromyography; NAM = Non-Audible Murmur; US = Ultrasound; VID = Video; UWB = Ultra-Wideband; MEG = Magnetoencephalography.

Ultrasound (US) imaging is another technology widely considered in SSI research, as it allows observing tongue movement sequences during the speech production process. Some recent works include those of Chen et al. [30] and Xu et al. [41], in which, respectively, a new technique for representing speech articulation resorting to an ultrasound-driven finite element model of the tongue is presented, and a novel sequential feature extraction approach for SSI systems is explored. Considering US studies, recognition results are typically disregarded, as they are mainly focused on synthesizing speech that is further subject to subjective assessment regarding how natural they sound [33,34]. Nevertheless, while US images yield good spatial and temporal resolutions, the images have relatively low quality due to the presence of speckle noise [42].

Video imaging can be used to capture visible speech articulators. By resorting to different models and algorithms, which allow extracting the articulators of interest from each frame, it is possible to obtain accuracy rates as high as 87.00% for a corpus of eight consonants [43] and 98.00% for 20 commands limited to usage context with slight over-articulation to increase the extent of lip movement [35]. Although generally being low-cost, this technology is susceptible to raise privacy concerns and are typically strongly affected by ambient illumination.

Regarding contactless radar-based silent speech recognition, and to the best of our knowledge, not much has yet been explored apart from a recent work by Shin et al. [8]. In this study, a recognition rate of 85.00% was reported for a corpus comprising 10 isolated words considering a dynamic time warping (DTW) approach. However, as stated by the authors, some limitations resided in the fact that distance and correlation amplitude were the only considered features, and that there was recognition degradation due to slight head movements of the participants throughout the acquisition sessions, something that would require additional methods to mitigate.

Concerning SSI for European Portuguese (EP), several technologies have been researched (e.g., EEG, sEMG, Ultrasonic Doppler (UD), Video, and Depth), also including multimodal approaches [7]. Freitas et al. [44] proposed Visual Speech Recognition (VSR) and Acoustic Doppler Sensors (ADS) for silent speech recognition, resorting to Dynamic Time Warping (DTW), achieving a 91.40% accuracy rate. Later, in 2013, the same author [45] selected four non-invasive modalities (Visual data from Video and Depth, sEMG, and UD) and proposed a system that explores their synchronous combination into a multimodal SSI. The same corpus as the one explored in this article was considered, and DTW and KNN were used. It was verified that the combination of multiple modalities presented a better recognition performance (93.80%), while subsets of the modalities produced lower results, such as 71.40% for the Video and Depth combination. In a more recent work, Teixeira et al. [46] proposed a VSR approach to enable real-time control of a media player, having achieved an accuracy of 81.30% for a corpus comprising eight control commands.

3. Method

In line with our research goals, and regarding its acquisition settings, an approach was established in which there was a compromise between keeping some aspects closer to real scenarios (e.g., no chin rest during acquisitions) and establishing some controlled conditions (e.g., frontal head orientation and fixed approximate distance from the radar). Such considerations, while not compromising the study's central purpose, should reduce, at this stage, some of the complexity of its data acquisition and post-processing phases.

In this section, all steps that are inherent to this project's methodology are described and the key stages are illustrated in Figure 1.

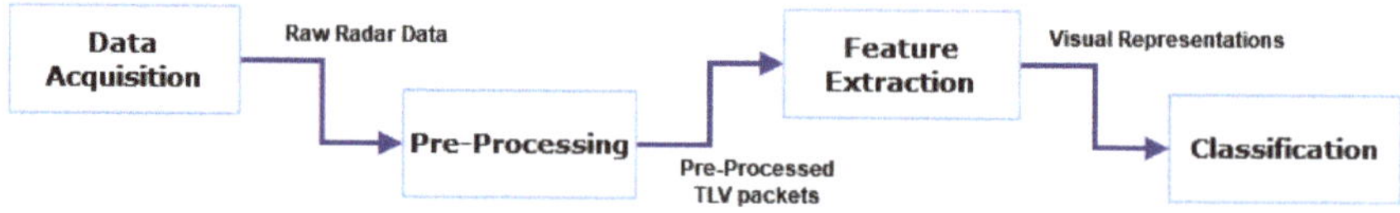

Figure 1. Acquisition and classification pipeline. From left to right, the respective processing steps are data acquisition, preprocessing, feature extraction, and classification.

3.1. Experimental Setup

The board considered for this investigation was the AWR1642BOOST-EVM from Texas Instruments, Dalas, TX, USA, an evaluation board for the AWR1642 FMCW radar sensor. This board is currently used at our research institute for a plethora of different purposes, ranging from robot navigation and human detection [47] to biosignal measurement [48].

The room selected for the acquisition sessions was free from any moving objects other than the participant, ensuring that no interference would negatively impact the acquired data.

In addition to the room settings, it was also established that, throughout the acquisition sessions, the participants would directly face the radar and sit at an approximate distance of 15 cm from it, a similar setting to how speakers are positioned in front of a tabletop microphone. However, and as we were aiming for reduced intrusiveness, we opted not to fix the position of the participants' heads, simply instructing them, without enforcing it, to try and keep the same relative position towards the radar.

3.2. Data Acquisition

For the data acquisition sessions, we resorted to Texas Instruments' DemoVisualizer application, a software that enables radar configuration, data acquisition, and data visualization. DemoVisualizer enabled testing different radar configurations while focusing on the acquisition aspects that most suited the particular research experiments. As previously explained, we privileged a less fixed head position, and this, as observed by the authors of [8], might yield added challenges in using distance to the radar as the data considered for recognition. Therefore, we opted for a configuration that prioritized the best possible velocity resolution, envisaging the acquisition of the participants' facial velocity dispersion patterns while they produced speech.

To automatically manage the data acquisition process, custom software was designed and developed. The participants were placed in front of the radar board, at approximately 15 cm from it (Figure 2). An LCD display, adequately positioned in the participant's line of sight, provided information about the word to be uttered. After signing the informed consent, speakers were asked to speak at a normal rhythm, and, subsequently, the acquisition procedure started. For each trial, a random corpus word would appear on the LCD, and a beep (along with a change in color of the screen) would signal that the speaker should utter it. After the beep, the acquisition software recorded radar data for two seconds.

It is also important to mention that, in addition to the first acquisition session, the realization of a second session was scheduled for three months later for those participants that could perform it to allow studying intra-speaker variability.

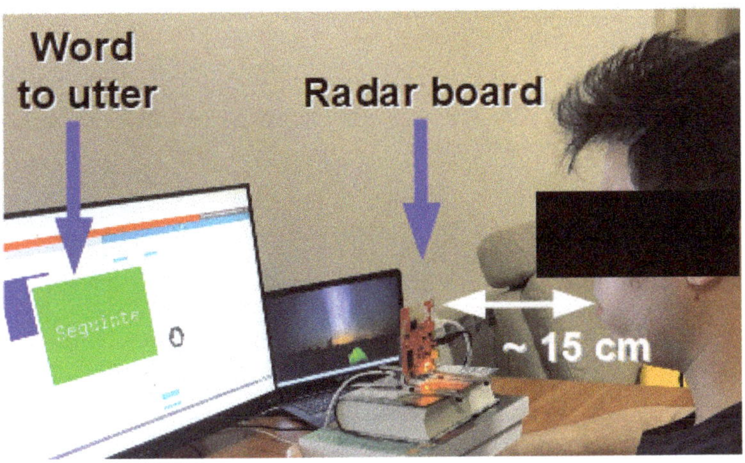

Figure 2. Radar setup for the data acquisition sessions. The participant is seated in front of the radar board while a monitor continuously displays the words to be uttered, turning the background green whenever the participant is asked to speak.

3.3. Corpus

Considering the team's body of work on Ambient Assisted Living (AAL) and its previous work in SSI research for these contexts, we adopted a previously considered corpus [45,49], presented in Table 2, containing a total of thirteen EP words. The words aggregated in the chosen corpus are mostly command instructions typically used in AAL contexts, such as when interacting with personal assistants (e.g., asking the personal assistant to turn something on, contact someone, or help in any way it can).

Table 2. Corpus considered for all radar data acquisition sessions, containing a total of thirteen European Portuguese words.

Ajuda	(help)	Anterior	(previous)
Calendário	(calendar)	Contactos	(contacts)
Email	(email)	Familia	(family)
Fotografias	(photographs)	Lembretes	(reminders)
Ligar	(turn on)	Mensagens	(messages)
Pesquisar	(search)	Seguinte	(next)
Vídeos	(videos)		

Four participants, all native EP speakers, enrolled in the data acquisition sessions: (a) one of the authors, an Engineering Ph.D. student, 26 years old, male; (b) a 24 years old female Psychologist; (c) a 50 years old female real estate manager; and (d) a 22 years old female Physiotherapist. Two of the speakers (one of them, Speaker 1, the first author, and Speaker 4) were asked to try and keep a more consistent speaking pattern throughout each acquisition session. This would inform a best-case scenario where a prospective user would be informed to be consistent in uttering the commands compared with speakers without such information.

For each of the 13 words present in the corpus, 60 validated repetitions were considered per participant. The validation stage mainly ensured the removal of the recordings in which no usable data was produced (e.g., participant missing the recording slot or data recording error).

3.4. Preprocessing

The preprocessing phase is mainly responsible for analyzing the produced data for each participant and adequately annotating each file with its corresponding class name and trial number. Besides organizing the data to a format convenient for the subsequent feature extraction phase, it also ensures the removal of any data inadequately acquired. The main difference between this processing step and the validation stage is that, while the validation stage consists of an empirical process where missed time slots or mispronunciations are removed through the observation of visual representations, this step removes the files at the earliest stage possible, verifying if the files, as soon as they are acquired by the radar board, are either corrupted or badly organized.

3.5. Feature Extraction

After the acquisition sessions, it was necessary to explore the data and define a set of features to be tested and used for classification. During data acquisition, the signals received by the RF front-end of the board are digitized and preprocessed by the built-in ADC and DSP, respectively, and the raw data thus obtained are assembled into several tag-length-value (TLV) packets. To process these packets, a parser was developed in Matlab to extract all the detected objects' relevant information (i.e., their Cartesian coordinates (X, Y, Z) and relative velocities expressed in the radar frame of reference). These data include the entire point cloud detected in the radar FoV, over the time acquisition windows, from which distinct subsets of points are clustered and associated, in real-time, by the firmware of the board, to represent different objects, or parts of a body, with dissimilar velocity measures. However, besides including static and moving objects, these readings may also contain "fake" target detections that often result from multiple reflections on walls or other surfaces, which led us to define that all data beyond 30 cm from the radar board would be filtered and excluded.

Regarding the visualization of the acquired data, for each word instance, two representations were created, one depicting distance variations over time and another the velocities dispersion over time (Figure 3).

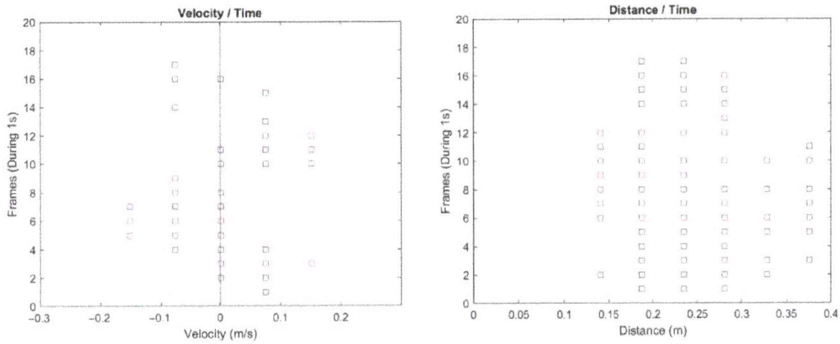

Figure 3. Illustrative visual representations for the data corresponding to one acquisition of the word "Ajuda" (Help): velocity dispersion pattern (**left**) and distance variation (**right**) over 1 second of acquisition frames (along the vertical axis). For this case study, although distance variations were acquired and presented, only the velocity representations were considered for model training and subsequent classification.

Although distance information could potentially provide pertinent data for classification purposes, in this work, they were disregarded, given the verification that slight distance differences between recordings of the same word would produce substantially different representations. That is especially true in this specific case, as no chin rest was used throughout the acquisition sessions. While additional postprocessing could help minimize this issue [8], such consideration was left for future work, and we ultimately opted for exploring the dispersion of velocity data associated with the users' facial motions.

Classifier Training and Testing

For testing the different classification approaches, the velocity dispersion data (as depicted in Figure 3) for each word instance (words being the classes) were provided to the machine learning algorithms. Additionally, and towards understanding the impact that different classifiers could have in the classification outcomes, several that have been commonly used in works pertaining SSI development and similar classification tasks were considered: Random Forests (RF), Linear Discriminant Analysis (LDA), Linear Regression (LR), Support Vector Machine (SVM), and Bagging (BAG). To implement them, Scikit-learning library was used with the different configuration parameters set to their default values. Regarding the classification process, a 5-fold cross-validation approach was adopted due to the limited size of the acquired data, ensuring that every observation from the original dataset had the chance of appearing in both training and testing. Finally, to assess the performance of the different classifiers, and due to the acquired dataset being class-balanced (i.e., no disparity between the number of instances belonging to each class), the Accuracy metric was adopted.

4. Results

As already mentioned, in this study, three experiments were conducted towards validating and assessing FMCW radar-based technology's silent speech recognition capabilities. Although the main objective was to understand if good classification results could be achieved, understanding the possibility of creating session-independent and speaker-independent models was also pivotal. This section is responsible for presenting and describing the achieved results.

4.1. Per-Session Speaker Performance

The first research experience of this study consisted of assessing FMCW radar technology's silent speech recognition capabilities. For this, the data acquired from both conducted acquisition sessions were considered. It is, however, worth mentioning that, although the initial plan was for all four speakers to participate in both sessions, one of them could not make it to the second due to personal life complications. In light of this circumstance, Table 3 presents the recognition results for the participants acquired on both sessions while excluding Speaker 3 from the recognition of the second session's data. In this experience, the classification process was pretty straightforward, as the models were independently created and tested with the data from each speaker per acquisition session.

Table 3. Mean (M), standard deviation (SD), and maximum (max) accuracy value for a specific k-fold iteration, and average recognition scores per speaker obtained for the different classifiers. For each column the highest recognition score is presented in bold face. All metrics were calculated and present values that are respective to a particular acquisition session.

Classif.	M		SD		max		SPK1		SPK2		SPK3		SPK4	
	1st	2nd	1st	2nd	1st	2nd	1st	2nd	1st	2nd	1st	2nd	1st	2nd
RF	83.50	86.00	7.30	3.70	93.60	91.00	88.10	88.20	**76.00**	-	80.60	82.00	89.40	88.00
BAG	**84.50**	86.20	8.70	4.00	95.50	91.70	**91.50**	87.60	74.00	-	**82.20**	81.90	90.10	89.10
LDA	81.40	86.30	9.80	5.40	94.20	93.60	88.70	89.70	71.70	-	74.50	79.80	90.80	89.60
LR	83.20	**88.30**	10.20	4.70	**96.10**	**96.10**	**91.50**	**92.20**	74.40	-	75.00	**83.10**	**92.20**	**89.80**
SVM	80.70	86.10	10.50	5.20	94.20	93.60	88.80	89.10	70.80	-	72.20	79.50	90.90	89.70

Through a careful analysis of the summarized classification results presented in Table 3, and part displayed in graphical form in Figure 4, it is possible to verify that all mean accuracy values across all participants, for all different classifiers, were superior to 80.00%. For session 1, the best mean accuracy value was obtained from the BAG classifier ($M = 84.50$; $SD = 8.70$), while for session 2, the best mean accuracy value was obtained from the LR classifier ($M = 88.3$; $SD = 4.70$). LR, however, produced the maximum accuracy values of 96.10% for a specific k-fold iteration on both conducted sessions. The classifiers which attained the lowest mean accuracy values were SVM ($M = 80.70$; $SD = 10.50$) for session 1, and RF ($M = 86.00$; $SD = 3.70$) for session 2.

Regarding the accuracy values obtained for each speaker and classifier, it is possible to verify that, for Speaker 1, both BAG and LR classifiers achieved the highest mean accuracy values for session 1 (91.50%), while, for session 2, LR produced the highest value (92.20%), for Speaker 2, which only enrolled in one of the acquisition sessions, the RF classifier presented the highest mean accuracy value (76.00%), for Speaker 3, the BAG classifier produced the highest mean accuracy value for session 1 (82.20%), while, for session 2, LR classifier produced the highest value (83.10%), and, for Speaker 4, LR produced the highest mean accuracy values for both session 1 (92.20%) and 2 (89.80%).

Considering the average accuracy results obtained from all classifiers, for all speakers, as depicted in Figure 4, it is possible to verify that Speaker 1 and Speaker 4 presented higher accuracy values than the remaining, which, in turn, obtained results similar between themselves.

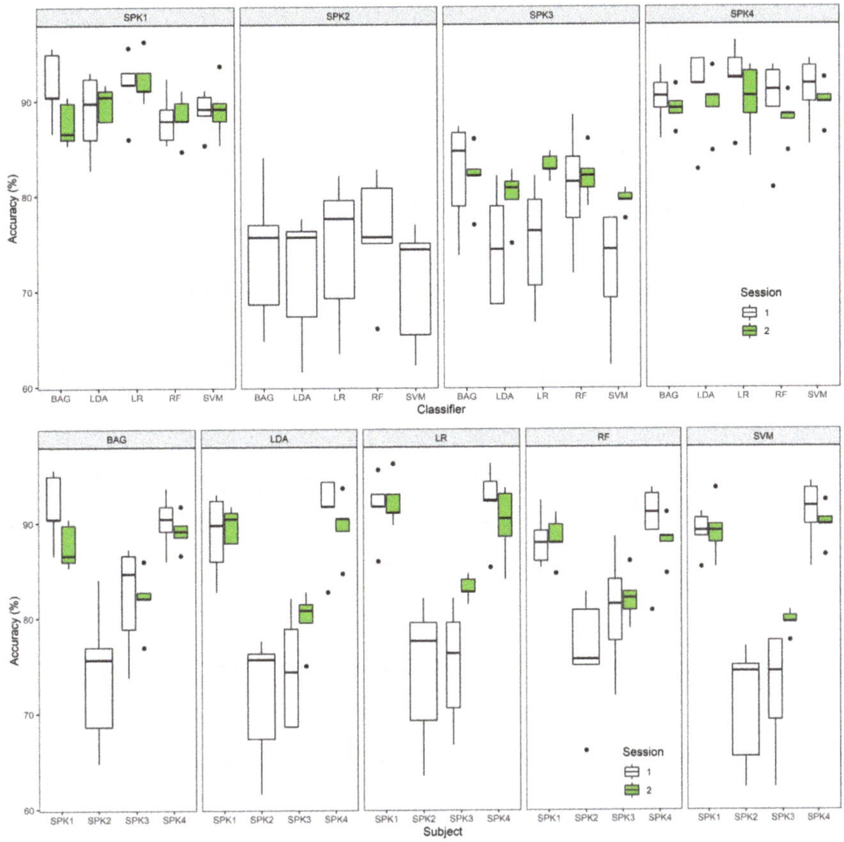

Figure 4. Boxplot representations depicting the average classification accuracies obtained for each acquisition session: average accuracies per classifier (**top**) and average accuracies per speaker (**bottom**). Speaker 2 only recorded one session.

To further understand the recognition results and identify the words that were most commonly mistaken, thus, negatively contributing to the accuracy outcomes, two Confusion Matrices were created (Figure 5). These matrices depict the average accuracy classifications across all participants while considering the best classifier from each session (i.e., BAG for the first session and LR for the second). By analyzing the first session's confusion matrix, it is possible to verify that the results were considerably worse for the words "Lembretes" (Reminders) and "Ligar" (Turn on) (with a correspondingly average accuracy, across all speakers, of 62.00% and 76.00%), typically being erroneously classified as "Email" and "Seguinte" (Next), respectively. As per the second session's confusion matrix, the only classification result that scored below 80.00% was the one corresponding to the word "Lembretes" (Reminders), having achieved an average accuracy of 74.00%, being, just like in the first session, typically confused with the word "Email".

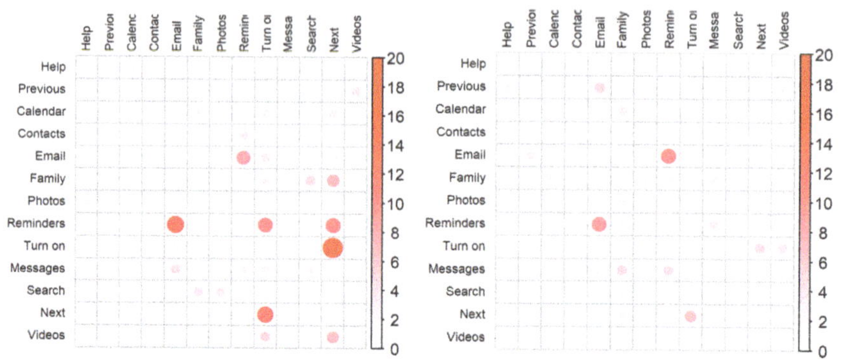

Figure 5. Confusion matrices for the best classifiers in each session illustrating the average recognition results across all participants. BAG classifier was considered for the first session (**left**), while LR classifier was considered for the second (**right**). The matrix rows represent the word instances submitted for recognition, while its columns represent the corresponding recognized words.

4.2. Session-Independence Performance

The possibility of achieving session-independent models is highly desirable across the several technologies considered for SSR purposes [27,28,32]. The main factor for such a necessity is that, for embracing such technologies in daily scenarios, it is mandatory that different usages, even if considerably distant between different points in time, do not deteriorate the recognition results due to slight variability in the acquired data [27]. One example of a technology that severely suffers from such changes in the captured data across different sessions is sEMG, as removing and reattaching the electrodes in-between sessions causes variations in the recorded EMG signals [27,28].

As already mentioned, two acquisition sessions were performed for this research project, being its central purpose to, while considering the same corpus and validated word repetitions, capture a dataset equal to the first session's one (besides the data acquired for the missing participant, Speaker 2). Although contributing towards all research experiences conducted in this study, its main intent was to allow performing this specific experiment, as it would allow understanding to what extent the creation of session-independent models is made possible by this technology.

After both datasets were acquired, the data belonging to the first session (session 1) was used for training, while the data from the second session (session 2) was considered for testing. The obtained recognition results are depicted in Table 4. In general, it is possible to verify that Speaker 1 obtained higher recognition results overall, being the best accuracy value obtained from the BAG classifier ($M = 81.79$). For Speaker 3, RF classifier produced the best accuracy value of 67.95% and, for Speaker 4, BAG classifier produced the best accuracy value of 71.79%.

Table 4. Mean accuracy values, produced by all classifiers, for the Intra-Speaker experiment. Acquired data from the first session (session 1) was used for training, while data from the second session (session 2) was considered for testing. Relative variances between session-independent results and per-session results are also presented for each speaker. The highest values in each column are highlighted using bold face.

	SPK 1 (S1) -> (S2)	R.Variance (S1)	R.Variance (S2)	SPK 3 (S1) -> (S2)	R.Variance (S1)	R.Variance (S2)	SPK 4 (S1) -> (S2)	R.Variance (S1)	R.Variance (S2)
RF	74.10	**14.00 (88.10)**	14.10 (88.20)	**67.95**	12.65 (80.60)	14.05 (82.00)	65.90	23.50 (89.40)	22.10 (88.00)
BAG	**81.79**	9.71 (91.50)	5.81 (87.60)	65.77	**16.43 (82.20)**	**16.13 (81.90)**	**71.79**	18.31 (90.10)	17.31 (89.10)
LDA	75.38	13.32 (88.70)	**14.32 (89.70)**	63.97	10.53 (74.50)	15.83 (79.80)	66.15	24.65 (90.80)	23.45 (89.60)
LR	78.08	13.42 (91.50)	14.12 (92.20)	67.17	7.83 (75.00)	15.93 (83.10)	63.33	**28.87 (92.20)**	**26.47 (89.80)**
SVM	76.53	12.27 (88.80)	12.57 (89.10)	66.28	5.92 (72.20)	13.22 (79.50)	68.46	22.44 (90.90)	21.24 (89.70)

Regarding the relative variance values between the session-independent and session-dependent models, per speaker, it is possible to verify that Speaker 4 was the one that presented a higher variability, having a relative variance, for the LR classifier, of 28.87% for the first session's data, and 26.47% for the second. Regarding Speaker 1, its session-independent model presented a maximum relative variance of 14.00% with RF classifier for session 1 and 14.32%, with LDA classifier, for session 1. For Speaker 3, its higher variances were 16.43% and 16.13%, both produced with the BAG classifier, for session 1 and 2 models, respectively.

4.3. Speaker-Independence Performance

Towards exploring and assessing FMCW radar-based technology's capabilities regarding the creation of speaker-independent models, a third research experiment ensued. In this experiment, recognition models were created with data belonging to several speakers, i.e., each speaker's data was left out for testing while the models were trained with the data belonging to the remaining speakers (n-1 models). The purpose of such consideration was to maximize the data from training, trying to achieve a more generalized recognition model, and approximate the case in which one intends to make a system that, after being trained, can be used by someone without the need to retrain the model with additional data.

The summarized classification results for the speaker-independent models are presented in Table 5, depicting the obtained mean accuracy values for all classifiers.

Table 5. Mean accuracy values, produced by all classifiers, for the Inter-Speaker experiment. The data belonging to each speaker were left out for testing, while the models considered for recognition were trained with the data belonging to the remaining speakers. The results were obtained considering the data for the first acquisition session, for each speaker. The highest mean accuracy obtained for each experiment is shown in bold face.

	SPK1			SPK2		
Model	**SPK2**	**SPK2 + SPK3**	**SPK2 + SPK3 + SPK4**	**SPK1**	**SPK1 + SPK3**	**SPK1 + SPK3 + SPK4**
RF	51.40	49.00	79.00	**40.10**	45.80	44.30
BAG	54.10	**54.00**	**80.50**	36.90	39.50	38.20
LDA	41.70	40.90	77.18	38.20	39.30	33.60
LR	52.80	50.00	67.90	38.30	46.20	43.80
SVM	**57.20**	52.20	77.60	39.70	**49.00**	**47.30**
	SPK3			SPK4		
Model	**SPK1**	**SPK1 + SPK2**	**SPK1 + SPK2 + SPK4**	**SPK1**	**SPK1 + SPK2**	**SPK1 + SPK2 + SPK3**
RF	32.60	43.80	42.30	74.70	75.60	74.70
BAG	**36.80**	44.70	42.90	**81.90**	77.80	**81.80**
LDA	32.90	43.30	33.80	79.40	77.40	72.60
LR	36.00	41.50	38.00	79.50	74.90	66.40
SVM	35.10	**45.30**	**43.40**	81.20	**78.50**	71.90

Regarding the recognition results for Speaker 1, it is possible to verify that the BAG classifier produced the best recognition result ($M = 80.50$) for the model trained with the data from all other speakers. Concerning Speaker 2, SVM achieved the best recognition accuracy ($M = 47.30$) for the model trained with the data from the remaining speakers, however, it produced a better classification score for when the data from Speaker 4 was not yet included ($M = 49.00$). Speaker 3 results share some similarities with the ones obtained from Speaker 2, with SVM also achieving the best recognition accuracy ($M = 43.40$) for the model trained with the data from the remaining speakers but produced a better classification score for when the data from Speaker 4 was not yet included ($M = 45.30$). Finally, for Speaker 4, BAG classifier produced the best recognition accuracy ($M = 81.80$) for the model trained with the data from all other speakers.

Figure 6 allows further verifying the mentioned recognition similarities between two different groups of speakers (i.e., speakers 1 and 4, and speakers 2 and 3). Speakers 1 and 4 achieved higher recognition accuracies when the models were trained with data containing each other's data, being their accuracy rates, with the models trained with all the remaining speakers' data, superior to the ones achieved for speakers 2 and 3. However, for speakers 2 and 3, higher recognition accuracies were obtained when the models had not yet been fed with speakers 1 and 4 data.

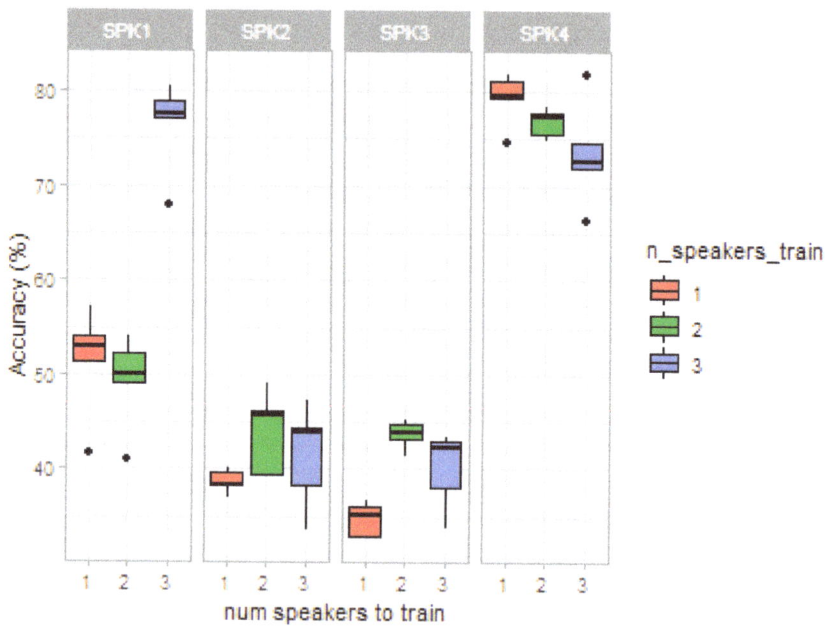

Figure 6. Study of speaker-independent models performance. Accuracy results, per speaker, when considering the remaining one, two, or three speakers for model training and subsequent classification).

5. Discussion

This work aimed to assert if, by resorting to FMCW radar, and expanding our preliminary work [13], three core aspects for SSI development could be tackled with this technology: (a) how capable is radar-based technology of successfully recognizing silent speech, (b) how discrepant can the results get when performing different acquisition sessions for the same participant), and (c) to what extent can inter-speaker models be created by resorting to this technology.

Concerning the per-speaker experiment, the average accuracy results of 84.50% using the BAG classifier and 88.30% using the LR classifier for both acquisition sessions translate into positive indications of FMCW radar-based technology SSR capabilities, particularly given that a set of thirteen words was considered. Through a careful analysis of the obtained results, it was possible to verify the positive impact that producing the words more consistently throughout the acquisition sessions can have in the establishment of more representative classification models. While the results were good for all speakers, this hints that such a simple instruction given to the speaker can potentially improve the recognition accuracies. Regarding the lower recognition accuracies for the words "Lembretes" (Reminders), sometimes recognized as "**Email**", and "Ligar" (Turn On), sometimes recognized as "Seguinte" (Next), we believe that this may be due to some notable articulatory similarity, e.g., at the beginning or middle of the word that, at some elocution speeds and, eventually, as a consequence of coarticulatory effects, may turn their velocity dispersion patterns more similar.

Comparing the obtained recognition results with previous works for the same AAL corpus allowed verifying that radar-based SSR either obtained comparable or superior accuracy values. In [49], an accuracy of 75.00% is reported for sEMG, and in [45], accuracies of 71.40%, 72.60%, and 83.00% were obtained for Video, Depth, and UDS technologies, respectively. Further comparing the obtained results with the ones in the existing literature for several other technologies, although such comparisons may not be necessarily fair given different considered corpus and technologies' nuances, allowed verifying that either comparable or superior results were also achieved. Two examples of such studies are the work by Sarmiento et al. [23], where the authors resort to EEG obtaining accuracies in the range of 67.78% and 72.67% for five syllables, and the work by Dash et al. [40], exploring Magnetoencephalography (MEG), and achieving an accuracy rate of 79.93% for five imagined phrases. Sun et al. [35] was capable of presenting an overall superior average recognition than the ones obtained in this study for a set of 20 commands. Nevertheless, the data considered resulted from asking the participants to over-articulate, something that is not considered in our work. Another representative study that managed to get a slightly superior recognition rate was the one by Kapur et al. [20], having achieved a 92.01% accuracy rate for a corpus of 15 words while resorting to sEMG.

Towards assessing the possibility of creating session-independent models from FMCW data, i.e., how a model created with data from one acquisition session can be used to perform recognition for another acquisition of the same speaker, a second research experiment ensued. Session recordings variability for the same participants is an aspect that limits several technologies considered towards SSR, typically requiring additional normalization algorithms across different sessions [27,28]. An analysis of the obtained results confirmed our initial hypothesis that whenever consistency is considered throughout the acquisition sessions it has a positive influence on the model's performance. Such aspect is clear from the outcomes as speakers asked to attempt being consistent were the one with higher recognition rates in the per-speaker experiment and also achieved higher performance between sessions.

Further comparing the acquired results with those presented in the literature for other technologies would be ideal. However, after an extensive review (Table 1), and to the best of our knowledge, most of the studies that mention session-independence aspects either focus their efforts on researching and developing normalization methods for tackling data variance between different acquisition sessions for the same participants [27,28] or highlight it as an aspect to explore in future work [32].

The final study's assessment—speaker-independence—aimed to assess the extent to which it was possible to create representative speaker-independent models by resorting to FMCW radar-based technology. By analyzing the obtained accuracy values, what stands out the most is that the results are significantly better when data from the two speakers asked to be consistent throughout the acquisition sessions (SPK1 and SPK4) is considered. Whenever one of these speakers is included in the training data, the accuracy for the other shows a strong improvement. Therefore, consistency seems to also play a pivotal role, here, enhancing the importance of particular patterns of articulation that are similar for these two speakers. Nevertheless, the nature of this advantage is yet to be established, particularly if it extends to more speakers observing a principle of consistency during the acquisition.

Another aspect worth noting concerns the higher (although to a smaller extent) recognition accuracies achieved for speakers 2 and 3 when the models had not yet been fed with data from speakers 1 and 4. The impact of the data from speaker 1 and 4 on the performance of models created considering speakers 2 and 3 may be sourced in a wide variety of factors: (1) first, we know that speakers 1 and 4 were instructed to be consistent and, therefore, it is conceivable that their data patterns have some degree of similarity and consistency, making the presence of one of them very favorable for the other, as our results show; (2) this consistency probably makes speakers 1 and 4 diverge more from speakers

2 and 3 and, thus, when these two 'kinds' of data are mixed, the resulting model is less capable of dealing with each of these latter speakers; and (3) naturally, this is potentiated by the moderate amount of speakers and data considered. In this context, future tests with a wider range of speakers may help clarify this aspect.

Still, the results obtained in this first experiment with speaker-independent models for FMCW radar are quite positive as, for the models created with three of the speakers, the worst case yielded 43.40% accuracy for a corpus of 13 words.

The possibility of creating speaker-independent models is an aspect that has been highly desired towards the development and integration of SSI in more ecological settings, allowing speakers to use the systems without requiring them to have previously contributed to the data used to create the model. Several studies, exploring different technologies (e.g., Video, US, sEMG, and MEG), have already stated the importance of considering such speaker-independent models; however, most either achieve poor recognition results or solely mention the topic and leave such consideration for future endeavors [14,20,30–32,36,38,40]. In this regard, two studies are worth mentioning. Sun et al. [35] explored video technology towards SSR and achieved an accuracy of 95.40% for 20 limited context-usage commands with over-articulation of the lips in speaker-independent settings, while Petridis et al. [38], also having explored video technology, achieved results ranging the 70.00% mark for two different corpus of respectively ten digits and ten phrases. One aspect, however, that highly contributes towards this technology's capability of creating speaker-independent models is that there already exist large volumes of data for several different speakers, something that does not happen for many other technologies. Nevertheless, video technology still remains dependent on ambient lighting and privacy concerns. Besides video technology, several other studies also mention how relevant achieving speaker-independent models would be but rarely explore it, typically leaving such consideration for future work [14,20,30–32].

Besides the promising results already presented, there are, nevertheless, some limitations of the present work that are worth noting. Regarding the acquisition sessions, the distance between the participant and the radar, although not being enforced as in similar studies in the literature, see, e.g., in [8], was kept at around 15cm with minor variations around this value. Although being less demanding and not obligating the speakers to be attached to the technology or remain immobile during the acquisitions, as it frequently happens in other technologies (e.g., US, EEG, and MEG), this is not yet the full extent of the capabilities we envisage for the technology to serve the considered scenarios.

Another limiting aspect resides in the fact that the considered corpus, although already comprising a word set comparable to the existing literature, contains inputs suited for specific interaction scenarios. Although capable of serving multiple domains, considering different interaction contexts usually requires different types of commands.

Finally, one last limitation concerns the number of speakers considered for acquisition. Promising results were achieved in all carried experiments with the four considered speakers. However, considering a higher number of speakers would allow establishing more generalized models by taking into account different individuals' anatomies and articulatory idiosyncrasies.

6. Conclusions

This paper proposes and demonstrates the consideration of a FMCW radar board to assess the plausibility of contactless radar-based technology towards SSR. Besides demonstrating its SSR capabilities, additional experiments were also performed to verify the possibility of creating session and speaker-independent models.

Regarding the per-session speaker experiment, based on velocity dispersion features, several classification models were trained and were subsequently capable of producing average recognition accuracies as good as 84.50% for the first acquisition session and 88.30% for the second one. Accuracies of 81.79% and 71.79% were also obtained for the session-independence experiment, suggesting that this technology may be resilient to variations in the recording data across different acquisition sessions for the same speakers. For the final carried experiment, Speaker-Independence, focusing on the possibility of training speaker-independent models, recognition accuracies as high as 80.50% and 81.80% were also achieved.

The obtained results, along with the inherent advantages of contactless radar-based technology (e.g., its non-invasive and privacy-preserving nature, its portability, and robustness against lighting conditions and environment noise), establish promising grounds for further exploring and more frequently considering this technology towards SSI development purposes.

Regarding future work, the team's focus will be on acquiring a larger amount of radar data from a wider range of speakers. This would allow us to consider other well-known classification models that require larger volumes of data (e.g., artificial neural networks (ANN) and convolutional neural networks (CNN)) and, most importantly, understand the impact that data from several participants has in the creation of the so needed speaker-independent models. Besides this central focus, the influence of speaker-to-radar distance and head orientation are other aspects that are deserving our attention. Initial assessments with different head orientations were already performed and allowed verifying that data from different orientations, when considered together with the frontal one, can improve recognition results, as the models learn how to more easily discern the word classes in which there is articulatory ambiguity. However, more experiments still need to be carried out including, e.g., multi-radar settings.

Author Contributions: Conceptualization, S.S. and A.T.; Formal analysis, D.F. and A.T.; Funding acquisition, S.S. and A.T.; Investigation, D.F., S.S. and A.T.; Methodology, D.F., S.S., F.C. and A.T.; Project administration, S.S. and A.T.; Resources, F.C. and A.T.; Software, D.F. and F.C.; Supervision, S.S., F.C. and A.T.; Validation, D.F.; Visualization, A.T.; Writing—original draft, D.F. and S.S.; Writing—review & editing, S.S. and A.T. All authors have read and agreed to the published version of the manuscript.

Funding: This work is partially funded by IEETA Research Unit funding (UID/CEC/00127/2019 UIDB/00127/2020), by Portugal 2020 under the Competitiveness and Internationalization Operational Program, and the European Regional Development Fund through project MEMNON (POCI-01-0145-FEDER-028976).

Institutional Review Board Statement: Not applicable.

Informed Consent Statement: Informed consent was obtained from all subjects involved in the study.

Data Availability Statement: The data presented in this study are available on request from the corresponding author. The data are not publicly available due to privacy concerns.

Conflicts of Interest: The authors declare no conflict of interest. The funders had no role in the design of the study; in the collection, analyses, or interpretation of data; in the writing of the manuscript, or in the decision to publish the results.

References

1. Kepuska, V.; Bohouta, G. Next-generation of virtual personal assistants (Microsoft Cortana, Apple Siri, Amazon Alexa and Google Home). In Proceedings of the 2018 IEEE 8th Annual Computing and Communication Workshop and Conference (CCWC), Las Vegas, NV, USA, 8–10 January 2018; pp. 99–103.
2. Denby, B.; Schultz, T.; Honda, K.; Hueber, T.; Gilbert, J.M.; Brumberg, J.S. Silent speech interfaces. *Speech Commun.* **2010**, *52*, 270–287. [CrossRef]
3. Levelt, W.J. *Speaking: From Intention to Articulation*; MIT Press: Cambridge, MA, USA, 1993; Volume 1.
4. Freitas, J.; Teixeira, A.; Dias, M.S.; Silva, S. SSI Modalities I: Behind the Scenes—From the Brain to the Muscles. In *An Introduction to Silent Speech Interfaces*; Springer: Berlin/Heidelberg, Germany, 2017; pp. 15–30.
5. Ahmed, S.; Cho, S.H. Hand gesture recognition using an IR-UWB radar with an inception module-based classifier. *Sensors* **2020**, *20*, 564. [CrossRef] [PubMed]
6. Hazra, S.; Santra, A. Short-range radar-based gesture recognition system using 3D CNN with triplet loss. *IEEE Access* **2019**, *7*, 125623–125633. [CrossRef]

7. Freitas, J.; Teixeira, A.; Dias, M.S.; Silva, S. Combining Modalities: Multimodal SSI. In *An Introduction to Silent Speech Interfaces*; Springer: Berlin/Heidelberg, Germany, 2017; pp. 51–71.
8. Shin, Y.H.; Seo, J. Towards contactless silent speech recognition based on detection of active and visible articulators using IR-UWB radar. *Sensors* **2016**, *16*, 1812. [CrossRef] [PubMed]
9. Rohling, H.; Meinecke, M.M. Waveform design principles for automotive radar systems. In Proceedings of the 2001 CIE International Conference on Radar Proceedings (Cat No. 01TH8559), Beijing, China, 15–18 October 2001; pp. 1–4.
10. Winkler, V. Range Doppler detection for automotive FMCW radars. In Proceedings of the 2007 European Radar Conference, Munich, Germany, 10–12 October 2007; pp. 166–169.
11. Vivet, D.; Checchin, P.; Chapuis, R. Localization and mapping using only a rotating FMCW radar sensor. *Sensors* **2013**, *13*, 4527–4552. [CrossRef] [PubMed]
12. van Delden, M.; Guzy, C.; Musch, T. Investigation on a System for Positioning of Industrial Robots Based on Ultra-Broadband Millimeter Wave FMCW Radar. In Proceedings of the 2019 IEEE Asia-Pacific Microwave Conference (APMC), Singapore, 10–13 December 2019; pp. 744–746.
13. Ferreira, D.; Silva, S.; Curado, F.; Teixeira, A. RaSSpeR: Radar-Based Silent Speech Recognition. *Proc. Interspeech* **2021**, *2021*, 646–650.
14. Meltzner, G.S.; Heaton, J.T.; Deng, Y.; De Luca, G.; Roy, S.H.; Kline, J.C. Development of sEMG sensors and algorithms for silent speech recognition. *J. Neural Eng.* **2018**, *15*, 046031. [CrossRef] [PubMed]
15. Dong, W.; Zhang, H.; Liu, H.; Chen, T.; Sun, L. A Super-Flexible and High-Sensitive Epidermal sEMG Electrode Patch for Silent Speech Recognition. In Proceedings of the 2019 IEEE 32nd International Conference on Micro Electro Mechanical Systems (MEMS), Seoul, Korea, 27–31 January 2019; pp. 565–568.
16. Liu, H.; Dong, W.; Li, Y.; Li, F.; Geng, J.; Zhu, M.; Chen, T.; Zhang, H.; Sun, L.; Lee, C. An epidermal sEMG tattoo-like patch as a new human–machine interface for patients with loss of voice. *Microsyst. Nanoeng.* **2020**, *6*, 1–13. [CrossRef] [PubMed]
17. Ruiz-Olaya, A.F.; López-Delis, A. Surface EMG signal analysis based on the empirical mode decomposition for human-robot interaction. In Proceedings of the Symposium of Signals, Images and Artificial Vision-2013: STSIVA-2013, Bogota, Colombia, 11–13 September 2013; pp. 1–4.
18. Diener, L.; Umesh, T.; Schultz, T. Improving fundamental frequency generation in emg-to-speech conversion using a quantization approach. In Proceedings of the 2019 IEEE Automatic Speech Recognition and Understanding Workshop (ASRU), Singapore, 14–18 December 2019; pp. 682–689.
19. Joy, J.E.; Yadukrishnan, H.A.; Poojith, V.; Prathap, J. Work-in-Progress: Silent Speech Recognition Interface for the Differently Abled. In Proceedings of the International Conference on Remote Engineering and Virtual Instrumentation, Bangalore, India, 3–6 February 2019; pp. 805–813.
20. Kapur, A.; Kapur, S.; Maes, P. Alterego: A personalized wearable silent speech interface. In Proceedings of the 23rd International Conference on Intelligent User Interfaces, Tokyo, Japan, 7–11 March 2018; pp. 43–53.
21. Merletti, R.; Parker, P.J. *Electromyography: Physiology, Engineering, and Non-Invasive Applications*; John Wiley & Sons: Hoboken, NJ, USA, 2004; Volume 11.
22. Shah, N.; Shah, N.J.; Patil, H.A. Effectiveness of Generative Adversarial Network for Non-Audible Murmur-to-Whisper Speech Conversion. In Proceedings of the INTERSPEECH 2018, Hyderabad, India, 2–6 September 2018; pp. 3157–3161.
23. Sarmiento, L.; Rodríguez, J.B.; López, O.; Villamizar, S.; Guevara, R.; Cortes-Rodriguez, C. Recognition of silent speech syllables for Brain-Computer Interfaces. In Proceedings of the 2019 IEEE International Conference on E-health Networking, Application & Services (HealthCom), Bogota, Colombia, 14–16 October 2019; pp. 1–5.
24. Morooka, T.; Ishizuka, K.; Kobayashi, N. Electroencephalographic Analysis of Auditory Imagination to Realize Silent Speech BCI. In Proceedings of the 2018 IEEE 7th Global Conference on Consumer Electronics (GCCE), Nara, Japan, 9–12 October 2018; pp. 683–686.
25. Ma, S.; Jin, D.; Zhang, M.; Zhang, B.; Wang, Y.; Li, G.; Yang, M. Silent Speech Recognition Based on Surface Electromyography. In Proceedings of the 2019 Chinese Automation Congress (CAC), Hangzhou, China, 22–24 November 2019; pp. 4497–4501.
26. Rameau, A. Pilot study for a novel and personalized voice restoration device for patients with laryngectomy. *Head Neck* **2020**, *42*, 839–845. [CrossRef] [PubMed]
27. Proroković, K.; Wand, M.; Schultz, T.; Schmidhuber, J. Adaptation of an EMG-Based Speech Recognizer via Meta-Learning. In Proceedings of the 2019 IEEE Global Conference on Signal and Information Processing (GlobalSIP), Ottawa, ON, Canada, 11–14 November 2019; pp. 1–5.
28. Wand, M.; Schultz, T.; Schmidhuber, J. Domain-Adversarial Training for Session Independent EMG-based Speech Recognition. In Proceedings of the INTERSPEECH 2018, Hyderabad, India, 2–6 September 2018; pp. 3167–3171.
29. Fernandes, R.; Huang, L.; Vejarano, G. Non-Audible Speech Classification Using Deep Learning Approaches. In Proceedings of the 2019 International Conference on Computational Science and Computational Intelligence (CSCI), Las Vegas, NV, USA, 5–7 December 2019; pp. 630–634.
30. Chen, S.; Zheng, Y.; Wu, C.; Sheng, G.; Roussel, P.; Denby, B. Direct, Near Real Time Animation of a 3D Tongue Model Using Non-Invasive Ultrasound Images. In Proceedings of the 2018 IEEE International Conference on Acoustics, Speech and Signal Processing (ICASSP), Calgary, AB, Canada , 15–20 April 2018; pp. 4994–4998.

31. Zhao, C.; Zhang, P.; Zhu, J.; Wu, C.; Wang, H.; Xu, K. Predicting tongue motion in unlabeled ultrasound videos using convolutional LSTM neural networks. In Proceedings of the IEEE International Conference on Acoustics, Speech and Signal Processing (ICASSP), Brighton, UK, 12–17 May 2019; pp. 5926–5930.
32. Gosztolya, G.; Pintér, Á.; Tóth, L.; Grósz, T.; Markó, A.; Csapó, T.G. Autoencoder-based articulatory-to-acoustic mapping for ultrasound silent speech interfaces. In Proceedings of the 2019 International Joint Conference on Neural Networks (IJCNN), Budapest, Hungary, 14–19 July 2019; pp. 1–8.
33. Kimura, N.; Kono, M.; Rekimoto, J. SottoVoce: An ultrasound imaging-based silent speech interaction using deep neural networks. In Proceedings of the 2019 CHI Conference on Human Factors in Computing Systems, Glasgow, UK, 4–9 May 2019; pp. 1–11.
34. Csapó, T.G.; Al-Radhi, M.S.; Németh, G.; Gosztolya, G.; Grósz, T.; Tóth, L.; Markó, A. Ultrasound-based silent speech interface built on a continuous vocoder. *arXiv* **2019**, arXiv:1906.09885.
35. Sun, K.; Yu, C.; Shi, W.; Liu, L.; Shi, Y. Lip-interact: Improving mobile device interaction with silent speech commands. In Proceedings of the 31st Annual ACM Symposium on User Interface Software and Technology, Berlin, Germany, 14 October 2018; pp. 581–593.
36. Vougioukas, K.; Ma, P.; Petridis, S.; Pantic, M. Video-driven speech reconstruction using generative adversarial networks. *arXiv* **2019**, arXiv:1906.06301.
37. Uttam, S.; Kumar, Y.; Sahrawat, D.; Aggarwal, M.; Shah, R.R.; Mahata, D.; Stent, A. Hush-Hush Speak: Speech Reconstruction Using Silent Videos. In Proceedings of the INTERSPEECH, Graz, Austria, 15–19 September 2019; pp. 136–140.
38. Petridis, S.; Shen, J.; Cetin, D.; Pantic, M. Visual-only recognition of normal, whispered and silent speech. In Proceedings of the 2018 IEEE International Conference on Acoustics, Speech and Signal Processing (ICASSP), Calgary, AB, Canada, 15–20 April 2018; pp. 6219–6223.
39. Birkholz, P.; Stone, S.; Wolf, K.; Plettemeier, D. Non-invasive silent phoneme recognition using microwave signals. *IEEE/ACM Trans. Audio Speech Lang. Process.* **2018**, *26*, 2404–2411. [CrossRef]
40. Dash, D.; Wisler, A.; Ferrari, P.; Wang, J. Towards a Speaker Independent Speech-BCI Using Speaker Adaptation. In Proceedings of the INTERSPEECH, Graz, Austria, 15–19 September 2019; pp. 864–868.
41. Xu, K.; Wu, Y.; Gao, Z. Ultrasound-based silent speech interface using sequential convolutional auto-encoder. In Proceedings of the 27th ACM International Conference on Multimedia, Nice, France, 21–25 October 2019; pp. 2194–2195.
42. Schultz, T.; Wand, M.; Hueber, T.; Krusienski, D.J.; Herff, C.; Brumberg, J.S. Biosignal-based spoken communication: A survey. *IEEE/ACM Trans. Audio Speech Lang. Process.* **2017**, *25*, 2257–2271. [CrossRef]
43. Thein, T.; San, K.M. Lip localization technique towards an automatic lip reading approach for Myanmar consonants recognition. In Proceedings of the 2018 International Conference on Information and Computer Technologies (ICICT), DeKalb, IL, USA, 23–25 March 2018; pp. 123–127.
44. Freitas, J.; Teixeira, A.; Bastos, C.; Dias, M. Towards a Multimodal Silent Speech Interface for European Portuguese. In *Speech Technologies*; InTech: London, UK, 2011; pp. 125–149.
45. Freitas, J.; Teixeira, A.; Dias, M.S. Multimodal Silent Speech Interface based on Video, Depth, Surface Electromyography and Ultrasonic Doppler: Data Collection and First Recognition Results. In Proceedings of the Workshop on Speech Production in Automatic Speech Recognition, Lyon, France, 30 August 2013.
46. Teixeira, A.; Vitor, N.; Freitas, J.; Silva, S. Silent speech interaction for ambient assisted living scenarios. In Proceedings of the International Conference on Human Aspects of IT for the Aged Population, Vancouver, BC, Canada, 9–14 July 2017; pp. 369–387.
47. Albuquerque, D.F.; Gonçalves, E.S.; Pedrosa, E.F.; Teixeira, F.C.; Vieira, J.N. Robot Self Position based on Asynchronous Millimetre Wave Radar Interference. In Proceedings of the 2019 International Conference on Indoor Positioning and Indoor Navigation (IPIN), Pisa, Italy, 30 September–3 October 2019; pp. 1–6.
48. Gouveia, C.; Tomé, A.; Barros, F.; Soares, S.C.; Vieira, J.; Pinho, P. Study on the usage feasibility of continuous-wave radar for emotion recognition. *Biomed. Signal Process. Control.* **2020**, *58*, 101835. [CrossRef]
49. Freitas, J. Articulation in Multimodal Silent Speech Interface for European Portuguese. Ph.D. Thesis, University of Aveiro, Aveiro, Portugal, 2015.

sensors

MDPI

Article

Lipreading Architecture Based on Multiple Convolutional Neural Networks for Sentence-Level Visual Speech Recognition

Sanghun Jeon, Ahmed Elsharkawy and Mun Sang Kim *

Center for Healthcare Robotics, Gwangju Institute of Science and Technology (GIST), School of Integrated Technology, Gwangju 61005, Korea; jeon7887@gist.ac.kr (S.J.); elsharkawy@gm.gist.ac.kr (A.E.)
* Correspondence: munsang@gist.ac.kr; Tel.: +82-10-9126-4628

Abstract: In visual speech recognition (VSR), speech is transcribed using only visual information to interpret tongue and teeth movements. Recently, deep learning has shown outstanding performance in VSR, with accuracy exceeding that of lipreaders on benchmark datasets. However, several problems still exist when using VSR systems. A major challenge is the distinction of words with similar pronunciation, called homophones; these lead to word ambiguity. Another technical limitation of traditional VSR systems is that visual information does not provide sufficient data for learning words such as "a", "an", "eight", and "bin" because their lengths are shorter than 0.02 s. This report proposes a novel lipreading architecture that combines three different convolutional neural networks (CNNs; a 3D CNN, a densely connected 3D CNN, and a multi-layer feature fusion 3D CNN), which are followed by a two-layer bi-directional gated recurrent unit. The entire network was trained using connectionist temporal classification. The results of the standard automatic speech recognition evaluation metrics show that the proposed architecture reduced the character and word error rates of the baseline model by 5.681% and 11.282%, respectively, for the unseen-speaker dataset. Our proposed architecture exhibits improved performance even when visual ambiguity arises, thereby increasing VSR reliability for practical applications.

Keywords: 3D densely connected CNN; 3D multi-layer feature fusion CNN; convolutional neural network; deep learning; lipreading; speech recognition; visual speech recognition

Citation: Jeon, S.; Elsharkawy, A.; Kim, M.S. Lipreading Architecture Based on Multiple Convolutional Neural Networks for Sentence-Level Visual Speech Recognition. *Sensors* **2022**, *22*, 72. https://doi.org/10.3390/s22010072

Academic Editor: Bruce Denby

Received: 3 November 2021
Accepted: 21 December 2021
Published: 23 December 2021

1. Introduction

Speech is the most common form of communication between humans and involves the perception of both acoustic and visual information. In 1976, McGurk and McDonald demonstrated that speech perception is influenced by vision, which is called the McGurk effect [1]. This effect indicates the necessity of matching both auditory and visual phonemes to perceive pronounced phonemes correctly.

Vision plays a crucial role in speech understanding, and the importance of utilizing visual information to improve the performance and robustness of speech recognition has been demonstrated [2–4]. Although acoustic information is richer than visual information when speaking, most people rely on watching lip movements to fully understand speech [2]. Furthermore, people rely on visual information in noisy environments where receiving auditory information is challenging. Similarly, people with hearing impairments depend on visual information to perceive spoken words. However, comprehending oral language using visual information alone, especially in the absence of context, can be challenging because it is difficult to understand lipreading actuations such as lip, tongue, and teeth movements without context [3]. Hearing-impaired people using visual information have achieved an accuracy of 17 ± 12%, even for a small subset of 30 monosyllabic words, and 21 ± 11% for 30 compound words, according to Easton and Basala [4]. Chung et al. [5] showed that experienced professional lip-leaders achieved 26.2% accuracy with the BBC News benchmark dataset when they could watch an unlimited number of videos.

The development of a visual speech recognition (VSR) system has enormous potential for various practical applications, such as speech recognition in noisy environments, biometric identification for security, communication in underwater environments, and silent movie analysis, and it can positively affect patients with speech impairments [6,7]. Therefore, it is important to develop a VSR system that exclusively uses visual information.

Recently, several researchers have investigated the possibility of developing a VSR system by decoding speech using only visual information to mimic human lipreading capability [8–13]. Despite their efforts, VSR systems still exhibit low performance and accuracy compared to those of audio or audio-VSR systems. A major challenge is the distinction of words with similar pronunciation, called homophones [9]; these lead to ambiguity at the word level. For example, although some words, such as pack, back, and mac, differ in their sound, the characters (e.g., [p], [b], and [m]) produce almost identical lip movements, thereby making them difficult to distinguish. As such, word distinction is the most difficult task for humans and crucial for accurate lipreading [14]. Another technical limitation of traditional VSR systems is that visual information does not provide sufficient data for learning words such as "a", "an", "eight", and "bin" because their length is no longer than 0.02 s [15].

To address the challenges of similar pronunciation and insufficient visual information, this paper presents a novel lipreading architecture that exhibits superior performance compared to those of traditional and existing deep learning VSR systems. This architecture consists of two sub-networks using end-to-end neural networks: the visual feature extraction module is made of a 3D convolutional neural network (CNN), a 3D densely connected CNN for each time step by reducing model parameters, and a multi-layer feature fusion (MLFF) CNN for capturing multichannel information in the temporal dimension of the entire video and localizing effective objects. The sequence processing module uses a two-layer bi-directional gated recurrent unit (GRU), which is followed by a linear layer. After applying a SoftMax layer to all time steps to obtain the probabilities, the entire network is trained using the connectionist temporal classification (CTC) loss function.

In our experiment, we compared the accuracy and efficiency of our architecture with those of other visual feature extraction models with excellent performance on a benchmark dataset [16]. The models used for this comparison were LeNet-5 [17], VGG-F [18], ResNet-50 [19], DenseNet-121 [20], and LipNet [21] as the baseline model. Extensive evaluation results show that the proposed architecture achieves state-of-the-art results and remarkable efficiency compared to existing deep learning methods.

The contributions of our work can be summarized as follows:

- We developed a novel lipreading architecture based on end-to-end neural networks that relies exclusively on visual information;
- We compared the architecture of our proposed model with that of LipNet as the baseline and those of 3D LeNet-5, 3D VGG-F, 3D ResNet-50, and 3D DenseNet-121 to evaluate the reliability of our model for practical applications;
- We demonstrated improved accuracy and efficiency of the proposed architecture over existing deep learning architectures applied to VSR system implementation.

The remainder of this paper is organized as follows. Section 2 reviews related work on VSR systems and the traditional and existing deep learning approaches. Section 3 introduces the proposed architecture. Section 4 presents information on benchmark datasets, data processing, data augmentation, implementation, and performance evaluation metrics. Along with certain comparative experiments and public processes, this section presents the experimental results. Finally, Section 5 provides a discussion and our conclusions.

2. Related Work

This section summarizes the traditional and existing deep learning approaches for VSR systems. Figure 1 illustrates the VSR processes of the traditional and deep-learning-based methods.

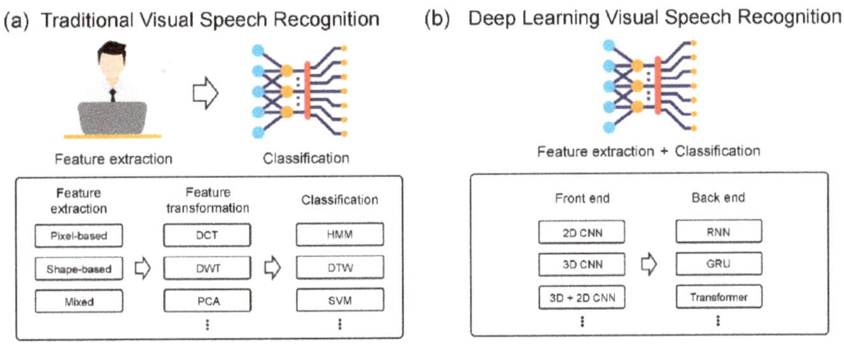

Figure 1. VSR process: (**a**) traditional step and (**b**) deep-learning step.

2.1. Traditional VSR

There are various traditional approaches for implementing VSR systems, such as pre-processing of extracted image features and temporal video feature detection, involving tasks such as optical flow, movement detection, and hand-worked vision pipelines. The traditional approach to implementing the VSR method can be split into two phases. The first phase involves the extraction of visual features from lip movements, and this process relies primarily on hand-labeled features of the geometric information of the lip, such as the lip contour. The existing visual feature extraction method for lip movement involves four steps, starting with the detection and extraction of the region of interest (ROI), the lip, from the video stream, followed by the extraction of lip features from the ROI. For reducing the dimension of the extracted features, a visual feature transformation is required as a complementary step during feature extraction. Different feature transformation algorithms have been developed and used for lipreading tasks, such as linear feature transformation (e.g., discrete cosine transformation (DCT) and discrete wavelet transform (DWT) [22]) and geometry-, motion- [23], and statistical model-based feature transformations [24,25]. The quality of these algorithms depends on the accuracy of training data that are hand-labeled, a task that requires significant amounts of time and effort.

The second phase involves text prediction using the dynamic visual features (classifier phase) and prediction of the words or sentences using a dynamic classifier such as the hidden Markov model (HMM). Using a limited dataset and the HMM model, Goldschen et al. [26] were the first to propose a visual-only sentence-level lipreading technique. They extracted visual features of the mouth region from codebook images to predict continuous sequences of tri-visemes. This study was followed by the development of multi-stream HMMs [27] and the creation of expanded datasets such as model audio and visual streams [22].

As the databases become increasingly complicated, issues such as a high number of speakers, variations in posture, and alterations in the conditions of lighting and background environment may arise. In addition, databases may possess other limitations such as high feature dimension and variations in image quality. Consequently, a complex lip feature extraction algorithm is required. Some classifiers run based on the conditional assumption and are not ideal for modeling long-term dependencies or for operating general classification tasks where several variables are merged.

2.2. Deep Learning VSR

In recent years, deep learning methods have been successfully applied to many fields, including VSR systems. Unlike traditional approaches, in which predictions are limited, deep learning methods attain high accuracy. For instance, when a CNN is combined with traditional methods, the trained classifier CNN architecture can distinguish between visemes, and an HMM framework is used to add temporal information after the CNN output [24,25]. Other researchers have combined long short-term memory (LSTM) with histograms of oriented gradients (HoGs) and input recognized short phrases from the GRID dataset [13,16]. Similarly, using the OuluVS and AVLetters datasets, a trained LSTM classifier with DCT and deep bottleneck features was employed to make word predictions [24].

The deep speech recognition architecture that reads the entire input sequence and then predicts the output sentence is called the sequence-to-sequence model (seq2seq). This model uses global information for longer sequences. Watch, listen, attend, and spell (WLAS) was the first seq2seq model to consist of both audio and visual modules, and it was used to recognize audio-visual speech from a real-world dataset [14,28].

The first suggested end-to-end model to deal with sentence-level lipreading and predict character sequences was LipNet [21]. This model combined spatial-temporal convolutions with Bi-GRU and was trained using the CTC loss function. A limited grammar and vocabulary dataset (GRID corpus) was used to evaluate the performance of the LipNet architecture: the word error rates in the overlapped and unseen-speaker databases were 4.8% and 11.4%, respectively, whereas the success rate of human lipreaders for the same database was 47.7%. Similar architectures have been introduced to investigate the convergence of audio-visual features, where digit sequences were predicted using a small subset of 18 phonemes and 11 terms, and a CTC cascading model was used [29–31]. Thus, the deep learning method can learn more deeply and extract more comprehensive features from the experimental data, demonstrating strong robustness for big data and visual ambiguity.

3. Architecture

This section describes a VSR deep learning architecture and proposes a novel visual feature extraction module (Figure 2c). The proposed module is compared with other visual feature extraction modules that exhibit outstanding feature extraction performance: (i) LipNet as the baseline module (Figure 2a) and (ii) four other comparative architectures with different visual feature extraction modules, namely, 3D LeNet-5, 3D VGG-F, 3D ResNet-50, and 3D DenseNet-121 (Figure 2b). Figure 3 and (Appendix A—Table A1) provide the detailed hyperparameters describing the proposed architecture.

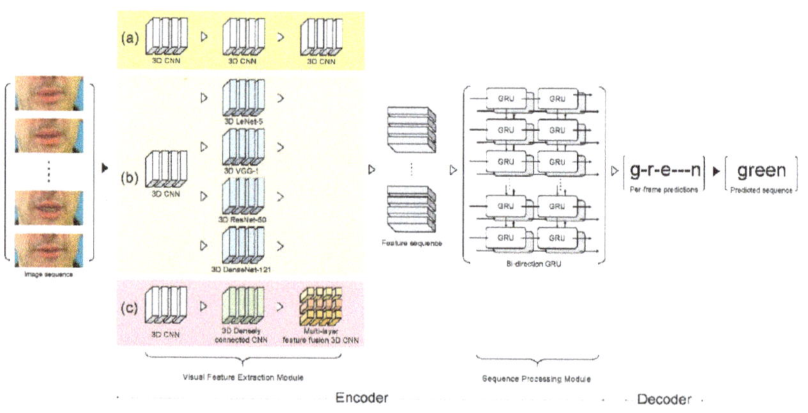

Figure 2. Schematic design of VSR architecture: (**a**) LipNet architecture (baseline), (**b**) four compared architectures, and (**c**) proposed architecture.

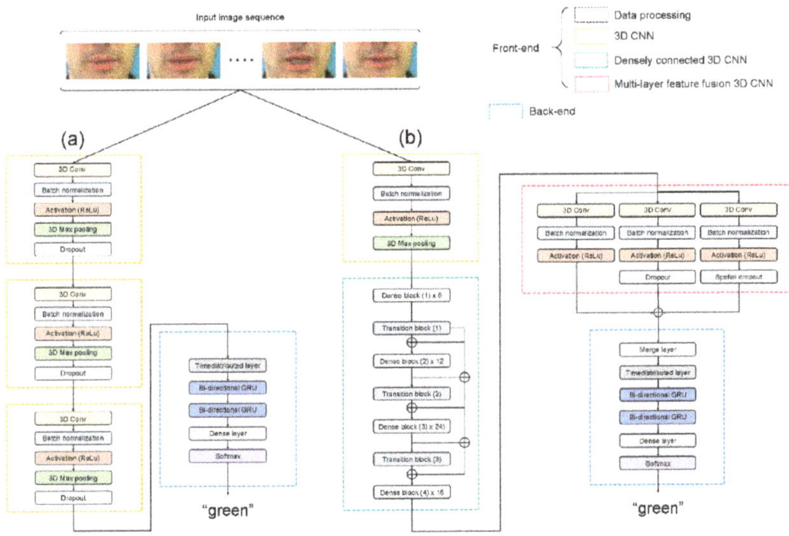

Figure 3. Detailed architecture: (**a**) baseline and (**b**) proposed architecture.

3.1. Spatial-Temporal CNN

CNNs directly use raw input data, thereby automating the feature construction process. When a 2D CNN is applied to an image recognition task, it captures the encoded information for a single image's data and then transfers the encoded information to compute features from the spatial dimensions using 2D feature maps. However, the application of a 2D CNN to a video identification task, where the motion information is encoded in multiple contiguous frames, is ineffective (Figure 4a). Therefore, we use a 3D CNN, which acts as a spatial-temporal CNN in the convolution process, to compute features of both the spatial and temporal dimensions and to capture different lipreading actuations, such as the movements of the lips, tongue, and teeth. This use of a 3D CNN is supported by studies that have shown that 3D CNNs are effective for the extraction of features from video frames, when spatial and temporal information encoded in subsequent frames is considered (Figure 4b) [21,25].

By transforming a single video frame into a cube by stacking several consecutive frames together, the spatial-temporal convolution uses a 3D kernel. In this construction, the feature maps of the convolutional layer are bound to several consecutive frames of the previous layer, which makes it possible to collect motion information during video analysis. Formally, $\tan\mathrm{h}(\cdot)$ is the hyperbolic tangent function, b_{ij} is the bias for this feature map, R_i is the size of the 3D kernel along the temporal dimension, w_{ijm}^{pqr} is the $(p,q,r)^{th}$ value of the kernel linked to the mth feature map in the previous layer, and the value at position $(x,\ y,\ z)$ on the jth feature map in the ith layer is given.

$$v_{ij}^{xyz} = \tan\mathrm{h}(b_{ij} + \sum_{m} \sum_{p=0}^{P_i-1} \sum_{q=0}^{Q_i-1} \sum_{r=0}^{R_i-1} w_{ijm}^{pqr} v_{(i-1)m}^{(x+p)(y+q)(z+r)}) \tag{1}$$

The heights and weights of the kernels are given by P_i and Q_i, respectively [25]. As 3D convolutional kernels replicate kernel weights around the entire cube in this construction, only one form of feature can be extracted from the frame cube.

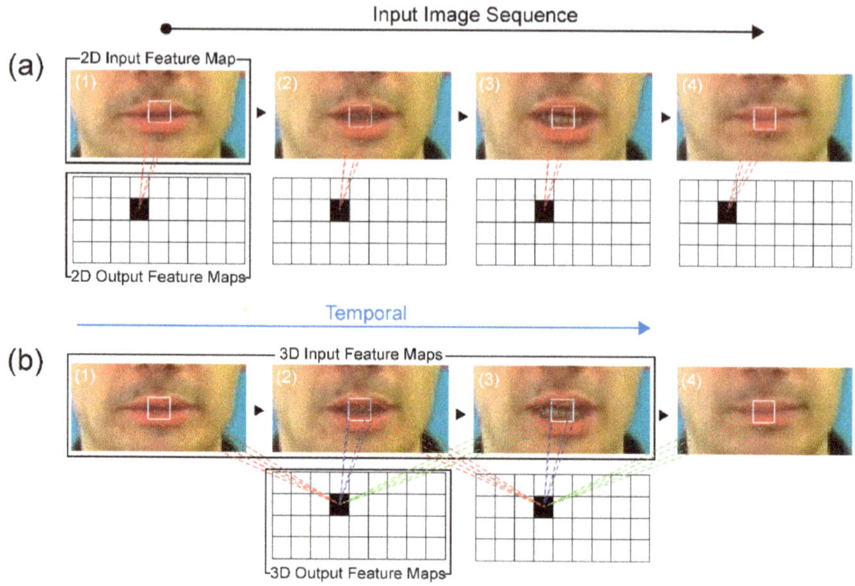

Figure 4. Comparison of (**a**) 2D and (**b**) 3D convolutions.

In our experiment, all the input video frames were fed into a spatial-temporal CNN to encode the visual information of the extracted lips. To be precise, we obtained spatial-temporal features using a 3D convolutional layer with 64 3D kernels of size $3 \times 7 \times 7$ in the input image extracted from multiple contiguous frames. We used a batch normalization (BN) layer to decrease the transformation of internal covariates and ReLU to speed up the training process. Then, to decrease the spatial scale of the 3D feature maps, a max-pooling 3D layer was added. Thus, the output shape was observed with $75 \times 50 \times 13 \times 64$ tensors for an input sequence of $75 \times 100 \times 50 \times 3$ (time/width/height/channel) frames.

3.2. 3D Densely Connected CNN

A densely connected CNN creates relationships between various layers of the connection, which helps enable full use of the features, reduces the gradient disappearance problem, and deepens the network. Before the convolution layer, the bottleneck layer reduces the input feature volumes. The multichannel feature volumes are then fused following the bottleneck layer process. As the preceding features remain, the next layer is only applied to a small set of feature volumes. In addition, with the hyperparameter theta regulating the degree of compression, transition layers are included to improve the model compactness further. Adopting a bottleneck layer, transition layer, and smaller growth rate results in a narrower network. This strategy reduces the model parameters, effectively suppresses overfitting, and saves computational power.

Although many researchers have used 2D CNNs and extracted visual information separately [28,29], 3D densely connected CNNs were used by adding temporal dimensions to densely connected convolution kernel and pooling layers. This approach was used because 2D CNNs often require complex pre-processing and post-processing to perform the same tasks. Therefore, we modified the 2D DenseNet-121 architecture into a 3D DenseNet-121 architecture to 35 maintain dense connectivity to enable deep feature extraction, and this architecture fully utilizes the information provided by the spatial-temporal CNN simultaneously. The dense block is a primary structure composed of densely connected composite functions in the 3D DenseNet-121 architecture, consisting of three sequential operations: BN, ReLU, and 3D convolution layers. The transition layers between different dense blocks contain a BN, ReLU, 3D convolution layer, and average 3D pooling layer.

We extended the strengths of 2D DenseNet-121 to 3D volumetric image processing tasks. The 2D DenseNet-121 network consists of l layers, where each layer represents a nonlinear transformation H_l. The output of the lth layer can then be written as x_l, defined as:

$$x_l = H_l([x_0, x_1, \ldots, x_{1-1}]) \quad (2)$$

where $x_0, x_1, \ldots, x_{1-1}$ are the volumes of the 3D features produced from the previous layers, and [...] refers to the operation of concatenation. Figure 5a illustrates a 3D densely connected CNN architecture consisting of four adjacent dense blocks and three transition layers. Dense block (1) was constructed using a BN layer, ReLU, 3 × 1 × 1 3D convolution layer, BN layer, ReLU, and 3 × 3 × 3 3D convolution layer (Figure 5b). The structures of dense blocks (2), (3), and (4) are similar to that of dense block (1). Figure 5c shows the transition layer composed of a BN layer, ReLU, 3 × 1 × 1 3D convolution layer, and 2 × 2 × 2 average 3D pooling layer.

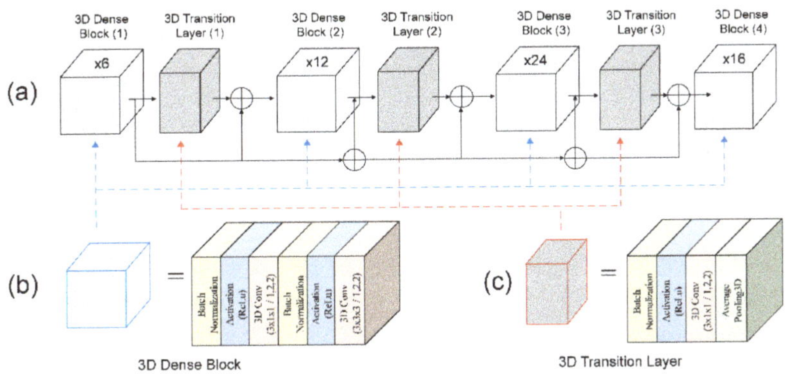

Figure 5. Densely connected 3D CNN architecture: (**a**) detailed densely connected 3D CNN; (**b**) 3D dense block structure; and (**c**) 3D transition layer structure.

3.3. MLFF 3D CNN

Currently, outstanding performance has been achieved for image classification problems using different CNN models. An example is the fusing of multiple CNNs for feature aggregation, where extracting various spatial and temporal features is possible by creating different structures and depths [30]. Different convolutional layers can extract features at various levels of abstraction for the MLFF 3D CNN training phase. Various features can also be derived from this training process with varying depths and filters of different sizes. Using this approach, some of the related features lost in the layered architecture can be selected, rendering the final feature richer.

The proposed MLFF 3D CNN architecture is shown in Figure 6. The first module (Figure 6a) consists of a 3D convolutional layer with 64 3D kernels of size 3 × 5 × 5 on a 3D densely connected convolution layer output feature, followed by a BN layer and ReLU layer. For the second module (Figure 6b), the structure of the first module is followed by a dropout layer to alleviate overfitting as the benchmark dataset used is not large compared with existing image datasets. The role of the dropout layer is to improve and generalize the performance by preventing the creation of strongly correlated activations, which solves overtraining and overfitting [31]. In the third module, the structure is similar to that of the second module, except that the dropout layer is replaced by a spatial dropout layer (Figure 6c), which is a method of dropping the entire feature map. Unlike the standard dropout method, which randomly drops pixels, this method exhibits excellent image classification using CNN models with strong spatial correlation [32]. Therefore, we applied a spatial dropout layer to effectively extract the shapes of the lips, teeth, and tongue with a strong spatial correlation that includes fine movements.

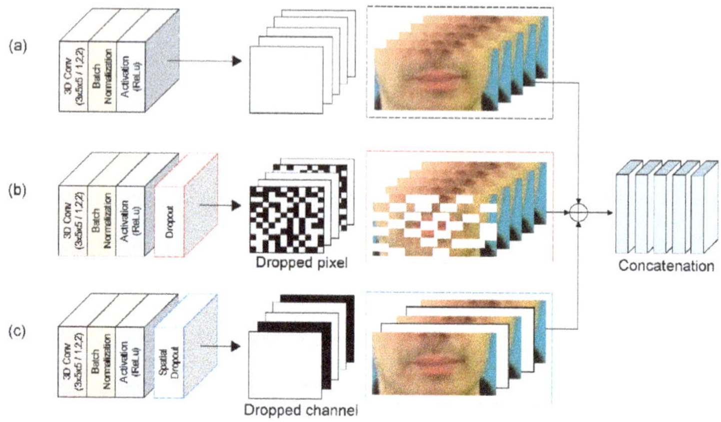

Figure 6. Detailed MLFF 3D CNN: (**a**) first architecture; (**b**) second architecture with dropout layer using dropped pixel; and (**c**) third architecture with spatial dropout layer using dropped channel.

3.4. GRU

The GRU is a recurrent neural network that learns to propagate and regulate the flow of information over more time stages [33]. The GRU can distinguish longer temporal contexts, which is helpful for discriminating ambiguity because 3D CNN captures only short viseme-level features. Moreover, the gradient vanishing problem can be solved by using a GRU, which uses an update gate and reset gate.

Our proposed architecture uses a two-layer bi-directional GRU as a sequence processing module (Figure 7a). Unlike the typical deployment of GRU, a two-layer bi-directional GRU is employed to present information in both forward and backward manners to two separate neural network architectures that are connected to the same output layer, such that both networks can obtain complete information regarding the input. The two-layer bi-directional GRU layer receives its input from the MLFF 3D CNN sequentially and then generates characters as output, as follows:

$$z_t = \sigma(W_z a_t + U_z h_{t-1} + b_z) \tag{3}$$

$$r_t = \sigma(W_r a_t + U_r h_{t-1} + b_r) \tag{4}$$

$$h_t = (1 - z_t) \circ h_{t-1} + z_t \circ \sigma_r(W_h a_t + U_h(r_t \circ h_{t-1}) + b_h) \tag{5}$$

The GRU consists of four components (x_t, z_t, r_t, and h_t) and a given sequence of image features a = $(a_1, a_2, \cdots, a_t)$. x_t is an input vector with its resulting weight parameter matrix and vector. z_t is an update gate vector with its resulting weight parameter matrix and vectors W_z, U_z, and b_z. r_t is a reset gate vector with its resulting weight parameter matrix and vectors W_r, U_r, and b_r. Finally, h_t is an output vector with its resulting weight parameter matrix and vectors W_h, U_h, and b_h. h_{t-1} is the previously hidden state output, which has the same structure as the current state. σ is the ReLU function, used as an activation function. represents the Hadamard product. To obtain an output with 75 × 512 tensors using the merge layer, we provided an input sequence of 75 × 3 × 1 × 192 (time/width/height/channel) frames in a bi-directional GRU.

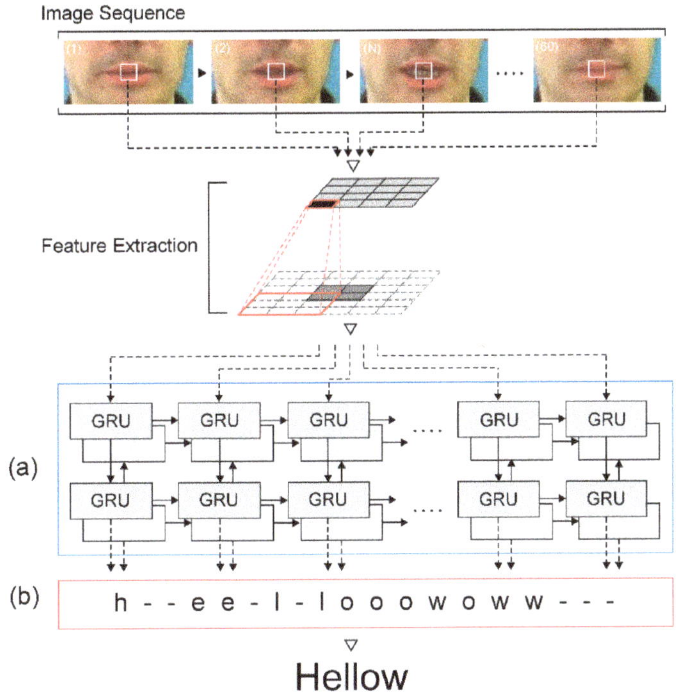

Figure 7. Sequence processing module: (**a**) a two-layer bi-directional GRU and (**b**) CTC.

3.5. Connectionist Temporal Classification (CTC)

We used the CTC approach and applied it to an end-to-end deep neural network. This approach uses a loss function to parameterize the distribution of a label token sequence without having to align the input sequence [34]. A single set of time step label tokens can be expressed as V by using CTC, where the series of size-T given by the temporal module is the output labeled by the blank symbols '⊔ and consecutive symbols are repeated (Figure 7b). We define a function B : $(V \cup \{\sqcup\})^* \rightarrow V^*$ to delete adjacent characters and to remove any blanks because the processed string may contain a blank token. It is possible to obtain the probability of observing a labeled sequence y by marginalizing this label, $p(y|x) = \sum_{u \in B^{-1}(y)} p(u_1|x) \cdots p(u_T|x)$, where x is the input video, for all possible alignments. The standard CTC loss L_{ctc} formula is defined as follows:

$$p_{ctc}(y|x) = \sum_{w \in B^{-1}(y)} p_{ctc}(w|x) = \sum_{w \in B^{-1}(y)} \prod_{t=1}^{T} q_{w_t}^{t} \tag{6}$$

$$L_{ctc} = -\ln p_{ctc}(y|x) \tag{7}$$

where T is the duration of the input sequence and $q_{(w_t)}^{t}$ represents the SoftMax probability of the output label w_t, where $w_t \in \{a, ai, an, ao, \cdots, zun, zuo, blank\}$ at frame t. The CTC path of a sequence is $w = (w_1, w_2, \cdots, w_T)$, and y is the sentence label (ground truth). $B^{-1}(y)$ is the set of all possible paths of the CTC that can be mapped to ground truth y. As CTC prohibits the use of autoregressive connections to control the inter-time-step dependencies of the label sequence, it is conditionally independent of the marginal distributions generated at each time step of the temporal module. Therefore, CTC models are typically decoded using a beam search procedure to restore the temporal dependency of the labels, which blends the probabilities of that language model.

4. Experiments and Results

This section describes the used dataset, data pre-processing, data augmentation, and implementation.

4.1. Dataset

The GRID audio-visual dataset is widely used for audio-visual recognition and VSR studies and is built on sentence-level audio video clips [16]. GRID is an openly available corpus containing an audio-visual database from 34 speakers with 1000 utterances per speaker and a total duration of 28 h. One sample has a single speaker clip, and each sample in this database lasts 3 s at 25 frames/s. The visual data for speaker number 21 are missing from the online available database corpus [16]. This database is sentence-level with a fixed grammar and is composed of "command (4) + color (4) + preposition (4) + letter (26) + digit (10) + adverb (4)". It has 51 unique words, and each sentence is a randomly chosen combination of these words.

To unify the test conditions for our experiment, we divided the training and validation sets as follows [21]. In unseen-speaker datasets that were not historically used in the literature, 3971 videos were used for the evaluation data relevant to male speakers (1 and 2) and female speakers (20 and 22). The remaining videos for the unseen speakers (28,775 videos) were utilized to train the models. Following this strategy, the models were evaluated using speakers that had not appeared in the training process, thus guaranteeing the generalized performance of the model. We employed sentence-level variants of segmentation for overlapped-speaker datasets, where 255 random sentences from each speaker were used for evaluation. For training, the leftover data from all speakers were pooled.

4.2. Data Pre-Processing and Augmentation

The data pre-processing stage detects the targeted face and mouth using a DLib face detector [35]. This detector utilized a HoG feature-based linear classifier [35]. The output is given as the (x, y) coordinates of the diagonal edges; these coordinates are used later to draw the bounding box around the mouth. Subsequently, the iBug tool was used with 68 landmarks coupled with an online Kalman filter as a face landmark predictor [36]. This tool is typically used to read lip movements and extract points on the lips, which correspond to those obtained from the trained dataset. These tools were employed to extract a mouth-centered area with dimensions of 100×50 pixels per frame using an affine transformation and to standardize the RGB channels over the entire training set to have zero mean and unit variance. For training data, we used the data augmentation process from [21] to prevent overfitting. We performed training with both regular and horizontally mirrored image sequences. As the dataset included start and end terms that acted as a timer for each "clip" sample, we augmented the training data at the sentence level using individual words as additional training instances. These instances had a decay rate of 0.925. Finally, if necessary, we detected the movement speed and duplicated the frames to prevent variation, and this process was conducted with a probability of 0.05/frame. All models were trained and tested under the same pre-processing and augmentation processes for the GRID dataset.

4.3. Implementation

All models were implemented using Keras with a TensorFlow backend and TensorFlow-CTC decoder to measure the character error rate (CER) and word error rate (WER) scores using CTC beam search. In Figure 3 and Table A1 (Appendix A), the detailed configuration and parameters used for each layer in the proposed architecture are summarized. The network parameters of all models were initialized via He initialization, except for the orthogonally initialized square GRU matrices and the default hyperparameters. The orthogonally initialized square GRU matrices were trained with mini batches of size 8 and used the optimizer ADAM [37] with a learning rate of 0.0001. The proposed model was trained utilizing channel-wise dropped pixels and the dropped channel using spatial dropout in the MLFF 3D CNN, where the proposed models included the baseline model trained on GRID until overfitting.

As mentioned earlier, LipNet is the baseline model in our study; therefore, we evaluated its performance for the categories of unseen and overlapped speakers. For the unseen-speaker category, 8.534% CER and 16.341% WER were achieved, versus 6.400% CER and 11.400% WER in the model paper [21]. In addition, for overlapped speakers, we obtained CER of 5.657% and WER of 14.779%, whereas 1.900% CER and 4.800% WER were mentioned in the model paper: our test results for the LipNet model are higher than those of the original LipNet model. Due to these variations in outcomes, video clips of individual words were not used for additional training instances or other defective operations in the training phase. As one of the contributions of this analysis is a feasibility test of the proposed model, to obtain the required CER and WER, we did not further subdivide the baseline model. However, we compared the results acquired in our environment with those obtained in the testing environments of the existing models to analyze both the existing models and the proposed model.

4.4. Performance Evaluation Metrics

We used standard automatic speech recognition evaluation metrics to assess the proposed model. The learning loss of all the models was measured to evaluate the learning state during the training process. To compare the performances and computational efficiencies of all models, we evaluated the parameters, epoch time, CER, and WER, of each model.

By calculating the total edit distance, the error rate metrics used for accuracy assessment were obtained and converted into percentages. It is necessary to compare the decoded text to the original text when assessing misclassifications. The equation is given, wherein N is the cumulative number of characters in the ground truth, S is the number of characters substituted for incorrect classifications, I is the number of characters inserted for non-picked characters, and D is the number of deletions that should not be present for decoded characters. Thus, the CER and WER are determined using Equations (8) and (9), where C and W denote characters and words, respectively.

$$\mathrm{CER}(\%) = \left(\frac{C_S + C_D + C_I}{C_N}\right) \times 100 \tag{8}$$

$$\mathrm{WER}(\%) = \left(\frac{W_S + W_D + W_I}{W_N}\right) \times 100 \tag{9}$$

We performed a CTC beam search using a TensorFlow-CTC decoder implementation to generate approximate maximum-probability predictions for all experimental models. We also compared the CER and WER with respect to the number of parameters and computational efficiency over the epoch time. To visualize the results, we used the phoneme-to-viseme mapping described in [38].

4.5. Training Process and Learning Loss

The training and validation losses during training on the GRID corpus are shown in Figures 8 and 9, and the definition of each tested model is presented in Table 1 for both the unseen and overlapped-speaker categories. For the former category in Figure 8, the gaps between training and validation in models A and B and the baseline model are similar, as illustrated in Figure 8a–c. Furthermore, our proposed model shows large gaps relative to those for the three models in Figure 8a–c, but slightly smaller gaps relative to those in Figure 8d.

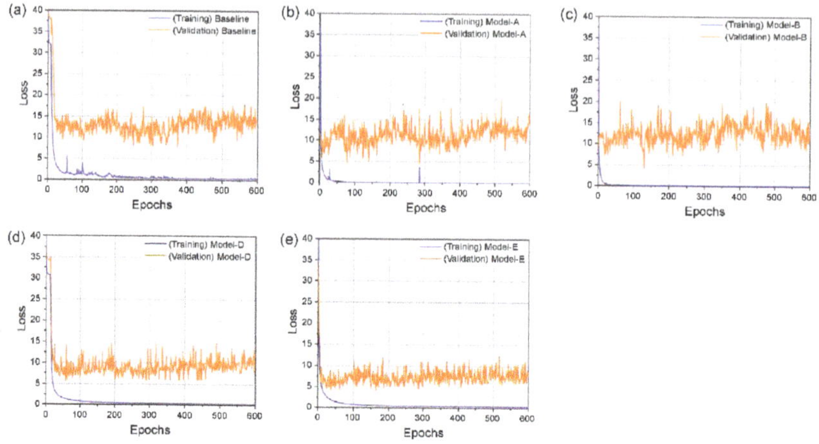

Figure 8. Training and validation loss of unseen speakers: (**a**) baseline; (**b**) model-A; (**c**) model-B; (**d**) model-D; (**e**) model-E.

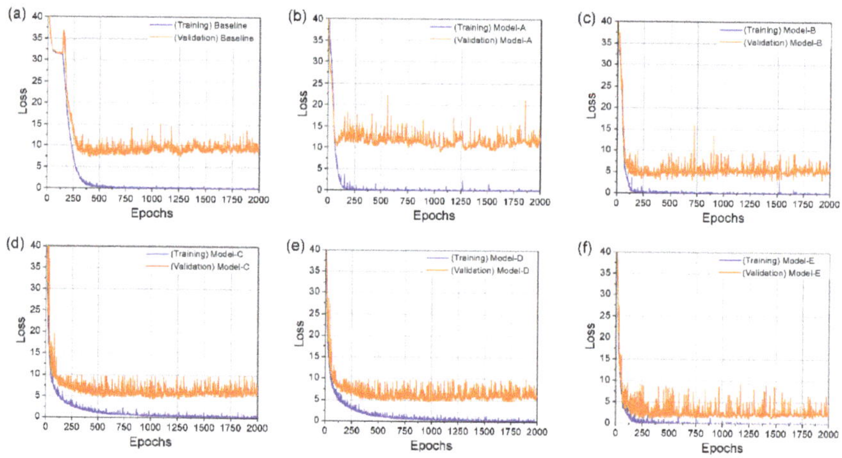

Figure 9. Training and validation loss of overlapped speakers: (**a**) baseline; (**b**) model-A; (**c**) model-B; (**d**) model-eC; (**e**) model-D; (**f**) model-E.

In the category of overlapped speakers, the training loss of each model started to decrease earlier than the baseline, as shown in Figure 9. The validation loss of the baseline and Figure 9b show similar tendencies, where the differences between the training and validation losses in Figure 9c–e are lower than the difference in the baseline model.

We performed the training process with the 3D ResNet-50 model; however, it did not perform as our experimental environment ran out of memory, and the results for unseen speakers were excluded. Finally, these results indicate that our proposed model for the two categories in the GRID dataset shows the lowest difference between the training and validation losses, which effectively prevents overfitting.

4.6. WER and CER

The results are reported for the unseen and overlapped speakers of the GRID dataset. The results of the comparison between our proposed model and the existing deep learning models are presented in Table 2, which reveals that our proposed model achieved state-of-the-art (SOTA) results: 2.853% CER and 5.059% WER for the unseen-speaker category and

1.004% CER and 1.011% WER for the overlapped-speaker category. These results exhibit decrease of all conditions over the current SOTA and baseline results.

Table 1. Number of parameters and epoch time of the proposed method compared to those of the baseline and different methods.

Model	Method		Parameters	Unseen Speakers	Overlapped Speakers
	Frontend	Backend		Epoch Time (s)	Epoch Time (s)
Baseline	3D CNN	Bi-GRU + CTC	45.7M	1152	131
Model-A	3D CNN + 3D LeNet-5	Bi-GRU + CTC	36.5M	1104	118
Model-B	3D CNN + 3D VGG-F	Bi-GRU + CTC	113.4M	1405	178
Model-C	3D CNN + 3D ResNet-50	Bi-GRU + CTC	667M	-	256
Model-D	3D CNN + 3D DenseNet-121	Bi-GRU + CTC	22.4M	1272	126
Model-E	Proposed architecture		34.5M	1286	127

Table 2. Performance of the proposed model compared to the baseline model and different existing models with unseen and overlapped speakers.

Year	Model		Unseen Speakers		Overlapped Speakers	
			CER (%)	WER (%)	CER (%)	WER (%)
	Hearing-impaired person (avg.) [21]		-	47.700	-	-
2016	Baseline-LSTM [21]		38.400	52.800	15.200	26.300
2016	Baseline-2D [21]		16.200	26.700	4.300	11.600
2017	LipNet-NoLM [21]		-	-	2.000	5.600
2017	LipNet [21]		6.400	11.400	1.900	4.800
2017	WAS [5]		-	-	-	3.300
2018	LCANet [39]		-	-	1.300	2.900
2018	LipNet + 3D-FPA [9]		7.246	14.178	-	-
2019	LRNeuNet [40]		6.100	9.500	1.200	2.700
2019	LipSound [41]		-	-	1.532	4.215
2020	PCPG [42]		-	11.200	-	-
2020	FastLR [43]		-	-	2.400	4.500
2020	LipNet + LipsID [44]		5.200	9.900	1.200	3.300
2020	TVSR-Net + SC-Block [45]		-	90.900	-	-
2020	DualLip [46]		-	-	1.600	2.710
2021	3D-ResNet50-TCN-CTC [47]		4.100	6.200	1.200	1.100
	Frontend	Backend	CER (%)	WER (%)	CER (%)	WER (%)
Baseline	3D CNN	Bi-GRU + CTC	8.534	16.341	5.657	14.779
Model-A	3D CNN + 3D LeNet-5	Bi-GRU + CTC	11.797	17.188	8.083	22.526
Model-B	3D CNN + 3D VGG-F	Bi-GRU + CTC	8.395	11.914	3.499	10.482
Model-C	3D CNN + 3D ResNet-50	Bi-GRU + CTC	-	-	3.089	8.203
Model-D	3D CNN + 3D DenseNet-121	Bi-GRU + CTC	5.314	10.286	3.165	8.529
Model-E	Proposed architecture		2.853	5.059	1.004	1.011

Although the accuracies of the models with 3D ResNet-50 and 3D DenseNet-121 architectures exceeded that of the baseline, no significant differences were detected (Table 2). In the case of unseen speaker's category (Figure 10a,b), Models A, B, D, and E exhibit an almost steady learning behavior until approximately 250 epochs; subsequently, the error rates show a continuous decrement. On the other hand, both baseline and Model A with a simple structure show higher error rates compared to others. Moreover, the smallest error rates are achieved by our proposed Model E with the same training steps. For the overlapped speaker category, Models B, C, and D show similar performance. Unlike the comparable performance of Model A for the unseen speaker category, an obvious degradation in performance is observed for the same model in the case of overlapped speaker's category (Figure 10c,d). For our proposed Model E, the error rate shows a noticeable decrement after approximately 130 epochs. Therefore, in terms of accuracy, our proposed model outperforms the existing models, including the baseline model, which can be attributed to the combination of multiple 3D CNN architectures. Figure 10 shows the training step with CER and WER on the GRID database.

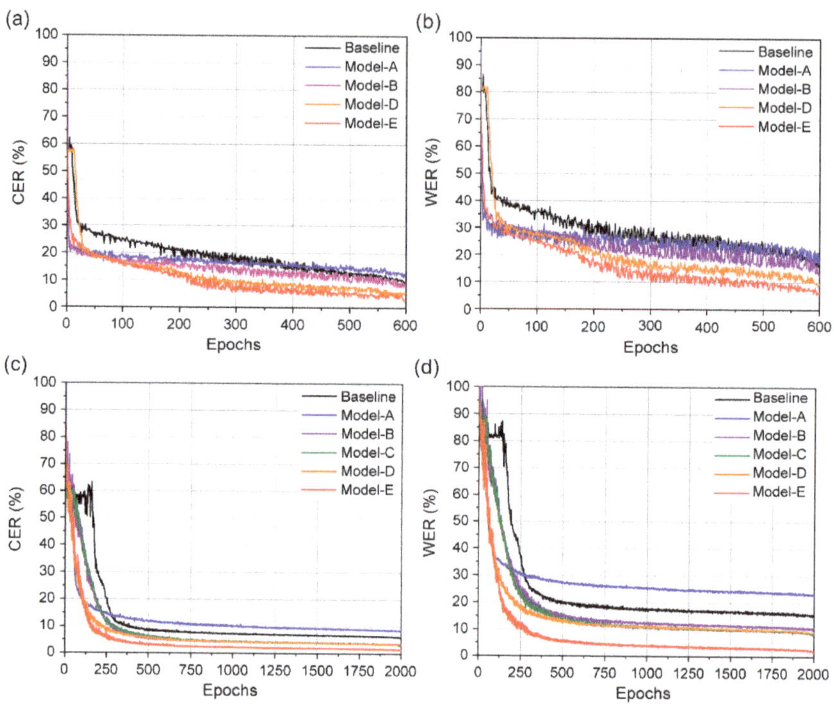

Figure 10. Training steps for CER and WER comparing our proposed model against the baseline and other models: (**a**) CER and (**b**) WER evaluated using unseen speakers, and (**c**) CER and (**d**) WER evaluated using overlapped speakers.

4.7. Model and Computational Efficiency

The major limitations of VSR systems in practical applications are their model size and computational efficiency. To evaluate the computational efficiencies of the models, we compared their accuracies with different numbers of trained parameters and epoch times (Figures 11 and 12). We summarized the number of parameters and epoch time for individual models for the two dataset categories in Table 1. Although our proposed model has an epoch time similar to that of the baseline model, lower CER and WER are seen due to 10 M fewer parameters used. In addition, our model shows a faster epoch time and lower number of parameters than those of the three other models except Model A with the GRID dataset. LeNet-5 has a gradient-based learning CNN structure, which is divided into an input layer, a convolution layer, a pooling layer, a fully connected layer, and an output layer, and the input layer is removed, and a total of seven layers are included. Model A is a simple structure using two convolution layers and pooling layers from LeNet-5. This simple structure has difficulty in processing high resolution images. Therefore, its accuracy will be significantly lowered when used for an application like lip reading, which requires delicate motion detection.

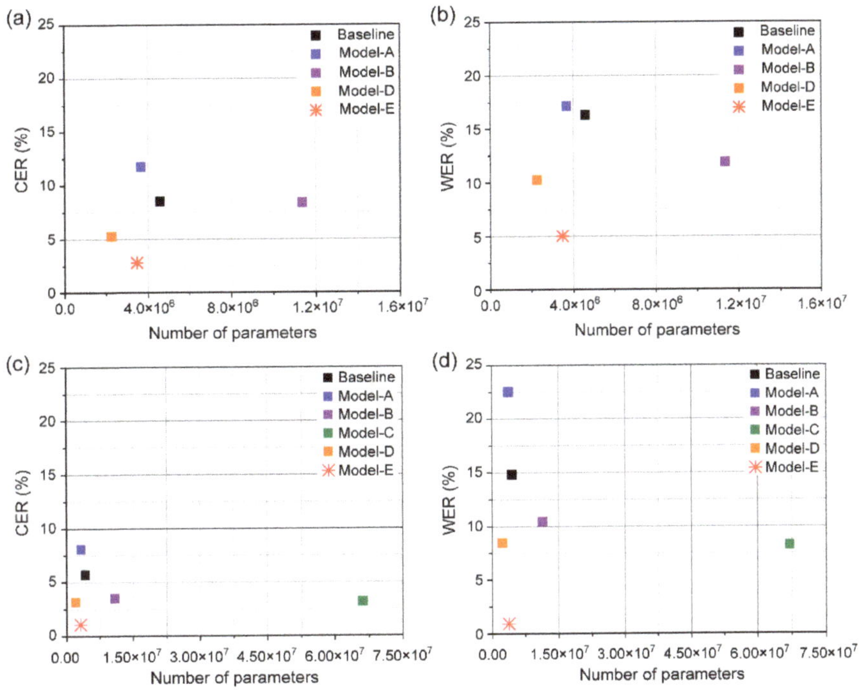

Figure 11. CER and WER of the baseline and different models according to the number of parameters, measured on two splits: (**a**) CER and (**b**) WER evaluated using unseen speakers, and (**c**) CER and (**d**) WER evaluated using overlapped speakers.

In conclusion, we obtained a lower error of GRID database while reducing the number of parameters by approximately 11.2 M compared to that of the baseline model; we also achieved a comparable epoch time.

4.8. Confusion Matrix

A mapping proposed by the IBM ViaVoice database was used for the visual analysis [38]. It consists of 43 phonemes grouped into 13 classes of visemes, including a silence class, vowels based on lip rounding (V), alveolar-semivowels (A), alveolar-fricatives (B), alveolar (C), palato-alveolar (D), bilabial (E), dental (F), labio-dental (G), and velar (H). For the bilabial viseme class, we plot a confusion matrix corresponding to the most confusing phoneme (Figure 13a), which represents the vowels based on lip rounding. In the experimental results, {/AE/, /IH/} is frequently misclassified during the text decoding process (Figure 13a). At a first glance, the confusion between /IH/ (a rather close vowel) and /AE/ (a very open vowel) is unexpected but only occurs in "at", a generally pronounced feature word with a shortened, weak vowel /AH/ in sample /AE/. This effect is due to the similar pronunciations of the {at, bin} text pair. Figure 13b represents the intra-viseme categorical confusion matrix. Distinguishing homophones is a major challenge, and we present the experimental results for [p], [b], and [m] in Figure 13c. Based on the evaluation of our model from different aspects, this model can help overcome the technical barriers for the practical implementation of VSR systems.

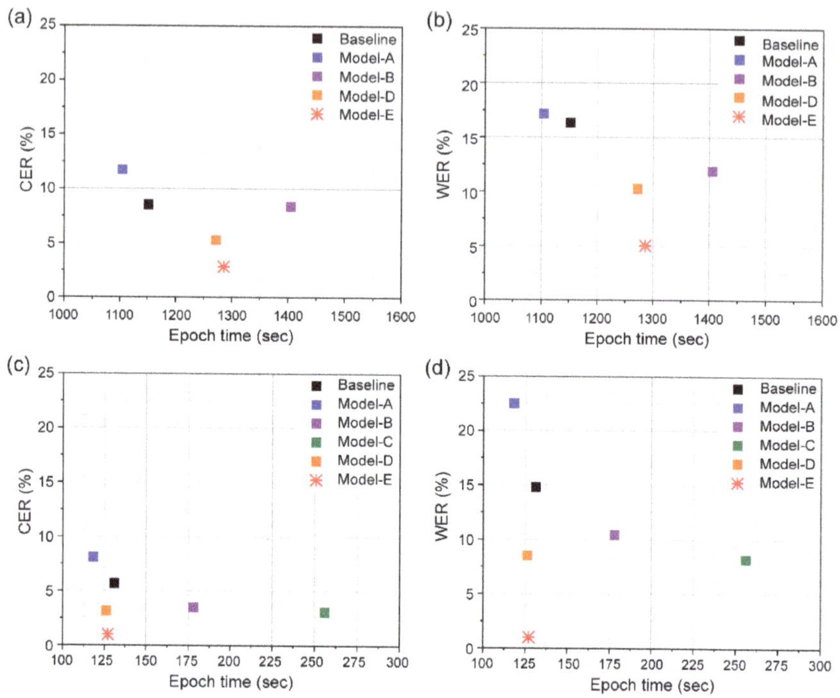

Figure 12. CER and WER of the baseline and different models according to epoch time, measured on two splits: (**a**) CER and (**b**) WER evaluated using unseen speakers, and (**c**) CER and (**d**) WER evaluated using overlapped speakers.

The confusion matrix in [21] shows the results for frequent misclassifying words in the text decoding process. Many errors occurred, but {/AA/, /AY/} and {/AE/, /IH/} accounted for the largest proportion among them. Furthermore, similar pronunciation between the text pairs of {r, i}, {at, bin}, and {four, two} causes frequent incorrect classification during the text decoding process for {/AA/, /AY /}, {/AE/, /IH/} and {/AO/, /UW/} in [39]. Although it was difficult to distinguish words of similar pronunciation for {/AE/, /IH/} in our proposed model, an enhancement can be noticed in Figure 13a for correctly classifying similar pronunciations for {/AA/, /AY/} and {/AO/, /UW/}, unlike their frequent misclassification in [21,39]. Additionally, we can determine the superiority of our model performance by comparing it with the bilabial and intra-visemes categorical confusion matrix in [21,39]. Therefore, our model showed excellent performance in distinguishing all similar pronunciations on bilabial-visemes and intra-visemes.

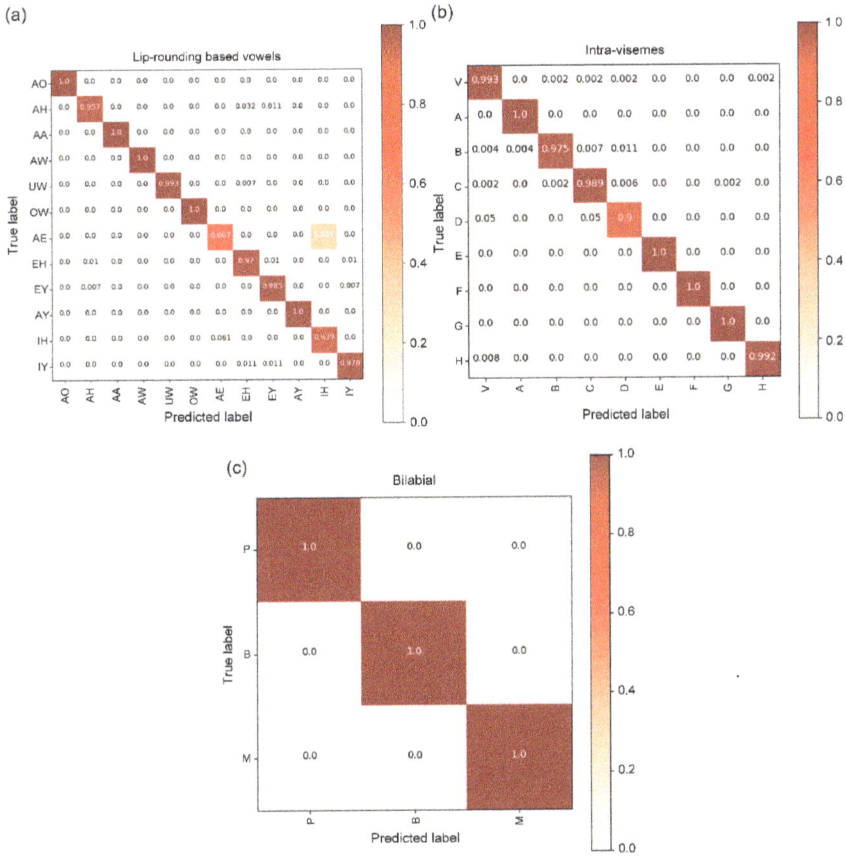

Figure 13. Detailed proposed architecture confusion matrices for the (**a**) lip-rounding based vowels; (**b**) intra-visemes; and (**c**) bilabial groups. The three groups with the most confusions were selected, as well as the confusions within viseme clusters.

5. Discussion and Conclusions

The primary reason lipreading is difficult is that much of the image in the video input remains unchanged—the movement of the lips is the biggest distinction. However, it is possible to perform action recognition, which is a type of video classification, from a single image. When lipreading, it is always important to derive the characteristics relevant to the speech content from a single image and to analyze the time relationship between the entire series of images to infer the content. The key problem with lipreading is visual ambiguity.

This paper presented a novel lipreading architecture for sentence-level VSR. By applying multiple visual feature extraction methods, we achieved accurate viseme prediction. To the best of our knowledge, this is the first time that a 3D CNN, 3D densely connected CNN, and MLFF 3D CNN have been used in combination to extract the features of lip movements as encoders. The strengths of each stage are as follows. The 3D CNN extracts features from multiple consecutive video frames efficiently. The 3D densely connected CNN helps in fully utilizing the features, effectively reducing the problem of gradient disappearance, and making the network deeper. In addition, the bottleneck layer, translation layer, and smaller growth rate make the network narrower, thereby reducing the number of model parameters, suppressing overfitting and saving computational power. Lastly, the MLFF 3D CNN with a dropout and spatial dropout layer avoids overfitting and effectively extracts shapes with strong spatial correlations with fine movements while exploring the context information of the movement in both the temporal and spatial domains.

We compared several deep learning models for predicting sentence sequences, and the results indicated that our proposed architecture achieves SOTA CER and WER values (Table 2). Smaller numbers of parameter and faster epoch times than those of the existing methods were realized using our proposed model. Moreover, the proposed architecture showed reduced CER and WER values than those of the baseline model for both the unseen-speaker and overlapped-speaker datasets.

It is important to develop a VSR system that exclusively uses visual information. This system has practical potential for various applications in speech recognition in noisy or underwater environments, biometric identification for security, and silent movie analysis; furthermore, it could be beneficial for patients with speech impairments. However, it remains difficult to perform automatic speech recognition using only VSR as speech uses acoustic and visual information. Thus, in future work, we will investigate a solution that can be directly applied to the loss function, because the loss function was not modified in our proposed model. Moreover, we intend to expand our concept to pursue performance enhancement and discover potential applications using both audio and visual information.

Author Contributions: Conceptualization, S.J.; methodology, S.J.; software, S.J.; validation, A.E.; formal analysis, S.J. and A.E.; investigation, S.J. and A.E.; data curation, S.J. and A.E.; writing—original draft preparation, S.J.; writing—review and editing, S.J., A.E. and M.S.K.; supervision, M.S.K.; funding acquisition, M.S.K. All authors have read and agreed to the published version of the manuscript.

Funding: This work was supported by the National Research Foundation of Korea (NRF) grant funded by the Korea government (MSIT) (NRF-2018X1A3A1069795).

Institutional Review Board Statement: Not applicable.

Informed Consent Statement: Not applicable.

Data Availability Statement: Not applicable.

Conflicts of Interest: The authors declare no conflict of interest. The funders had no role in the design of the study; in the collection, analyses, or interpretation of data; in the writing of the manuscript, or in the decision to publish the results.

Appendix A

The following table provides a detailed hyperparameters further describing the proposed end-to-end lip-reading architecture.

Table A1. Hyperparameters of proposed architecture.

Layers	Size/Strid/Pad	Output Size	Dimension Order
3D Conv	$[3 \times 5 \times 5]/(1,2,2)/(1,2,2)$	$75 \times 50 \times 25 \times 64$	$T \times C \times H \times W$
3D Max Pooling	$[1 \times 2 \times 2]/(1,2,2)$	$75 \times 50 \times 13 \times 64$	$T \times C \times H \times W$
3D Dense Block (1)	$[3 \times 1 \times 1]$ 3D Conv / $[3 \times 3 \times 3]$ 3D Conv $(\times 6)$	$75 \times 25 \times 13 \times 96$	$T \times C \times H \times W$
3D Transition Block (1)	$[3 \times 1 \times 1]$ 3D Conv / $[1 \times 2 \times 2]$ average pool$/(1 \times 2 \times 2)$	$75 \times 12 \times 6 \times 6$	$T \times C \times H \times W$
3D Dense Block (2)	$[3 \times 1 \times 1]$ 3D Conv / $[3 \times 3 \times 3]$ 3D Conv $(\times 12)$	$75 \times 12 \times 6 \times 38$	$T \times C \times H \times W$
3D Transition Block (2)	$[3 \times 1 \times 1]$ 3D Conv / $[1 \times 2 \times 2]$ average pool$/(1 \times 2 \times 2)$	$75 \times 6 \times 3 \times 3$	$T \times C \times H \times W$

Table A1. *Cont.*

Layers	Size/Strid/Pad		Output Size	Dimension Order
3D Dense Block (3)	[3 × 1 × 1] 3D Conv / [3 × 3 × 3] 3D Conv	(×24)	75 × 12 × 6 × 38	T × C × H × W
3D Transition Block (3)	[3 × 1 × 1] 3D Conv / [1 × 2 × 2] average pool/(1 × 2 × 2)		75 × 3 × 1 × 1	T × C × H × W
3D Dense Block (4)	[3 × 1 × 1] 3D Conv / [3 × 3 × 3] 3D Conv	(×16)	75 × 3 × 1 × 33	T × C × H × W
MLFF 3D CNN (1)	[3 × 5 × 5]/(1, 2, 2)/(1, 2, 2)		75 × 3 × 1 × 64	T × C × H × W
MLFF 3D CNN (2)	[3 × 5 × 5]/(1, 2, 2)/(1, 2, 2)		75 × 3 × 1 × 64	T × C × H × W
MLFF 3D CNN (3)	[3 × 5 × 5]/(1, 2, 2)/(1, 2, 2)		75 × 3 × 1 × 64	T × C × H × W
Bi-GRU (1)	256		75 × 512	T × F
Bi-GRU (2)	256		75 × 512	T × F
Linear	27 + blank		75 × 512	T × F
Softmax			75 × 28	T × V

References

1. McGurk, H.; MacDonald, J.J.N. Hearing lips and seeing voices. *Nature* **1976**, *264*, 746–748. [CrossRef] [PubMed]
2. Chitu, A.G.; Rothkrantz, L.J.M. Automatic visual speech recognition. In *Speech Enhancement, Modeling, Recognition—Algorithms, and Applications*; Ramakrishnan, S., Ed.; Intechopen: London, UK, 2012; p. 95.
3. Fisher, C.G. Confusions among visually perceived consonants. *J. Speech Hear. Res.* **1968**, *11*, 796–804. [CrossRef] [PubMed]
4. Easton, R.D.; Basala, M. Perceptual dominance during lipreading. *Atten. Percept. Psychophys.* **1982**, *32*, 562–570. [CrossRef] [PubMed]
5. Chung, J.S.; Senior, A.; Vinyals, O.; Zisserman, A. Lip reading sentences in the wild. In Proceedings of the 2017 IEEE Conference on Computer Vision and Pattern Recognition (CVPR), Honolulu, HI, USA, 21–26 July 2017; pp. 3444–3453.
6. Kastaniotis, D.; Tsourounis, D.; Fotopoulos, S. Lip Reading Modeling with Temporal Convolutional Networks for Medical Support applications. In *2020 13th International Congress on Image and Signal Processing, BioMedical Engineering and Informatics (CISP-BMEI)*; IEEE: Chengdu, China, 2020; pp. 366–371.
7. Zhao, H.; Zhang, B.; Yin, Z. Lip-Corrector: Application of BERT-based Model in Sentence-level Lipreading. *J. Phys. Conf. Ser.* **2021**, *1871*, 012146. [CrossRef]
8. Fernandez-Lopez, A.; Sukno, F.M. Survey on automatic lip-reading in the era of deep learning. *Image Vis. Comput.* **2018**, *78*, 53–72. [CrossRef]
9. Hao, M.; Mamut, M.; Yadikar, N.; Aysa, A.; Ubul, K. A survey of research on lipreading technology. *IEEE Access* **2020**, *8*, 204518–204544. [CrossRef]
10. Chen, X.; Du, J.; Zhang, H. Lipreading with DenseNet and resBi-LSTM. *Signal Image Video Process.* **2020**, *14*, 981–989. [CrossRef]
11. Tsourounis, D.; Kastaniotis, D.; Fotopoulos, S. Lip Reading by Alternating between Spatiotemporal and Spatial Convolutions. *J. Imaging* **2021**, *7*, 91. [CrossRef] [PubMed]
12. Fenghour, S.; Chen, D.; Guo, K.; Xiao, P. Lip Reading Sentences Using Deep Learning with Only Visual Cues. *IEEE Access* **2020**, *8*, 215516–215530. [CrossRef]
13. Ma, S.; Wang, S.; Lin, X. A Transformer-based Model for Sentence-Level Chinese Mandarin Lipreading. In *2020 IEEE Fifth International Conference on Data Science in Cyberspace (DSC)*; IEEE: San Francisco, CA, USA, 2020; pp. 78–81.
14. Zhou, Z.; Zhao, G.; Hong, X.; Pietikäinen, M. A review of recent advances in visual speech decoding. *Image Vis. Comput.* **2014**, *32*, 590–605. [CrossRef]
15. Xiao, J. 3D feature pyramid attention module for robust visual speech recognition. *arXiv* **2018**, arXiv:1810.06178.
16. Cooke, M.; Barker, J.; Cunningham, S.; Shao, X. An audio-visual corpus for speech perception and automatic speech recognition. *J. Acoust. Soc. Am.* **2006**, *120*, 2421–2424. [CrossRef]
17. LeCun, Y.; Bottou, L.; Bengio, Y.; Haffner, P. Gradient-based learning applied to document recognition. *Proc. IEEE* **1998**, *86*, 2278–2324. [CrossRef]
18. Chatfield, K.; Simonyan, K.; Vedaldi, A.; Zisserman, A. Return of the devil in the details: Delving deep into convolutional nets. *arXiv* **2014**, arXiv:1405.3531.
19. He, K.; Zhang, X.; Ren, S.; Sun, J. Deep residual learning for image recognition. In Proceedings of the 2016 IEEE Conference on Computer Vision and Pattern Recognition (CVPR), Las Vegas, NV, USA, 27–30 June 2016; pp. 770–778.
20. Huang, G.; Liu, Z.; Van Der Maaten, L.; Weinberger, K.Q. Densely connected convolutional networks. *arXiv* **2017**, arXiv:1608.06993.

21. Assael, Y.M.; Shillingford, B.; Whiteson, S.; De Freitas, N. Lipnet: End-to-end sentence-level lipreading. *arXiv* **2016**, arXiv:1611.01599.
22. Chu, S.M.; Huang, T.S. Bimodal speech recognition using coupled hidden Markov models. In Proceedings of the Sixth International Conference on Spoken Language Processing (ICSLP 2000), Beijing, China, 16–20 October 2000; pp. 747–750.
23. Wand, M.; Koutník, J.; Schmidhuber, J. Lipreading with long short-term memory. In Proceedings of the 2016 IEEE International Conference on Acoustics, Speech and Signal Processing (ICASSP), Shanghai, China, 20–25 March 2016; pp. 6115–6119.
24. Petridis, S.; Pantic, M. Deep complementary bottleneck features for visual speech recognition. In Proceedings of the 2016 IEEE International Conference on Acoustics, Speech and Signal Processing (ICASSP), Shanghai, China, 20–25 March 2016; pp. 2304–2308.
25. Ji, S.; Yang, M.; Yu, K. 3D convolutional neural networks for human action recognition. *IEEE Trans. Pattern Anal. Mach. Intell.* **2013**, *35*, 221–231. [CrossRef]
26. Goldschen, A.J.; Garcia, O.N.; Petajan, E.D. Continuous automatic speech recognition by lipreading. In *Motion-Based Recognition*; Springer: Berlin, Germany, 1997; pp. 321–343.
27. Potamianos, G.; Graf, H.P.; Cosatto, E. An image transform approach for HMM based automatic lipreading. In Proceedings of the 1998 International Conference on Image Processing. ICIP98 (Cat. No. 98CB36269), Chicago, IL, USA, 7 October 1998; pp. 173–177.
28. Noda, K.; Yamaguchi, Y.; Nakadai, K.; Okuno, H.G.; Ogata, T. Lipreading using convolutional neural network. In Proceedings of the Fifteenth Annual Conference of the International Speech Communication Association, Singapore, 14–18 September 2014; pp. 1149–1153.
29. Chung, J.S.; Zisserman, A. Learning to lip read words by watching videos. *Comput. Vis. Image Under.* **2018**, *173*, 76–85. [CrossRef]
30. Zhang, P.; Wang, D.; Lu, H.; Wang, H.; Ruan, X. Amulet: Aggregating multi-level convolutional features for salient object detection. *arXiv* **2017**, arXiv:1708.02001.
31. Hinton, G.E.; Srivastava, N.; Krizhevsky, A.; Sutskever, I.; Salakhutdinov, R.R. Improving neural networks by preventing co-adaptation of feature detectors. *arXiv* **2012**, arXiv:1207.0580.
32. Tompson, J.; Goroshin, R.; Jain, A.; LeCun, Y.; Bregler, C. Efficient object localization using convolutional networks. In Proceedings of the 2015 IEEE Conference on Computer Vision and Pattern Recognition (CVPR), Boston, MA, USA, 7–12 June 2015; pp. 648–656.
33. Chung, J.; Gulcehre, C.; Cho, K.; Bengio, Y. Empirical evaluation of gated recurrent neural networks on sequence modeling. *arXiv* **2014**, arXiv:1412.3555.
34. Graves, A.; Fernández, S.; Gomez, F.; Schmidhuber, J. Connectionist temporal classification: Labelling unsegmented sequence data with recurrent neural networks. In Proceedings of the 23rd International Conference on Machine Learning, Pittsburgh, PA, USA, 25–29 June 2006; pp. 369–376.
35. King, D.E. Dlib-ml: A machine learning toolkit. *J. Mach. Lean. Res.* **2009**, *10*, 1755–1758.
36. Sagonas, C.; Tzimiropoulos, G.; Zafeiriou, S.; Pantic, M. 300 faces in-the-wild challenge: The first facial landmark localization challenge. In Proceedings of the 2013 IEEE International Conference on Computer Vision Workshops, Sydney, Australia, 2–8 December 2013; pp. 397–403.
37. Kingma, D.P.; Ba, J. Adam: A method for stochastic optimization. *arXiv* **2014**, arXiv:1412.6980.
38. Neti, C.; Potamianos, G.; Luettin, J.; Matthews, I.; Glotin, H.; Vergyri, D.; Sison, J.; Mashari, A.; Zhou, J. *Audio-Visual Speech Recognition*; Final Workshop 2000 Report; Center for Language and Speech Processing, The Johns Hopkins University: Baltimore, MD, USA, October 2000.
39. Xu, K.; Li, D.; Cassimatis, N.; Wang, X. LCANet: End-to-end lipreading with Cascaded Attention-CTC. In Proceedings of the 13th IEEE International Conference on Automatic Face & Gesture Recognition, Xi'an, China, 15–19 May 2018.
40. Rastogi, A.; Agarwal, R.; Gupta, V.; Dhar, J.; Bhattacharya, M. LRNeuNet: An attention based deep architecture for lipreading from multitudinous sized videos. In Proceedings of the 2019 International Conference on Computing, Power and Communication, New Delhi, India, 27–28 September 2019.
41. Qu, L.; Weber, C.; Wermter, S. LipSound: Neural mel-spectrogram reconstruction for lip reading. In Proceedings of the INTERSPEECH 2019, Graz, Austria, 15–19 September 2019; pp. 2768–2772.
42. Luo, M.; Yang, S.; Shan, S.; Chen, X.J. Pseudo-convolutional policy gradient for sequence-to-sequence lip-reading. In Proceedings of the 15th IEEE International Conference on Automatic Face and Gesture Recognition (FG 2020), Buenos Aires, Argentina, 16–20 November 2020; pp. 273–280.
43. Liu, J.; Ren, Y.; Zhao, Z.; Zhang, C.; Huai, B.; Yuan, J. FastLR. In Proceedings of the 28th ACM International Conference on Multimedia, Seattle, WA, USA, 12–16 October 2020.
44. Hlaváč, M.; Gruber, I.; Železný, M.; Karpov, A. Lipreading with LipsID. In Proceedings of the International Conference on Speech and Computer, St. Petersburgh, Russia, 7–9 October 2020; pp. 176–183.
45. Yang, C.; Wang, S.; Zhang, X.; Zhu, Y. Speaker-independent lipreading with limited data. In Proceedings of the 2020 IEEE International Conference on Image Processing (ICIP), Abu Dhabi, United Arab Emirates, 25–28 October 2020; pp. 2181–2185.
46. Chen, W.; Tan, X.; Xia, Y.; Qin, T.; Wang, Y.; Liu, T.-Y. DualLip: A system for joint lip reading and generation. In Proceedings of the 28th ACM International Conference on Multimedia, Seattle, WA, USA, 12–16 October 2020; pp. 1985–1993.
47. Zhang, T.; He, L.; Li, X.; Feng, G. Efficient end-to-end sentence level lipreading with temporal convolutional network. *Appl. Sci.* **2021**, *11*, 6975. [CrossRef]

sensors

MDPI

Article

Beyond the Edge: Markerless Pose Estimation of Speech Articulators from Ultrasound and Camera Images Using DeepLabCut

Alan Wrench [1,2,*] and Jonathan Balch-Tomes [2]

[1] Clinical Audiology, Speech and Language Research Centre, Queen Margaret University, Musselburgh EH21 6UU, UK
[2] Articulate Instruments Ltd., Musselburgh EH21 6UU, UK; jbalchtomes@articulateinstruments.com
* Correspondence: awrench@articulateinstruments.com; Tel.: +44-131-474-0000

Abstract: Automatic feature extraction from images of speech articulators is currently achieved by detecting edges. Here, we investigate the use of pose estimation deep neural nets with transfer learning to perform markerless estimation of speech articulator keypoints using only a few hundred hand-labelled images as training input. Midsagittal ultrasound images of the tongue, jaw, and hyoid and camera images of the lips were hand-labelled with keypoints, trained using DeepLabCut and evaluated on unseen speakers and systems. Tongue surface contours interpolated from estimated and hand-labelled keypoints produced an average mean sum of distances (MSD) of 0.93, s.d. 0.46 mm, compared with 0.96, s.d. 0.39 mm, for two human labellers, and 2.3, s.d. 1.5 mm, for the best performing edge detection algorithm. A pilot set of simultaneous electromagnetic articulography (EMA) and ultrasound recordings demonstrated partial correlation among three physical sensor positions and the corresponding estimated keypoints and requires further investigation. The accuracy of the estimating lip aperture from a camera video was high, with a mean MSD of 0.70, s.d. 0.56 mm compared with 0.57, s.d. 0.48 mm for two human labellers. DeepLabCut was found to be a fast, accurate and fully automatic method of providing unique kinematic data for tongue, hyoid, jaw, and lips.

Keywords: multimodal speech; lip reading; ultrasound tongue imaging; pose estimation; speech kinematics; keypoints; landmarks

Citation: Wrench, A.; Balch-Tomes, J. Beyond the Edge: Markerless Pose Estimation of Speech Articulators from Ultrasound and Camera Images Using DeepLabCut. *Sensors* **2022**, *22*, 1133. https://doi.org/10.3390/s22031133

Academic Editors: Bruce Denby, Tamás Gábor Csapó and Michael Wand

Received: 29 November 2021
Accepted: 28 January 2022
Published: 2 February 2022

1. Introduction

In speech science, kinematic analysis of speech articulators is a key methodology in the quantification of speech production [1]. It can be used to relate movement to muscle activation and the timing of neural control signals. Biomechanical engineers can evaluate their models, sociophoneticians can quantify changes in articulatory gestures, clinical phoneticians can assess progress after intervention for speech disorders, and speech technologists can use the objective measures as input for silent speech recognition or lip-reading.

Electromagnetic articulography (EMA) is an important method for measuring the kinematics of speech articulators in 3D space. It has an advantage over image-based techniques because it generates movement coordinates of keypoints on articulators, such as tongue tip, blade and dorsum, lips, and jaw. It is the preferred technique for kinematic speech studies and, since the decommissioning of X-ray microbeam facilities, unique in providing intraoral keypoint data. Limitations on where the 2 mm $\times$ 3 mm electromagnetic sensors can be attached means that movement of the posterior tongue surface and hyoid cannot be monitored.

Ultrasound tongue imaging and camera video of the lips and face are instrumental techniques within the budget of most speech laboratories and have become popular as a source of articulatory speech data. They are non-invasive, convenient, and suitable for field work. Dynamic MRI of the vocal tract is another fast-evolving imaging technique with the important ability to image all the structures in the vocal tract although with significant disadvantages in cost, access, temporal/spatial resolution, unnatural recording conditions and severe acoustic noise. All of the aforementioned imaging techniques provide data, which must be postprocessed to extract measurable dynamic and static features of the vocal tract. Postprocessing of speech articulator image data has almost exclusively taken the form of edge detection or boundary segmentation. Accurate boundaries are useful for estimating vocal tract area functions but not so useful for measuring kinematics of articulators. It is also the case that edge detectors are sometimes fooled by imaging artefacts.

Recent advances in computer vision and machine learning offer an alternative approach, learning the mapping between an entire articulatory image and keypoints, labelled by experts, which need not be related to an edge or boundary. This paper investigates the potential of such pose estimation deep neural nets. We show that pose estimation can estimate the position and shape of articulatory structures within an image to an accuracy matching that of a human labeller. The movement of estimated keypoints partially correlates with that of EMA sensors. Further, more rigorous investigation is required to establish the limits of pose estimation in this regard.

1.1. State-of-the-Art in Ultrasound Tongue Contour Estimation

In order to determine which edge detection methods to compare with pose estimation we will review the state-of-the-art. Early attempts to extract a tongue surface contour from a midsagittal ultrasound image of the oral cavity were based on active contours (aka snakes) [2]. The most frequently referenced technique is *EdgeTrak* [3], where a spline with up to 100 control points is iteratively attracted to contiguous edge features in the image. The technique must be "seeded" with a contour close to the desired edge. To avoid the need to seed by hand every frame in a movie sequence, it is common to hand-label the first image and proceed through the movie by seeding each following frame with the estimated position of the contour in the preceding frame. This process leads to a tendency for the estimated contour to drift away from the tongue contour over time and become longer or shorter [4]. This approach is also, by design, bound to find an "edge" (continuous line where pixels are brighter above than below or vice versa). It cannot estimate the position of the tongue where there is no edge. *SLURP* [5] forms the most recent and successful development of the active contour approach. It incorporates a particle filter to generate multiple tongue configuration hypotheses. These hypotheses are used as seeding for the active contour to avoid the problem of drift. It also employs an active shape model, trained on a small number of tongue contour samples, for the purpose of constraining the shape and iteratively driving the snake optimization.

Machine learning was first used to estimate ultrasound tongue contours by Fasel and Berry [6]. They report a mean sum of distances (MSD, see Appendix A) accuracy of 0.7 ± 0.02 mm for their deep belief network, *AutoTrace*, which is remarkable given there were only 646 inputs to the network, meaning each image was resized to 19×34 pixels. The high accuracy score can be explained by the holdout method commonly used for testing network performance whereby a small percentage of images are randomly selected from the same dataset used for training and isolated for testing. Due to the slow rate of change of tongue movement with respect to sampling frequency, many test images are therefore almost identical to images seen in the training set. This 'holdout' method of testing produces accuracy scores that are not representative of how the estimation network would perform on data from unseen speakers and recording conditions. This was demonstrated by Fabre et al. [7] who showed their MSD accuracy of 1.9 mm diminished to 4.1 mm when no image frames from a test speaker were used in training, even when the recording conditions (ultrasound model, probe geometry, depth, field of view and contrast) were the same. Fasel and Berry [6] used a 20% holdout. Xu et al. [8] used a holdout of 8% of their hand-labelled frames reporting a MSD accuracy of 0.4 mm. More recent work by Mozzaffari [9] used a 5% holdout. Akgul and Aslan [10] used a 44% holdout and reported an MSD of 0.28 mm.

Many previous attempts to use deep networks for tongue contour estimation (*BowNet* [9], *MTracker* [11], and *DeepEdge* [12]) have adopted the U-net architecture [13] or U-net-like architecture (*IrisNet* [9]). U-Net is a convolutional neural network, developed for biomedical image segmentation with an architecture designed to work with a few thousand training images. This network classifies pixels with a probability of them belonging to a learned boundary. *TongueNet* [9] constitutes the only previous report on using a network for landmark feature identification as opposed to image segmentation. The authors indicate that about 10 keypoints along the tongue surface is optimal for accurate performance (see Table 1).

AutoTrace [6] was re-evaluated along with *TongueTrack* [14] and *SLURP* by Laporte [5]. The MSD accuracy scores are summarised in Table 1.

Table 1. Mean sum of distances (MSD) error scores reported in the literature for estimated vs. hand traced contours quoted by authors for speaker independent tests.

Algorithm	MSD (Mean/s.d.) mm
EdgeTrak [1]	6.8/3.9
SLURP [1]	1.7/1.1
TongueTrack [1]	3.5/1.5
AutoTrace [1,2]	2.6/1.2
DeepEdge (NN + Snake) [3]	1.4/1.4
MTracker [4]	1.4/0.7
BowNet [5]	3.9/-
TongueNet [5]	3.1/-
IrisNet [5]	2.7/-
Human-human	0.9 [6]/-, 1.3 [7]

[1] MSD values taken from Laporte et al. [5]. [2] when trained on the first 1000 frames of the test set. [3] MSD values taken from Chen et al. [12] 5.7 pixels at 0.25 mm/pixel. [4] MSD values taken from Zhu et al. [11]. [5] MSD values taken from Mozzafari et al. [9]. Value in mm estimated based on 128 × 128 images of 80 mm depth = 0.638 mm/pixel. Trained and tested on the same dataset with 5% test holdout. [6] Reported MSD between two hand-labellers Jaumard-Hakoun et al. [15]. [7] Reported RMSE standard deviation of 7 labellers Csapo and Lulich [4].

From the contour estimation algorithms listed in Table 1, *SLURP*, *DeepEdge*, and *MTracker* report the best performance with MSD values of 1.7, 1.4, and 1.4 mm, respectively. The authors of these methods also provide code that can be freely downloaded. These algorithms are therefore selected for further investigation and comparison with DeepLabCut pose estimation.

1.2. Lip Contour Estimation

Estimating lip contours from video of the face has a similar history to ultrasound tongue contour estimation. Early attempts used Snakes [16] and Active Shape Models [17]. Kaucic et al. [18] used Kalman filters to track the mucosal (inner) and vermillion (outer) borders of the lips. There is a need for lip feature extraction for the speechreading/lipreading application. Since 2011, with the development of convolutional neural networks (CNNs), this approach has dominated. However, the CNN lip feature encoders form part of a larger network for speech recognition and are embedded with no means to extract the lip features.

In the speech science field, the most often referenced technique, and one currently still in use for estimating lip contours for gestural speech research, is from a 1991 PhD by Lalouche [19]. This requires the participant's lips to be coloured blue. All blue pixels are then extracted from the image by chroma key, and post-processing is carried out to estimate the mucosal and vermillion borders. In a similar approach, but without the requirement for blue lips, King and Ferragne [20] have used the *semanticseg* function in the MATLAB deep learning toolbox to extract lip boundaries and postprocessed by fitting an ellipse to the boundary shape to give an estimate of width and height of the vermillion border.

The Lalouche chroma key method cannot operate on greyscale images so evaluation of DeepLabCut is compared here only with hand-labelling.

2. Pose Estimation

Recent advances in computer vision and machine learning have dramatically improved the speed and accuracy of markerless body-pose estimation [21]. There are a growing number of freely available toolkits that apply deep neural networks to the estimation of human and animal limb and joint positions from 2D videos. These include DeepLabCut [22], DeepPoseKit [23], and SLEAP [24]. These software packages all use Google's open-source TensorFlow platform to build and deploy convolutional deep neural network models. The DeepLabCut toolkit (DLC) [21,22,25] has a broad user base and has continuing support so was selected here for evaluation of pose estimation in the speech domain.

DeepLabCut

Once installed, the Python-based DeepLabCut toolbox is run using a simple graphical user interface (GUI) requiring no programming skills. The GUI makes it easy for users to label keypoints, train the convolutional neural network, apply the resulting model to identify pixel coordinates of keypoints in images, and output them in a simple comma separated text file. The processes for training and estimating pose with DeepLabCut are outlined in Figure 1. Auxiliary tools, for visualizing and assessing the accuracy of the tracked keypoints are also available within the DLC graphical user interface. Deep learning approaches require very large amounts of labelled data for training. Large, labelled corpora are publicly available for classical problems, such as facial landmark detection and body pose estimation [26,27], but not for ultrasound tongue image contouring. It is not practical to hand-label tens of thousands of ultrasound images, but it is possible to leverage existing networks trained on large datasets in one domain, and transfer learning to a new domain using only a few hundred frames. DLC applies transfer learning from object recognition and human pose estimation. With only a small number of training images and a few hours of machine learning, the resultant network can perform to within human-level labelling accuracy [22]. Here, we evaluate that performance claim on the domains of ultrasound tongue imaging and lip camera imaging.

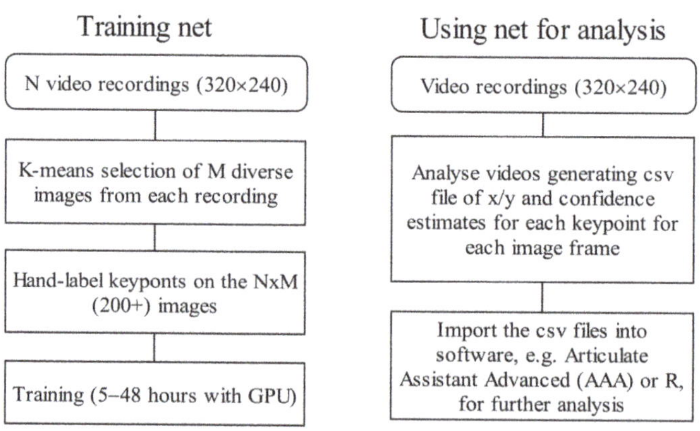

Figure 1. DeepLabCut training and analysis processes.

Deep net architectures designed for markerless pose estimations are typically composed of a backbone network (encoder), which functions as a feature extractor, integrated with body part detectors (decoders). DeepLabCut provides a choice of encoders (MobileNetV2 [28], ResNet [29], or EfficientNet [30]), all with weights pretrained on the ImageNet corpus that consists of 1.4 million images labelled according to the objects they contain. The body part detector algorithms are taken from a state-of-the art human pose estimation network called DeeperCut [31] from which it takes its name. DeeperCut is in turn trained on the Max-Planck-Institut für Informatik human pose dataset [27], consisting of 25,000 images containing over 40,000 people with annotated body joints. The "Lab" nomenclature references the ability of DeepLabCut to transfer learning from human pose estimation to other domains, such as animals or medical images, using only a few labelled images, making the process manageable for a single research laboratory.

Encoders are continuously being redesigned to improve the speed and accuracy of object recognition, and it has been shown that this improved encoder performance feeds directly through to improve pose estimation [25], particularly with respect to:

1. Shorter training times.
2. Less hand-labelled data required for training.
3. Robustness on unseen data.

For each labelled keypoint, the decoder produces a corresponding output layer whose activity represents a probability score-map (aka heat-map). These score-maps represent the probability that a body part is at a particular pixel [31]. During training, a score-map is generated with unity probability for pixels within a 'pos_dist_threshold' (default = 17) pixel radius of the labelled keypoint pixel and zero elsewhere. Mathis et al. recommend a 17-pixel distance threshold after experimenting on different threshold values for a 1062 $\times$ 720-pixel resolution video input.

It is possible to use features of the score-maps such as the amplitude and width, or heuristics such as the continuity of body part trajectories, to identify images for which the decoder might make large errors. Images with insufficient automatic labelling performance that are identified in this way can then be manually labelled to increase the training set and iteratively improve the feature detectors.

DeepLabCut can use deep, residual networks, with either 50, 101, or 152 layers (ResNet). The optional MobileNetV2 is faster for both training and analysis and make analysis with CPU (as opposed to GPU) feasible. EfficientNet encoders are also available.

Labelling the training set with multiple related anatomical keypoints improves the accuracy of individual keypoint estimates. Mathis [22] shows a network, trained with all body part labels simultaneously, outperforms networks trained on only one or two body parts by nearly twofold. DeepLabCut applies image augmentation to artificially expand the training set by modifying the base set with images transformed by scaling, rotating, mirroring, contrast equalization, etc. In this paper, only scaling and rotation were applied.

3. Materials and Methods

3.1. Ultrasound Data Preparation

3.1.1. Training Data

Ultrasound images were downsampled to fit in an image of 320 $\times$ 240 pixels. Where the original image was not 4:3 aspect ratio, it was letterboxed to lie centrally, and a black background added. The images were encoded using H.264 (greyscale, rate factor 23, zero latency, and YUV_4_2_0 palate) to provide a compact data storage with minimal loss. The original images had vertical heights of 80, 90, or 100 mm, and after letterboxing, the output images had vertical heights of 100–125 mm, leading to a pixel resolution of approximately 0.4–0.5 smm per pixel. It is worth noting that the axial resolution in mm of a 3 MHz 3-cycle ultrasonic pulse is $3 \times 0.5 \times 1540/3{,}000{,}000 \times 1000 = 0.77$ mm so our 320 $\times$ 240 image has better resolution than the underlying data. Preliminary tests indicated that, compared to a 320 $\times$ 240 video, a 800 $\times$ 600 video took 5$\times$ longer to analyse and a 1200 $\times$ 900 video took 12$\times$ longer. Therefore, 320 $\times$ 240 was determined to be a practical resolution.

The tongue contour may be partly obscured by mandible or hyoid shadows or otherwise indistinct. The labeller then has a choice either to omit keypoints in these regions or to estimate their position based on clues elsewhere in the image or audio. For this paper, we mainly adopted the latter approach.

Hand-labelling was carried out on 20 frames each, from 26 recordings. The frames were selected by k-means clustering using the DLC labelling tool so that they were distinct from each other. A few frames were rejected if they had no discernible features leaving a total of 520 test frames. The recordings comprised:

- A total of 10 recordings from 6 TaL Corpus [32] adult speakers (Micro system, 90° FOV, 64-element 3 MHz, 20 mm radius convex depth 80, 81 fps). These recordings were the first few recordings from the corpus and not specially selected.
- A total of 4 recordings from 4 UltraSuite corpus [33] Ultrax typically developing children (Ultrasonix RP system, 135° FOV, 128 element 5 MHz 10 mm radius microconvex, depth 80 mm, 121 fps). These were randomly selected. 10 recordings of the authors, using the Micro system with 64-element, 20 mm radius convex probe, and with different field of view and contrast settings
- A total of 2 recordings by Strycharczuk et al. [34] using an EchoB system with a 128-element, 20 mm radius convex probe. These data are from an ultrasound machine not represented in the test set and included to generalize the model.

3.1.2. Test Data

Hand-labelling was carried out on 40 k-means selected frames from 25 recordings using the DLC labelling tool to generate a total of 1000 test frames. Each recording was from a different speaker and were taken from several publicly available corpora:

- A total of 10 TaL corpus adult speakers (Articulate Instruments Micro system, 90 FOV, 64-element 3 MHz, 20 mm radius convex depth 80, 81 fps).
- A total of 6 UltraSuite Ultrax typically developing children (Ultrasonix RP system, 135° FOV, 128 element 5 MHz 10 mm radius microconvex, depth 80 mm, 121 fps).
- A total of 2 UltraSuite Ultrax speech sound disordered children (recorded as previous).
- A total of 2 UltraSuite UltraPhonix children with speech sound disorders (SSD) (recorded as previous).
- A total of 2 UltraSuite children with cleft palate repair. Ultrasound (Articulate instruments Micro system, 133° FOV, 64-element 5 MHz, 10 mm radius microconvex, depth 90, 91 fps.
- A total of 3 UltraspeechDataset2017 [35] adults. Ultrasound images (Terason t3000 system, 140° FOV, 128-element, 3–5 MHz 15 mm radius microconvex, depth 70 mm, 60 fps).

None of the test speakers were used in the training set. The hand-labelling was conducted by the first author with the same protocol used to train the DLC model (see Section 3.1.3). The second author also hand-labelled 25% of the same test frames (every fourth frame) for the purpose of comparing hand-labelling similarity.

3.1.3. Ultrasound Keypoint Labelling

Eleven points were selected along the upper surface of the tongue: vallecula, root1, root2, back1, back2, dorsum1, dorsum2, blade1, blade2, tip1, tip2. This number is sufficient to describe the shape of the surface. Separation between the consecutive blade and tip points were approximately half that of other points in order to better represent the flexibility of that part of the tongue. An attempt was made to maintain consistency in placement relative to the tongue surface, even when these points were obscured by hyoid or mandible shadow. This approach to labelling differs from traditional labelling, which is limited in length to the extent of the bright edge visible in the image. In addition, keypoints were labelled on the hyoid and on the mandible at its base and at the mental spine where the short tendon attaches. The latter point is important as it forms the insertion point for the fanned genioglossus muscle fibres. These fibres principally control the midsagittal shape of the tongue body. Figure 2 shows the location of the labelled keypoints.

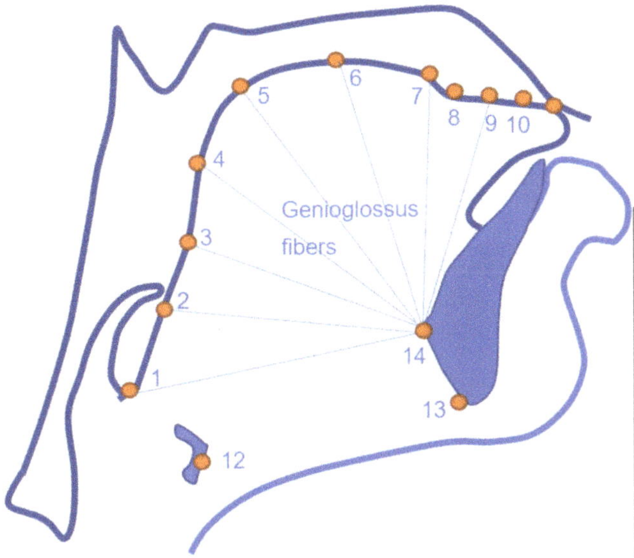

1	Vallecula
2	Root1
3	Root2
4	Body1
5	Body2
6	Dorsum1
7	Dorsum2
8	Blade1
9	Blade2
10	Tip1
11	Tip2
12	Hyoid
13	Mandible
14	Short tendon

Figure 2. Outline of midsagittal tongue contour and the labelled keypoints.

The bright surface of the epiglottis and any saliva bridge between the tip of the epiglottis and the tongue root is often traced as part of the tongue (see Figure 3A for an example). This may be an appropriate contour to trace if the boundary of the oral cavity is being assessed, but for studies investigating tongue root retraction and for the sake of consistently modelling the tongue surface, in this study, we elected to follow the surface of the tongue to the vallecula rather than the visible surface of the epiglottis.

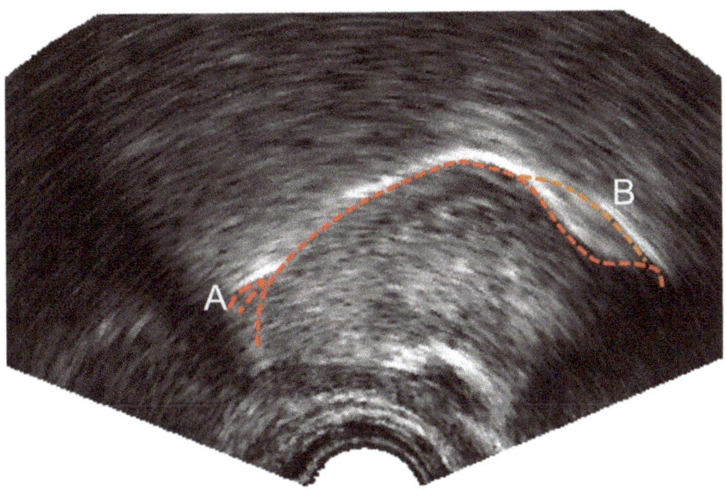

Figure 3. Ultrasound image showing (**A**) bright reflection from tip of epiglottis (**B**) double reflection parasagittal surface (upper) and midsagittal surface (lower) of the tongue blade.

It is sometimes the case that there are two apparent edges (e.g., Figure 3B). This most often occurs at the tongue blade when it is grooved to produce an (s) sound. In this study, we hand-labelled the lower of the two edges even when it was less distinct, as this represents the contour of the midline of the tongue.

Where possible, the tongue tip position was estimated even when there was no bright contour. In particular, the bright artefact often generated by the tip raising gesture was not labelled as part of the tongue contour. This bright artefact is due to the ultrasound beam reflecting off the surface of the tongue to the underside of the blade and back along the same path (Figure 4). Again, this means that the hand labels do not strictly follow the brightest edge.

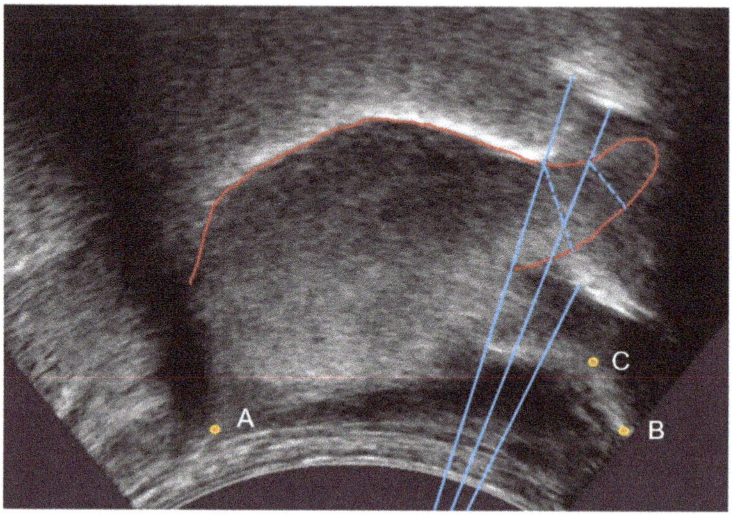

Figure 4. Ultrasound image with beam tracing (blue) showing actual path of ultrasound beam (dotted) and the resulting bright artefacts based on the equivalent time of travel in the direction of the transmitted beam (solid). A—hyoid; B—mandible base; C—short tendon base.

Hyoid, mandible base, and short tendon base are indicated in Figure 4 as points A, B, and C, respectively.

3.2. *EMA-Ultrasound Test Data*

Simultaneously recorded ultrasound and EMA data are rare. Some pilot data generously made available by Manchester and Lancaster Universities (UKRI grant AH/S011900/1) were used here to evaluate the ability of DLC output to emulate EMA sensor movement. EMA data were recorded using the Carstens 501 system (Carstens Medizinelektronik GmbH, Bovenden, Germany) with three coils placed on the tongue tip (1 cm from apex), tongue blade and dorsum (approximately 15 mm separation between each sensor). The corpus consisted of three carrier phases "She said X clearly", "She said X", and "She said X again". Unfortunately, in many of the ultrasound recordings there was loss of tip information as the probe failed to make contact with the chin. As a result of the restricted vocabulary and missing tip images, only five recordings were used to evaluate the ability of DeepLabCut to estimate EMA sensor movement. These were all the phrase "She said X clearly" with X = Bide, Bart, Bore, Bead, Bee.

3.3. *Lip Camera Data Preparation*

3.3.1. Training Data

The TaL sample corpus was used for training. A total of 24 recordings, one from each of 24 speakers, were selected at random. Moreover, 7 or 8 frames were selected by k-means clustering from each recording, providing 207 training frames.

3.3.2. Test Data

Testing was carried out on 10 TaL corpus speakers who were not included in the training set. A total of 40 frames were labelled from each speaker, providing 400 test frames. These speakers were selected to represent a range of age, sex, ethnicity, and facial hair. A second labeller re-labelled every fifth hand-labelled frame (20% of the test set).

3.3.3. Lip Keypoint Labelling

The lip video was taken from the TaL corpus [32], which uses a front facing camera. This presents the opportunity to label the commissures (corners), midsagittal point on the upper and lower mucosal boundary, and points midway between the commissures and midpoints, as indicated in Figure 5.

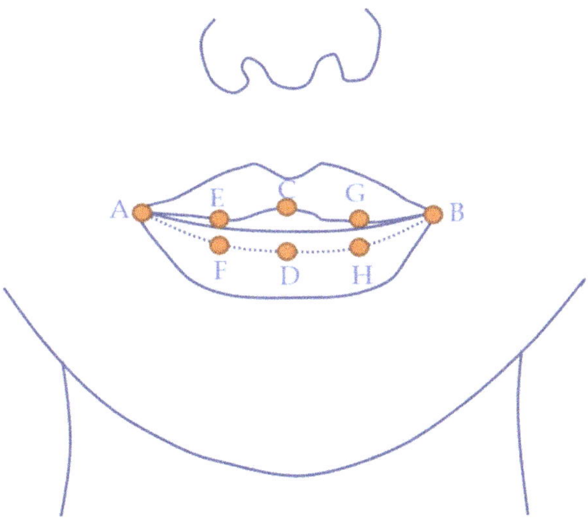

Figure 5. Keypoints labelled on the lip video. Dotted line indicates mucosal border on the lower lip that makes contact with the upper lip when the mouth is closed. A and B—commissures; C and D—centre of the upper and lower lip, respectively (defined by philtrum midpoint and not necessarily equidistant from A and B); E—equidistant from A and C; F—equidistant from A and D; G—equidistant from B and C; H—equidistant from B and D.

3.4. Accuracy Measures

To assess the accuracy of an estimated tongue contour, Jaumard-Hakoun et al. [15] proposed a mean sum of distances (MSD). For every point on the hand-labelled curve, the distance to the closest point on the estimated curve was calculated, and for every point on the estimated curve, the distance to the closest point on the hand-labelled curve was calculated. The total sum of these distances was then divided by the total number of distance measures. This per-frame mean sum of distances was then averaged across all frames to give a single score. This curve similarity measure can give an identical score for the case where two curves match perfectly but one is longer than the other, and the case where the curves are the same length but do not match. We chose to separate these two factors. We used the standard MSD calculation, but when considering the endpoints of spline B, only one distance to a point on Spline A was included in the calculation: the shortest one. This means that the MSD is not affected by the relative lengths of the splines. Instead, we report a separate spline length difference measure (length hand-labelled spline—length estimated spline). Each spline was cubically interpolated prior to performing this MSD calculation so that it had 100 points regardless of its length.

DeepLabCut includes a root mean square error (RMSE) for the distance between hand-labelled points and estimated points. This measure is only applicable to DeepLabCut, since it is the only point estimation algorithm being considered.

For lip analysis, the area bounded by the upper and lower lip contours was measured in mm^2. The width between commissures was measured. The MSD of the upper and lower lips were reported separately.

For comparison between EMA tongue sensors and the estimated tongue tip, blade, and dorsum keypoints, a Pearson correlation coefficient was calculated.

3.5. *Ultrasound Tongue Contour Estimation Methods*

The configuration of DeepLabCut, *SLURP*, *MTracker*, and *DeepEdge* are described in the following sections and Table 2 summarises their training and analysis rates.

Table 2. Comparison of ultrasound contour tracking algorithms showing the analysis frame rate, image size, training frame rate, and time for the network training to converge.

Algorithm	Frames per Second [1] (GPU/CPU)	Image Size	Training Data/Time (Frames/Hours)
SLURP [2,3]	NA/8.5	data	N/A
DeepEdge (NN + Snake)	2.7/NA	64 × 64	2700/2
DeepEdge (NN only)	3.0/NA	64 × 64	2700/2
MTracker	27/NA	128 × 128	35,160/2
DeepLabCut (MobNetV2_1.0)	287/7.3 [4]	320 × 240	520/7.5
DeepLabCut (ResNet50)	157/4.0 [4]	320 × 240	520/16
DeepLabCut (ResNet101)	105/2.6 [4]	320 × 240	520/30
DeepLabCut (EfficientNet B6)	27/1.7 [4]	320 × 240	520/48

[1] Using Windows laptop PC with Core i7-10750H 16GB RAM and NVIDIA RTX 2060 MaxQ. [2] SLURP is the only algorithm tested here that does not use the NVIDIA GPU. [3] SLURP requires the first frame of each recording to be manually seeded with at least 6 points using GetContours MATLAB GUI. The timing recorded here excludes this manual labelling step. [4] Analysing using batch size 8.

3.5.1. DLC Ultrasound

For body tongue/hyoid/mandible inference, we used DeepLabCut (version 2.1.10.0) [22,36]. We used a MobileNetV2-1.0 [25] based neural network with default parameters *. We also compared this with ResNet50, ResNet101, and EfficientNetB6 [25]. A total of 0.8 million iterations were used for training after preliminary testing (see Appendix B) showed convergence occurred with this amount of training. We validated with one held-out folder of the 1000 hand-labelled test frames. The image size was 320 × 240; ~0.5 mm/pixel. We then used a p-cut-off of 0.6 to determine root mean square error scores. This network was then used to analyse each of the test videos generating csv files of keypoints with associated confidence values, which were imported into the AAA software package (version 219_06, 2021, Articulate Instruments Ltd., Musselburgh, UK). The 11 tongue keypoints were converted into a single cubic tongue spline with 11 control points. The pixel to mm scale was calculated separately for each recording. MSD and length measures in millimetres were then made with respect to the hand-labelled keypoints similarly imported.

* ImgAug with ±25° rotation and random scaling in the range 0.5–1.25 (40% of the original dataset); pos_dist_threshold of 17.

3.5.2. SLURP

The GetContours GUI [37] implemented in MATLAB was used to run the *SLURP* edge detection function. *SLURP* employs tongue-shape models but does not provide tools for in-domain training. Retraining the shape models on the training data used in this study was thought unlikely to make a substantial difference to the performance. Two different shape models provided by the author were tested and the one that gave the best results was selected. Increasing the minimum number of particles did not substantially improve the performance. The resulting reduction in the analysis rate was therefore not justified and the default settings were used:

Colormap = "gray", Sigma = 5.0, Delta = 2.0, Band Penalty = 2.0, Alpha = 0.80, Lambda = 0.95, Adaptive Sampling = Enabled, Particles = Min 10, Max 1000.

Each frame was seeded by hand with a 15-point spline. The tracker ran at 8.5 fps producing 100 edge contour points. These spline points were downsampled to 50 by removing every alternate point and imported into AAA software for MSD and length analysis.

3.5.3. MTracker

A region of interest was defined as the area between the coordinates [50,50] and [200,300] relative to the top-left of the image. Dense U-Net model "dense_aug.hdf5" was used. This model has 50% of the training data with image augmentation. The tracker ran at 27 fps producing 100 points or fewer when the confidence threshold of 50% was not reached. These spline points were downsampled to 50 by removing every alternate point when imported into the AAA software for MSD and length analysis.

3.5.4. DeepEdge

DeepEdge version 1.5 ran under MATLAB R2021a with deep learning toolbox, image processing toolbox, and computer vision toolbox. Three optional models are provided, each trained on different datasets. A model trained on the same ultrasound system and probe used in our 6 Ultrax TD child test recordings was tried first. However, this model performed more poorly on the test set than another of the models. The best performing model was ("DpEdg_CGM-OPUS5100_CLA651_21JUL2021"), and this was the model used for this study. All videos were mirrored, such that the tongue was pointing to the left, then after running *DeepEdge* and exporting the data, the results were then mirrored again to face tongue tip right before importing into AAA. The tracker ran at 3 fps producing 20 edge contour points. These contour points were imported into AAA software for MSD and length analysis.

3.6. Method for Comparing EMA Position Sensors to DLC Keypoints

The ultrasound data were analysed using the DLC ResNet50 model trained as per Section 3.5.1. The estimated tip1, blade1, and dorsum1 keypoints were picked as close matches to the tip, blade, and dorsum EMA sensors. Sections of the five recordings corresponding to the spoken utterances were selected. The sensor positions were compared to the estimated keypoints for every ultrasound frame timepoint, and Pearson correlation values recorded.

3.7. Method for Evaluating DLC Performance on Lip Camera Data

The same DeepLabCut configuration used for ultrasound images was used to train lip images. Only MobileNetV2_1.0 and ResNet50 encoders were tested. Keypoints were imported into the AAA analysis package (version 219_06, 2021, Articulate Instruments Ltd., Musselburgh, UK) as an upper lip spline and lower lip spline. Both splines shared the commissure keypoints as endpoints. MSD values were calculated for the upper lip and lower lip separately. A width (distance between commissure keypoints) and aperture (area enclosed by the upper and lower lip splines) were also recorded, as these are measures that speech scientists are interested in.

4. Results

4.1. Ultrasound Contour Tracking

We evaluated DLC with the MobileNetV2_1.0 encoder by training on 100% (twice), 75%, 50%, and 25% of the 520 hand-labelled frames. A small difference in MSD scores occurred between models generated in two separate training runs with 100% of the training data. This is likely due to the random selection of frames for image augmentation and the random amounts of augmented scaling and rotation. Table 3 shows that, compared to using 100% of the data, using 75% (390 frames) did not reduce performance significantly. ResNet50 backbone also produced no significant difference when using 75% compared to 100% of the training data. Using 50% of the total available hand-labelled frames, i.e., 260 hand-labelled frames with distinct tongue shapes extracted from 26 recordings, gave marginally poorer performance ($p = 0.03$). Performance was reduced when the number of frames used to transfer learning from human pose estimation to ultrasound tongue images was limited to 130. For this paper, we used models trained on all 520 hand-labelled frames. While it is possible that generalization to a very different scanner and probe might require the model to be retrained with supplementary frames from that domain, very few additional images would be needed. Certainly, no more than 260 and likely less than 50. The test set used here included recordings from a Terason scanner and probe unseen in the training set and performed very well (mean MSD 1.00 SD 0.30 for speaker TH c.f. mean MSD 1.06 SD 0.71 for all 25 speakers.

Table 3. Error scores vs. hand contoured for 20, 15, 10, and 5 frames per recording used for training.

MobileNetV2 Training Data	MSD (Mean, s.d., Median)	MSD p Value [1]	%Length Diff (Mean, s.d., Median)
conf 80% 520 frames	1.06, 0.59, 0.90	1.00	+1.8, 7.0, +2.0
	1.06, 0.71, 0.89	0.89	+1.8, 9.1, +1.8
conf 80% 390 frames	1.12, 0.86, 0.91	0.09	+2.8, 10.7, +2.5
conf 80% 260 frames	1.13, 0.71, 0.94	0.03	+1.9, 9.7, +1.7
conf 80% 130 frames	1.17, 0.79, 0.94	<0.001	+3.5, 8.1, +3.2

[1] Two tailed t-test assuming equal variance with reference to the MSD data generated by the model corresponding to the first row MSD distribution.

MSD mean and standard deviation values reported in Table 4 show *SLURP*, *MTracker*, and *DeepEdge* all performed less well on the test set used for this study than previously reported (1.7, 1.1 c.f. 2.3, 1.5) (1.4, 0.7 c.f. 3.2, 5.8) (1.4, 1.4 c.f. 2.7, 3.1). DeepLabCut still performed better than the originally reported MSD values for these other methods. 0.9 mm vs. 1.4–1.7 mm.

Table 4. Error scores vs. hand contoured (including regions where hand labels had to be guessed at tip and vallecula.

Algorithm	MSD (Mean, s.d., Median)	MSD p Values [1]	%Length Diff (Mean, s.d., Median)
SLURP	2.3, 1.5, 1.9	<0.001	−3.8, 14.4, −4.6
DeepEdge (NN only)	2.8, 3.1, 1.9	<0.001	−27.5, 25.3, −26.0
MTracker	3.2, 5.8, 1.5	<0.001	−49.0, 28.7, −44.4
DLC (MobileNetV2_1.0 conf 80%)	1.06, 0.59, 0.90	0.04	+1.8, 7.0, +2.0
DLC (ResNet50 conf 80%)	0.93, 0.46, 0.82	0.29	+1.6, 8.8, +2.2
DLC (ResNet101 conf 80%)	0.96, 0.67, 0.81	0.80	+1.8, 9.1, +1.8
Inter-labeller	0.96, 0.39, 0.88	1.0	−4.3, 6.2, −4.8

[1] Two tailed t-test assuming unequal variance with reference to the MSD data generated by the inter-labelling.

The quality of the ultrasound images may have been poorer in this test set than the original *SLURP*, *MTracker*, and *DeepEdge* studies, partly explaining the reduction in performance. It may also be the case that the training and test sets in the original studies were closely matched and the trackers have a limited ability to generalise to unseen data. In particular, *DeepEdge* comes with three models, each trained on a different system rather than one general model. If *DeepEdge* were trained on the same dataset that DeepLabCut was trained on it may have performed better, but DeepEdge requires at least 4× the available hand-labelled frames to train successfully and no training software is provided. Furthermore, the hand-labelled contours, used here as ground truth, follow the tongue contour and not necessarily the brightest edge. The original studies may have been evaluated against hand-labels of the brightest edge.

Table 4 shows that DLC with a ResNet50 encoder provided MSD scores equivalent to the MSD between the two hand-labellers in this study. The inter-labeller mean of 0.96 mm is close to the inter-labeller MSD of 0.9 mm reported by Jaumard-Hakoun [15]. It also indicates that, while the second hand-labeller tended to assign tongue contours 4% shorter than the first labeller, DLC was closer in length producing contours that were on average only 2% longer. The slightly poorer performance of ResNet101 compared with ResNet50 may be due to overtraining or variance in performance vs. number of training iterations (see Appendix B).

Table 5 shows the mean root mean square error scores across all keypoints with a confidence greater than 0.6 (60%). For example, if a tongue tip keypoint is obscured by the mandible shadow, then the network might generate a low confidence in its position and this point would be ignored. Using MobileNetV2 with 520 training samples as a baseline, the RMSE pixel accuracy is shown to decrease by up to 3.5% when less training data are used and increase by up to 3% when using a ResNet encoder. Interestingly ResNets are less accurate when all keypoints are considered but more accurate when unconfident points are ignored. EfficientNetB6 performed poorly, perhaps because the amount of training data were insufficient for such a large encoder network.

Table 5. Root mean square error scores on test set and times for training.

Network	RMSE Test ($p > 0.6$) Pixels	Train Time [1] 0.8 Million Iterations	Analyse Time [1] Frames/s
DLC (MobileNetV2_1.0)	6.15	7.5 h	190
DLC (MobileNetV2_1.0)	6.17	7.5 h	190
DLC (MobileNetV2_1.0) 75%	6.28	7.5 h	190
DLC (MobileNetV2_1.0) 50%	6.39	7.5 h	190
DLC (MobileNetV2_1.0) 25%	6.38	7.5 h	190
DLC (ResNet50)	6.07	16 h	100
DLC (ResNet101)	5.99	30 h	46
DLC (EfficientNet b6)	11.55	48 h	14

[1] Time measured using a GTX 1060 GPU (slower than the GPU used for timings in Table 2).

Figure 6 shows two frames evaluated within DLC. Image (a) shows that, although the 11 tongue keypoints hug the tongue surface, producing a low MSD value, they sometimes do not match the hand-labelled locations along that surface. This leads to RMSE scores of 6 pixels (~3 mm) compared to only 1 mm for MSD. Figures 7 and 8 show how the overall results in Table 4 break down across test speakers. Speakers 01F_BL1 and PB both have very poor image quality, with DLC ignoring keypoints in some frames, resulting in shorter length estimates.

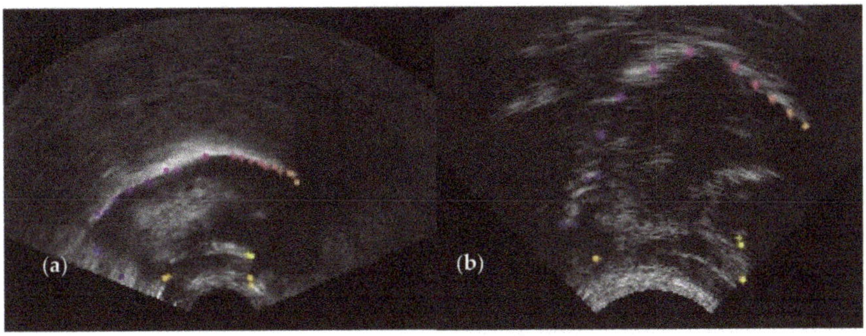

Figure 6. (**a**) Shows points estimated to lie on the tongue surface but distributed differently to the hand labels; (**b**) an example where the positions are estimated accurately. '+' indicates the estimated position.

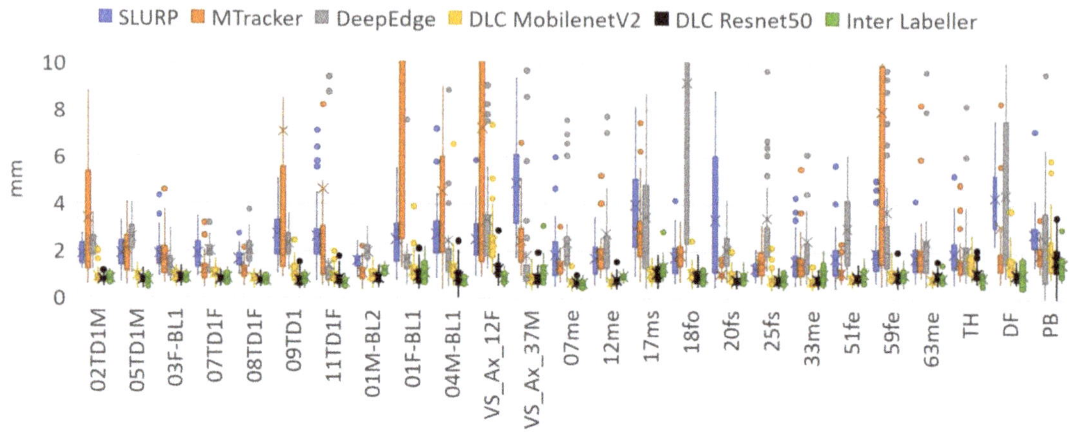

Figure 7. Mean sum of distances (MSDs) between hand-labelled randomly selected frames (40 frames for each speaker) and each of the assessed methods (DLC using MobileNetV2_1.0 encoder).

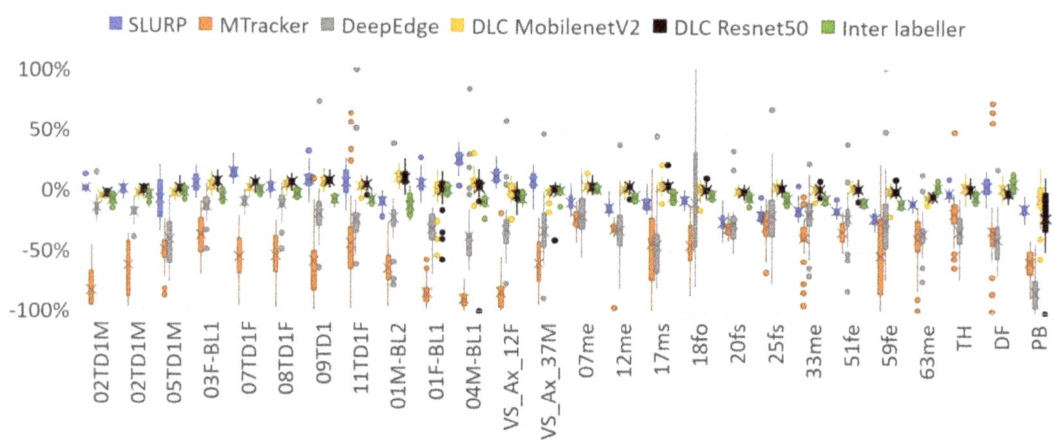

Figure 8. Relative length distance between hand contoured randomly selected frames (40 frames for each participant) and each of the assessed methods (DLC using MobileNetV2_1.0 encoder).

Figure 9 shows plots of x-axis = MSD vs. y-axis = %length difference for every test frame that generate the overall results in Table 4. An ideal estimator would have all points at (0,0) (see Appendix A for why an MSD of 0.0 is unlikely). Of the three previously reported estimators, *SLURP* is the most robust when MSD and length are considered. The tighter cluster for DLC more closely matches the inter-labeller plot.

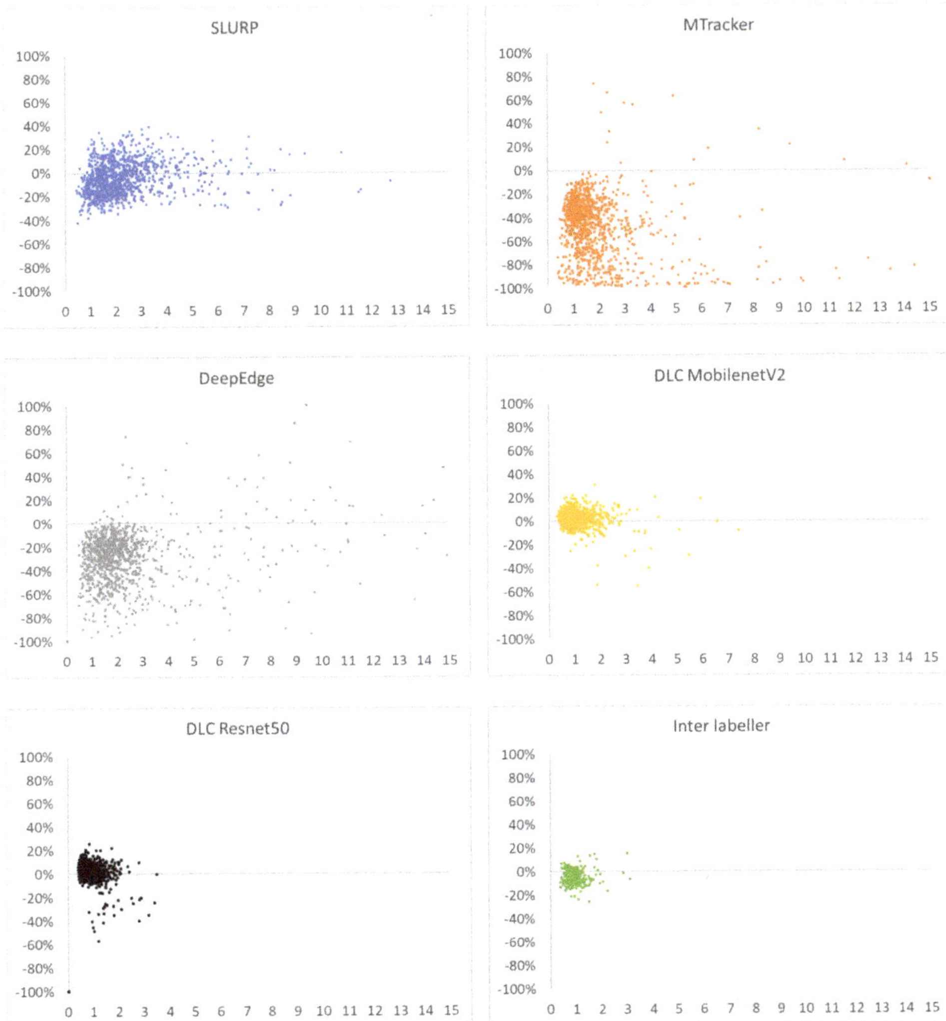

Figure 9. MSD vs. %length difference (hand-label estimator) for SLURP, MTracker, DeepEdge, DLC MobileNetV2, DLC ResNet50, and second labeller.

Figures 10–12 show every fifth hand-labelled test frame for the speakers 17ms (from the TaL corpus), 03F_BL1 (from the UltraSuite UXSSD corpus) and DF (from the UltraSpeech corpus). *SLURP* (green) has pretrained shape models, which restrict the shape of the contour. In Figure 10, a plausible tongue shape does not always match the underlying data. The flick upwards at the root of the tongue in Figure 11 may be as a result of how *SLURP*'s shape model was trained. *MTracker* (yellow) fits the tongue surface quite well, but because the length is controlled by a 50% confidence threshold, it very often omits the more difficult tip and root sections of the tongue contour. When we raised the threshold, *MTracker* performed very poorly in these regions. *DeepEdge* (pink) tended to underestimate the length. The option to postprocess by applying *EdgeTrak* to the neural net output produced poorer results, and so is not reported here. DLC ResNet50 (cyan) matches the hand-labelled contour (blue) so well that, in many frames, it sits directly on top. Disagreements mainly occur at the root where the hand-labelled contour is often speculative.

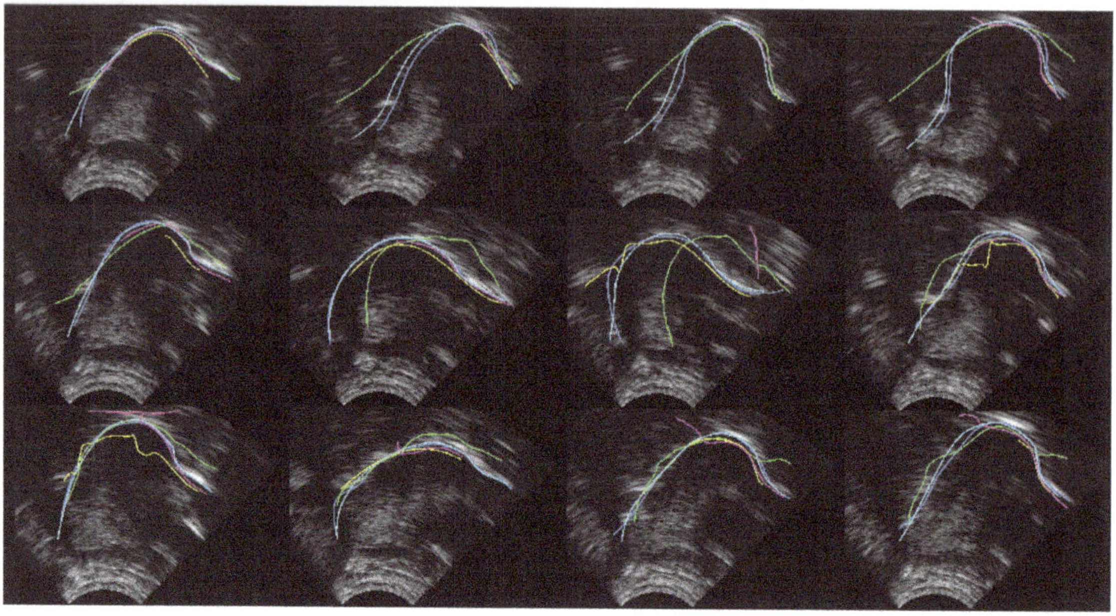

Figure 10. Speaker TaL17ms; blue—hand-label; green—SLURP; yellow—MTracker; pink—DeepEdge; cyan—DLC_ResNet50.

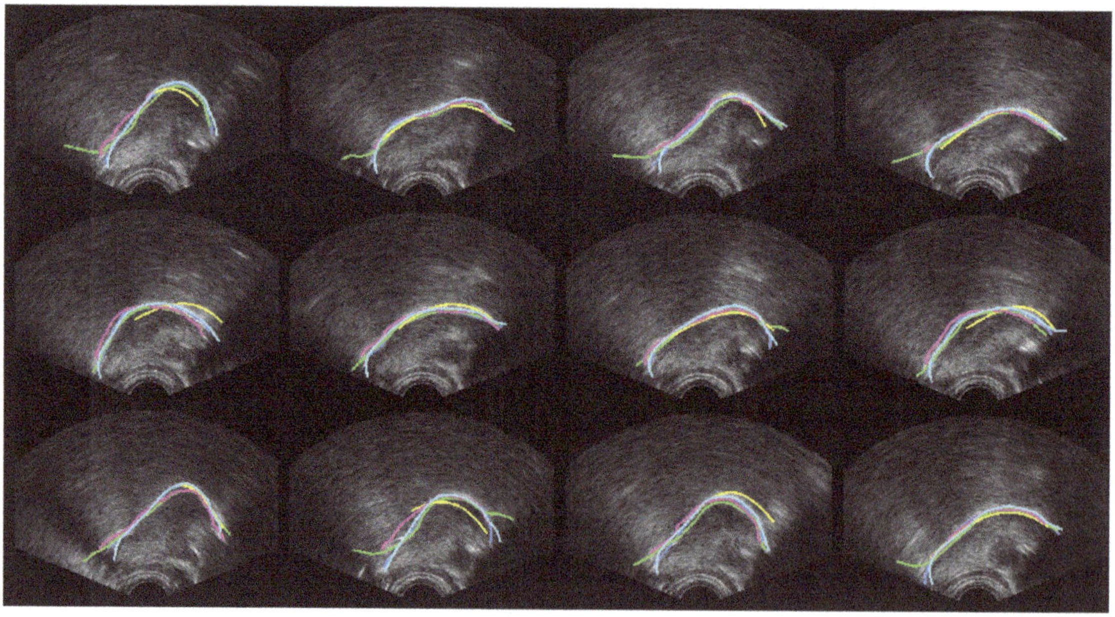

Figure 11. Speaker UXSSD03F; blue—hand-label; green—SLURP; yellow—MTracker; pink—DeepEdge; cyan—DLC_ResNet50.

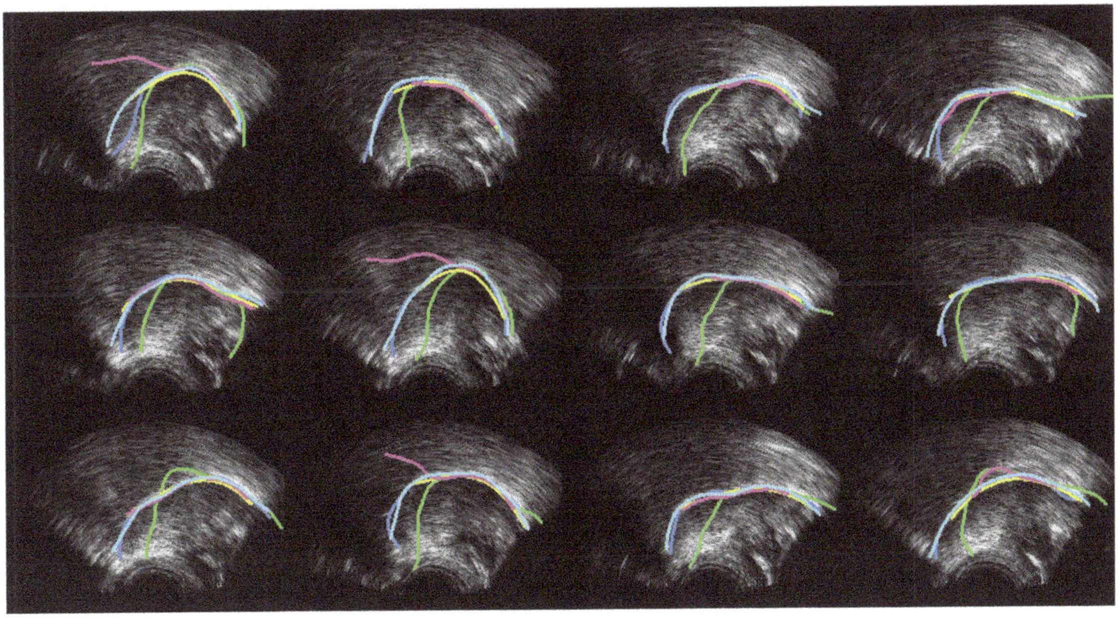

Figure 12. Speaker UltraSpeechDF; blue—hand-label; green—SLURP; yellow—MTracker; pink—DeepEdge; Cyan—DLC_ResNet50.

4.2. Ultrasound-EMA Point Tracking

The splines were scaled in mm and consisted of the 11 tongue-surface keypoints. The TT1, TBl1, and TD1 keypoints were selected as being close to the positions of the three EMA sensors on the tip blade and dorsum, respectively. The bite plane [38] was recorded in both the EMA and ultrasound data (see Figure 13) and both sets of data were rotated so that the bite plane formed the *x*-axis.

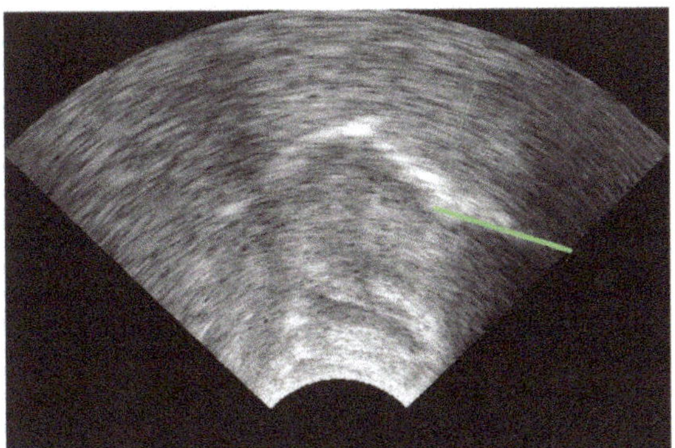

Figure 13. Image of the tongue pressed against a bite-plate and a green fiducial line superimposed. All coordinates were rotated so that the green line formed the horizontal axis.

Figure 14 shows the comparison of x and y EMA sensor positions (red) with the positions estimated by DLC ResNet50 (black) as the phrase "She said bead clearly" is spoken. It is apparent that there is very little correlation in the *x*-axis, while there is a modest correlation in the *y*-axis.

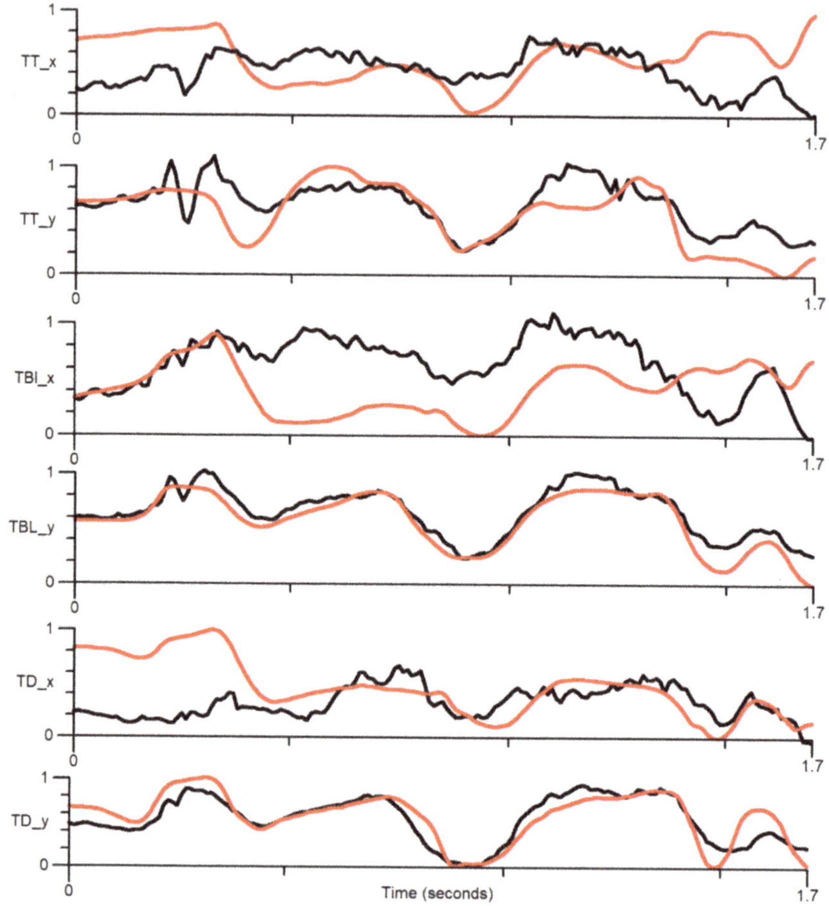

Figure 14. The phrase 'She said "bead" clearly' showing x and y position against time for the tongue tip (TT), blade (TBl), and dorsum TD. Red—EMA sensor; black—DLC estimated position. The *y*-axis has no units.

Figure 15 plots, each EMA coordinate vs. the corresponding DLC estimated the coordinate for every ultrasonic frame of the five simultaneous EMA/ultrasound recordings. Again, correlation is only good for the y-coordinates of the tip and blade. Table 6 shows that the Pearson correlation coefficients calculated across all ultrasound frames for the five recordings confirm the visual findings.

Table 6. Pearson correlation values for each sensor coordinate calculated over the five recordings.

Sensor Coordinate	Pearson Correlation Coefficient
Tongue tip x	0.37
Tongue tip y	0.88
Tongue blade x	0.39
Tongue blade y	0.93
Tongue dorsum x	−0.03
Tongue dorsum y	0.44

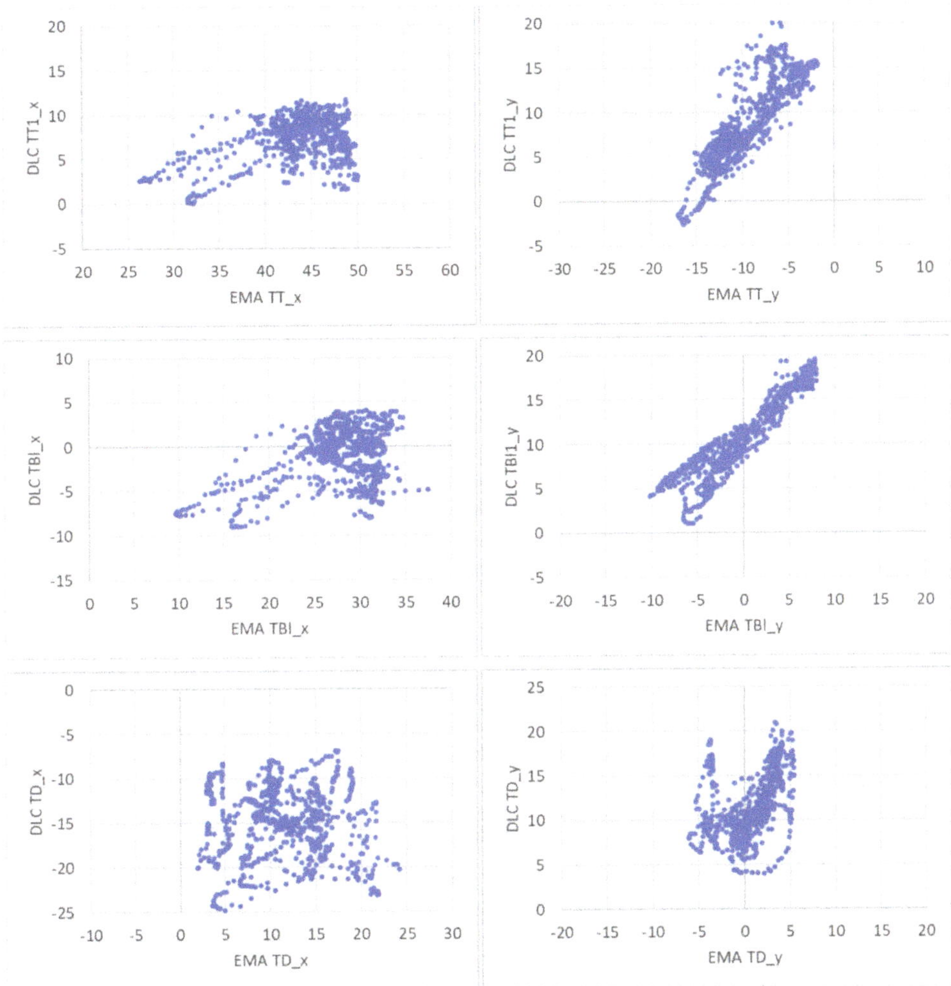

Figure 15. EMA coordinate vs. DLC estimated coordinate plotted for every ultrasonic frame of the 5 simultaneous EMA/ultrasound recordings. TT—tongue tip; TBl—tongue blade; TD—tongue dorsum.

4.3. Camera Lip Tracking

Table 7 shows RMSE scores for all lip keypoints. Unexpectedly, ResNet50 does not significantly outperform MobileNetV2. It may be that although experiments on the ultrasound images showed that 260 frames were adequate for good MobileNetV2 performance, the 207 training frames used here were insufficient for ResNet50 to reach its full potential. We used fewer training frames because the training set was homogeneous as each speaker was recorded under identical conditions.

Table 7. Root mean square error (average for all lip keypoints).

Network	RMSE Test ($p > 0.6$) Pixels
DLC (MobileNetV2_1.0)	3.79
DLC (ResNet50)	3.74

Table 8 shows a comparison of MSD for upper and lower lips, lip aperture and lip width for the two DLC encoders and the second labeller. As with RMSE, these performance indicators reveal that unlike for ultrasound images, DLC does not quite match the inter-labeller lip performance. More training data might improve the performance. Lip contouring does not follow a bright edge. Indeed, the lower lip contour labelling criterion was to mark where it would meet the upper lip rather than the visible boundary of the lip and oral cavity. As shown by the smaller average area estimates, DLC tends to mark this visible boundary of the lips and oral cavity. This is also apparent in the labelled images shown in Figures 16–18. Estimation of the commissures and, therefore, of the width of the mouth is, however, as accurate as the inter-labeller score.

Table 8. MSD, aperture difference, and width difference comparing hand labels to DLC (MobileNetV2_1.0).

Lip Measure	Inter Labeller Mean/s.d./Median	DLC MobileNetV2_1.0 Mean/s.d./Median (p Value) [1]	DLC ResNet50 Mean/s.d./Median (p Value)
MSD upper lip (mm)	0.41/0.23/0.36	0.59/0.29/0.54 (<0.001)	0.59/0.40/0.47 (=0.001)
MSD lower lip (mm)	0.73/0.71/0.55	0.86/0.75/0.64 (0.17)	0.82/0.67/0.64 (0.65)
Lip aperture (mm^2)	4.6/54/6.2	−23/61/−10	−19/48/−11
Lip width (mm)	−0.1/3.6/-0.5	0.8/2.4/0.7	−0.2/3.7/0.4

[1] Two tailed t-test assuming unequal variance with reference to the MSD data generated by the DLC inter-labeller distribution. Not applicable to aperture and width.

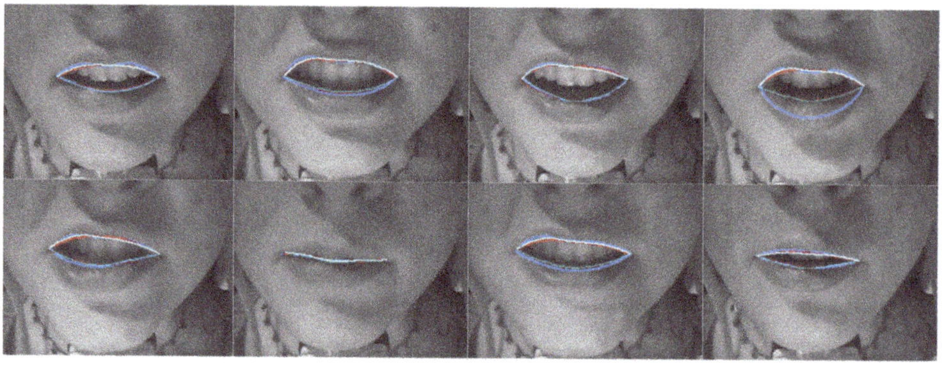

Figure 16. Speaker 25fs; blue—hand-label; red—MobileNetV2; cyan—ResNet50.

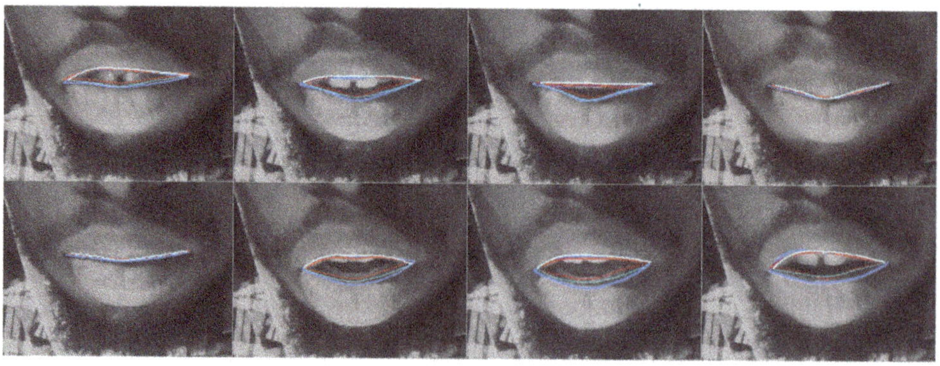

Figure 17. Speaker 12me; blue—hand-label; red—MobileNetV2; cyan—ResNet50.

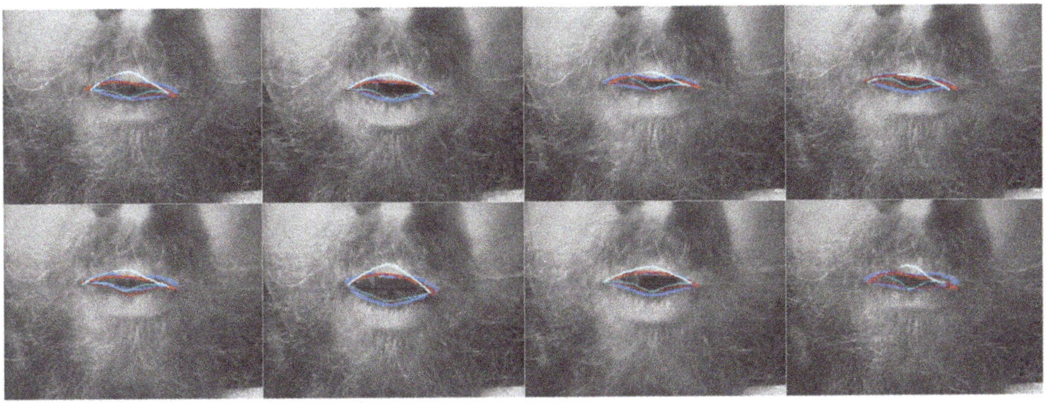

Figure 18. Speaker 17ms; blue—hand-label; red—MobileNetV2; cyan—ResNet50.

Figure 19 shows the MSD values for lower and upper lips, comparing DLC MobileNetV2, DLC ResNet50, and inter-labeller. Of note, ResNet50 improves the estimates of speaker 12me, but it performs more poorly on speaker 17ms (their lips are partially obscured by a moustache). Example frames from these two speakers are shown in Figures 17 and 18, respectively. MobileNetV2 estimates the lower lip closer to the lip edge for speaker 12me than the labeller. For speaker 17ms, ResNet50 can be seen to perform very poorly and with low confidence.

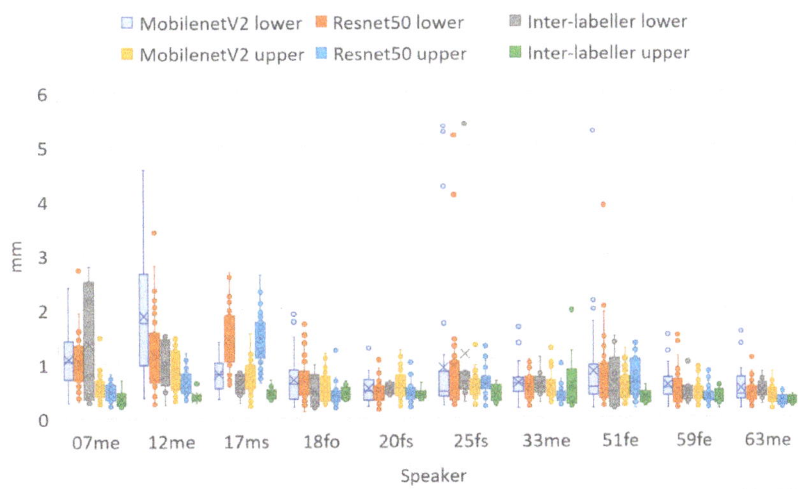

Figure 19. Mean sum of distances (MSDs) between hand-labelled randomly selected frames (40 frames for each speaker) and DLC using MobileNetV2_1.0 and ResNet50 encoders. Upper lip and lower lip shown separately.

Figures 20 and 21 show the lip aperture and width respectively, comparing DLC MobileNetV2, DLC ResNet50 and inter-labeller for each speaker. As can be observed in Figure 19, ResNet50 had trouble identifying the lip commissures for speaker 17ms, but MobileNetV2 did surprisingly well. These figures also show that the second human labeller had trouble following the instructions for the positioning of the lower lip boundary for speakers 07me and 12me. They also overestimated the width in speaker 12me, where whiskers obscured the commissure positions. High Pearson correlation values of 0.95 for hand-labelled vs. DLC lip aperture and 0.89 for hand-labelled vs. DLC lip width were recorded for DLC MobileNetV2.

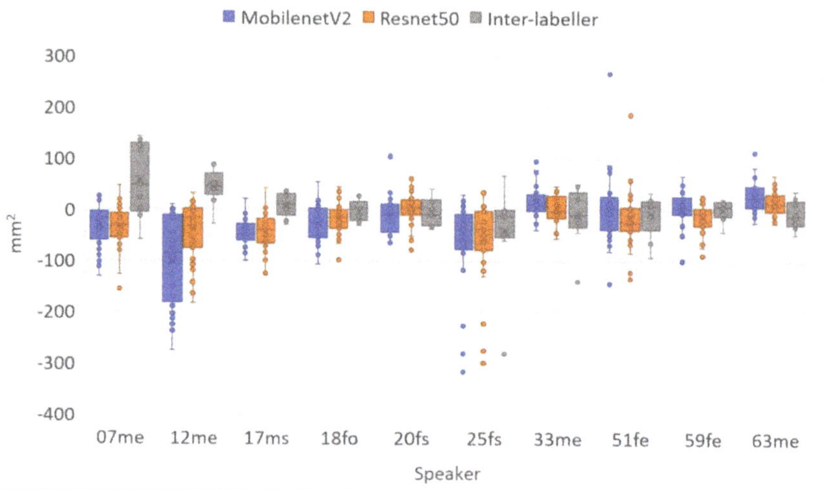

Figure 20. Difference in lip area between hand-labelled randomly selected frames (40 frames for each speaker) and DLC MobileNetV2_1.0, DLC ResNet50 and a second hand-labeller.

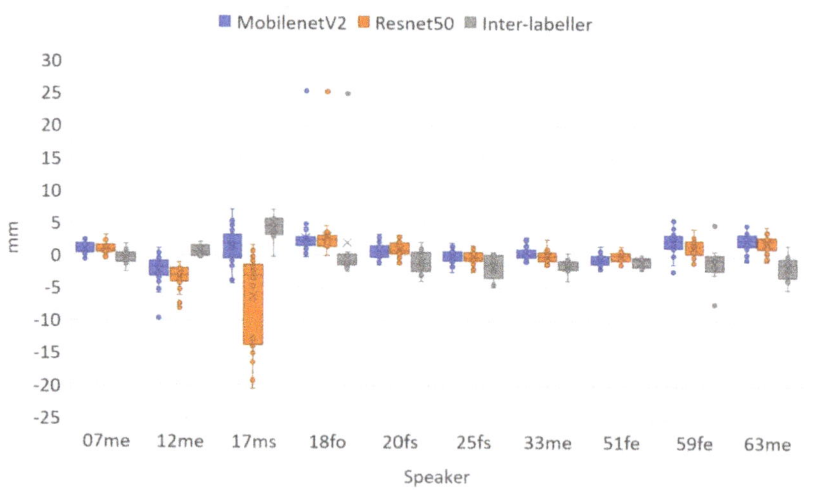

Figure 21. Difference in lip width between hand-labelled randomly selected frames (40 frames for each speaker) and DLC MobileNetV2_1.0, DLC ResNet50, and a second hand-labeller.

5. Discussion

In this study, we investigated an open tool for pose estimation applied to speech articulator image data. By leveraging existing networks pretrained on general object recognition and human body pose estimation, relatively small amounts of speech articulator training data result in a model capable of achieving human-level accuracy on unseen data.

In a field dominated by segmentation and edge-detection methods, we show the following for the first time:

- Pose estimation is capable of learning how to label features that do not necessarily correspond to edges.
- Pose estimation can estimate feature positions to the same level of accuracy as a human labeller.

The hand-labels used here, as in similar studies, were subjectively determined. However, labelling, adhering to the prescribed guidance (see Section 3.1.3), could be learnt by both another human and the DLC network equally well. From this, we can be encouraged that if a more principled ground truth were to be established, perhaps by mapping points from MRI or EMA onto the ultrasound images, then that ground truth would be learnt.

It is possible that the performance of *SLURP*, *MTracker*, and *DeepEdge* could be improved if trained on the same data as DLC. However, Zhu et al. [11] test *MTracker* on two of the test sets used here; namely, the Ultrax Child corpus and the French UltraSpeech corpus. Looking at Figures 7 and 8, speakers from those datasets (02TD1M, 05TD1M, 07TD1F, 08TD1F, 09TD1, 11TD1F, TH, DF, PB) do not show broadly better performance by *MTracker*. Conversely, the UltraSpeech corpus is not represented in the DLC training data and Figures 7 and 8 do not show worse performance by DLC on TH, DF, and PB than on other speakers. Neither *DeepEdge* nor *MTracker* make a training package publicly available. If users could train these models on their data, it is questionable whether they would choose to hand-label the 2000–35,000 frames needed to train these networks. By contrast, the DLC model trained in this study appears to generalise well. DLC includes a simple training package and the training data used in this study is available online. Thus, if the model did perform poorly on a user's dataset, a few (<100) hand-labelled frames from that dataset could be added to the existing training data and the model retrained. The small number of images required for training also permits time for more careful, consistent, and expert labelling.

Outside the scope of this study, DeepLabCut can also estimate the position of the hyoid, and jaw if these features lie within the image. These are point structures rather than edges and cannot easily be estimated by segmentation or edge detection algorithms.

Where speed is a consideration, DLC MobileNetV2 and ResNet50 perform faster (with a GPU) than real time even with ultrasound frame rates of 119 fps. DLC could therefore perform tongue contour estimation on live ultrasound and lip images. Real-time performance is important for live lipreading or visual feedback of tongue for speech therapy. For offline analysis, DLC MobileNetV2 performs at 7 fps using a CPU and can process a batch of recordings at this rate. It does not require manual intervention for each recording so can be left to run overnight if necessary.

DeepLabCut analyses each frame independently. No frame-to-frame continuity is applied. Given that it tracks so well, the absence of temporal continuity constraints can be seen as an advantage because problems of "drift" in contour position cannot occur. Frame-to-frame jitter in keypoint position can be filtered out in post-processing if the frame rate is significantly faster than the articulator movement.

Pose estimation offers the possibility of tracking keypoint positions. Whether it is possible to track points on the tongue remains an open question. Results from our pilot investigation comparing the EMA sensor position to DLC estimates, and high RMSE values (~3 mm) w.r.t. MSD values (~1 mm), both indicated poor estimation of the sensor position along the tongue surface. This is likely due, in part, to inconsistency of training keypoint placement by the human labeller, despite an effort in this study to try to label as if the keypoints were attached to a specific flesh point. A further multi-speaker study where simultaneous EMA and ultrasound is used to train and to evaluate the estimation of the sensor positions is required.

Edge tracked partial tongue contours provide no indication of which part of the contour corresponds to root, body, or tip. This has dictated what kind of further analysis can be performed on estimated lip and tongue contours. The intersection of the tongue contour with a fixed measurement axis is often used to assess raising or lowering of a part of the tongue. Measuring tongue tip movement in this way runs into problems when the tip is retracted, and the contour no longer crosses the axis. Pose estimation opens the possibility of measuring the contraction of the root, body, dorsum, and blade with respect to the anatomically defined position of the short tendon where bundles of the genioglossus muscle attach to the mandible (see Figure 2). This provides a measure independent of probe rotation. Lip rounding can be identified not only using the overall width and aperture measures but also the relative height of the midpoints compared to the parasagittal points. With reference to Figure 5, a measure for lip rounding can be formulated as:

$$(\text{abs}\,(C - D) - 0.5\,(\text{abs}\,(E - F) + \text{abs}\,(G - H)))\ /\ \text{abs}\,(A - B)$$

Pose estimation has recently been applied to sustained speech articulations recorded using MRI of the vocal tract [39]. A total of 256 × 256-pixel images with 1 pixel/mm resolution were analysed and RMSE accuracy results of 3.6 mm reported. These results are similar to the RMSE scores reported here for ultrasound. Beyond the scope of the current study, we piloted articulatory keypoint estimation using dynamic MRI of the vocal tract taken from a public multi-speaker dataset [40]. The data consisted of 84 × 84-pixel images (83 Hz) and perhaps because of the low spatial resolution, the method was less successful. A larger amount of training data were perhaps required, and this would be something to be investigated further.

DeepLabCut provides a package for estimating 3D positions using multicamera data and could be applied to form a richer feature set for lip movement. A side-facing camera would capture lip protrusion information. DeepLabCut could also be investigated as a means for tracking other expressive facial features, such as eyebrows, or for monitoring head movement. DeepLabCut also provides a package to run on a live video stream and work is underway to implement this for live ultrasound input.

In summary, the combination of transfer learning and pose estimation, evaluated here using DeepLabCut, provides a ground-breaking level of efficiency, practicality, and accuracy when applied to feature labelling of speech articulatory image data. The models generated by this study have been made available in Supplementary Materials for use and further evaluation by other research groups.

Supplementary Materials: The following supporting information can be downloaded at: https://www.mdpi.com/article/10.3390/s22031133/s1. Videos of all test recordings with DLC MobileNetV2 contour estimates superimposed; an AAA project freely accessible by downloading the AAA software where all of the estimation methods can be compared.

Author Contributions: Conceptualization, A.W.; methodology, A.W.; software, A.W. and J.B.-T.; validation, A.W. and J.B.-T.; formal analysis, A.W.; data curation, A.W. and J.B.-T.; writing—original draft preparation, A.W.; writing—review and editing, A.W. and J.B.-T.; visualization, A.W. All authors have read and agreed to the published version of the manuscript.

Funding: This research received no external funding.

Institutional Review Board Statement: The study was conducted according to the guidelines of the Declaration of Helsinki and approved by the Research Ethics Panel of the Division of Speech and Hearing Sciences, Queen Margaret University (protocol code 2020.07.01.AW dated 1 July 2020).

Informed Consent Statement: Not applicable. Use of publicly available datasets.

Data Availability Statement: All files and folders used for DeepLabCut training and testing are available at https://github.com/articulateinstruments/DeepLabCut-for-Speech-Production (accessed on 28 November 2021).

Acknowledgments: Thanks go out to Wei-rong Chen for the discussions on how to optimise DeepEdge, to Mark Tiede and Cathy Laporte for discussions on how to optimise SLURP, and to Jian Zhu for discussions on how to optimise MTracker. The simultaneous EMA-ultrasound data were collected by Sam Kirkham and Patrycja Strycharczuk with funding from the UK Arts and Humanities Research Council (AH/S011900/1).

Conflicts of Interest: The authors declare no conflict of interest.

Appendix A

The mean sum of distances (MSDs) between two contours A and B each consisting of n equally spaced x/y coordinates is defined by Jaumard-Hakoun et al. [15] as:

$$MSD(A,B) = \left(\sum_{i=1}^{n} minj(\|Bi - Aj\| + \sum_{j=1}^{n} mini(\|Ai - Bj\|\right)/2n$$

The MSD first considers each defining x/y coordinate of contour A and its distance to the closest point on contour B, and then considers each point along contour B and its distance to the closest point on contour A. It then averages the resulting $2n$ distances. Using this formulation, the two sums in the MSD are considerably different in magnitude when the length of one contour is significantly different from the length of the other.

In this paper we modified the above formula so that the MSD is unaffected by difference in length between the two contours. While measuring distances between the first point on contour A and all points on contour B only the distance to the closest point on contour B is accumulated and n is incremented only for this distance. The same process applies to the last point on contour A. Then the process is repeated when taking points along contour B and comparing to contour A. In this way, distances corresponding to disparate endpoints of the contours are not counted. If contour B is very much shorter than contour A but the two contours match exactly, then the MSD score will be close to zero.

Note: the mean tongue contour length is 86 mm. For this paper $n = 100$ so the distance between contour vertices is on average 0.86 mm. If two contours sit perfectly on top of each other but the vertices are offset by 0.43 mm then the MSD score would be 0.43 mm, not zero. Our MSD results show very few image frames where the score was less than 0.4 mm.

Appendix B

Typical training loss and corresponding root mean square error (RMSE) for ultrasound keypoint estimation showing hand-labelled data vs. test data.

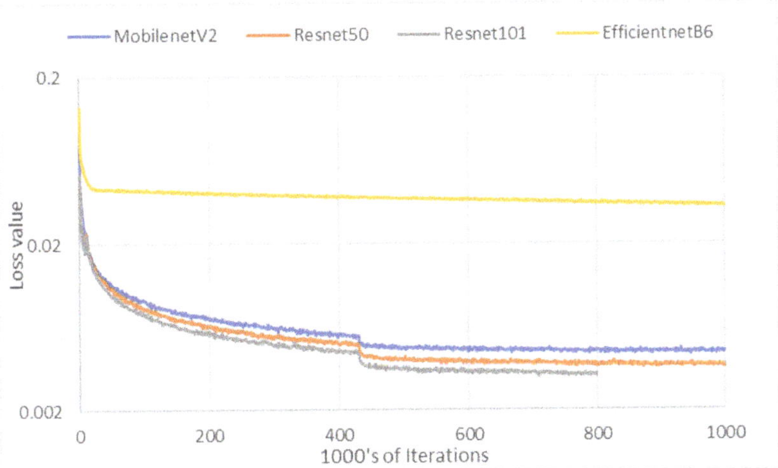

Figure A1. Training iteration loss.

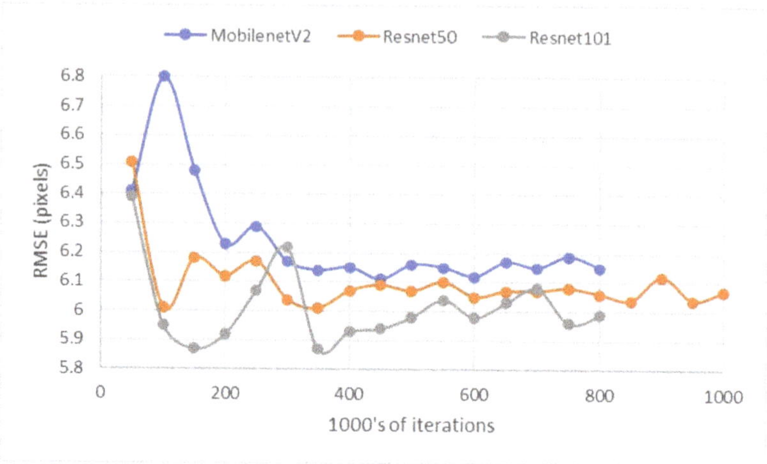

Figure A2. RMSE for ultrasound test set (excludes points with confidence less than 0.6).

References

1. Fuchs, S.; Perrier, P. On the complex nature of speech kinematics. *ZAS Pap. Linguist.* **2005**, *42*, 137–165. [CrossRef]
2. Kass, M.; Witkin, A.; Terzopoulos, D. Snakes: Active contour models. *Int. J. Comput. Vis.* **1988**, *1*, 321–331. [CrossRef]
3. Li, M.; Kambhamettu, C.; Stone, M. Automatic contour tracking in ultrasound images. *Clin. Linguist. Phon.* **2005**, *19*, 545–554. [CrossRef] [PubMed]
4. Csapó, T.G.; Lulich, S.M. Error analysis of extracted tongue contours from 2D ultrasound images. In Proceedings of the Sixteenth Annual Conference of the International Speech Communication Association, Dresden, Germany, 6–10 September 2015.
5. Laporte, C.; Ménard, L. Multi-hypothesis tracking of the tongue surface in ultrasound video recordings of normal and impaired speech. *Med. Image Anal.* **2018**, *44*, 98–114. [CrossRef]
6. Fasel, I.; Berry, J. Deep belief networks for real-time extraction of tongue contours from ultrasound during speech. In Proceedings of the 2010 20th International Conference on Pattern Recognition, Istanbul, Turkey, 23–26 August 2010; pp. 1493–1496.
7. Fabre, D.; Hueber, T.; Bocquelet, F.; Badin, P. Tongue tracking in ultrasound images using eigentongue decomposition and artificial neural networks. In Proceedings of the 16th Annual Conference of the International Speech Communication Association (Interspeech 2015), Dresden, Germany, 6–10 September 2015.
8. Xu, K.; Gábor Csapó, T.; Roussel, P.; Denby, B. A comparative study on the contour tracking algorithms in ultrasound tongue images with automatic re-initialization. *J. Acoust. Soc. Am.* **2016**, *139*, EL154–EL160. [CrossRef]
9. Mozaffari, M.H.; Yamane, N.; Lee, W. Deep Learning for Automatic Tracking of Tongue Surface in Real-time Ultrasound Videos, Landmarks instead of Contours. In Proceedings of the 2020 IEEE International Conference on Bioinformatics and Biomedicine (BIBM), Seoul, Korea, 16–19 December 2020; pp. 2785–2792.
10. Aslan, E.; Akgul, Y.S. Tongue Contour Tracking in Ultrasound Images with Spatiotemporal LSTM Networks. In Proceedings of the German Conference on Pattern Recognition, Dortmund, Germany, 10–13 September 2019; pp. 513–521.
11. Zhu, J.; Styler, W.; Calloway, I. A CNN-based tool for automatic tongue contour tracking in ultrasound images. *arXiv*, 2019; arXiv:1907.10210.
12. Chen, W.; Tiede, M.; Whalen, D.H. DeepEdge: Automatic Ultrasound Tongue Contouring Combining a Deep Neural Network and an Edge Detection Algorithm. 2020. Available online: https://issp2020.yale.edu/S05/chen_05_16_161_poster.pdf (accessed on 5 February 2021).
13. Ronneberger, O.; Fischer, P.; Brox, T. U-net: Convolutional networks for biomedical image segmentation. In Proceedings of the International Conference on Medical Image Computing and Computer-Assisted Intervention, Munich, Germany, 5–9 October 2015; pp. 234–241.
14. Tang, L.; Bressmann, T.; Hamarneh, G. Tongue contour tracking in dynamic ultrasound via higher-order MRFs and efficient fusion moves. *Med. Image Anal.* **2012**, *16*, 1503–1520. [CrossRef]
15. Jaumard-Hakoun, A.; Xu, K.; Roussel-Ragot, P.; Dreyfus, G.; Denby, B. Tongue contour extraction from ultrasound images based on deep neural network. *arXiv*, 2016; arXiv:1605.05912.
16. Chiou, G.I.; Hwang, J. Lipreading from color video. *IEEE Trans. Image Process.* **1997**, *6*, 1192–1195. [CrossRef]
17. Luettin, J.; Thacker, N.A.; Beet, S.W. Speechreading using shape and intensity information. In Proceedings of the Fourth International Conference on Spoken Language Processing, ICSLP'96, Philadelphia, PA, USA, 3–6 October 1996; pp. 58–61.
18. Kaucic, R.; Dalton, B.; Blake, A. Real-time lip tracking for audio-visual speech recognition applications. In Proceedings of the European Conference on Computer Vision, Cambridge, UK, 15–18 April 1996; pp. 376–387.

19. Lallouache, M.T. Un Poste" Visage-Parole" Couleur: Acquisition et Traitement Automatique des Contours des Lèvres. Ph.D. Thesis, INPG, Grenoble, France, 1991.
20. King, H.; Ferragne, E. Labiodentals /r/ here to stay: Deep learning shows us why. *Anglophonia Fr. J. Engl. Linguist.* **2020**, *30*. [CrossRef]
21. Mathis, M.W.; Mathis, A. Deep learning tools for the measurement of animal behavior in neuroscience. *Curr. Opin. Neurobiol.* **2020**, *60*, 1–11. [CrossRef] [PubMed]
22. Mathis, A.; Mamidanna, P.; Cury, K.M.; Abe, T.; Murthy, V.N.; Mathis, M.W.; Bethge, M. DeepLabCut: Markerless pose estimation of user-defined body parts with deep learning. *Nat. Neurosci.* **2018**, *21*, 1281–1289. [CrossRef] [PubMed]
23. Graving, J.M.; Chae, D.; Naik, H.; Li, L.; Koger, B.; Costelloe, B.R.; Couzin, I.D. DeepPoseKit, a software toolkit for fast and robust animal pose estimation using deep learning. *eLife* **2019**, *8*, e47994. [CrossRef]
24. Pereira, T.D.; Tabris, N.; Li, J.; Ravindranath, S.; Papadoyannis, E.S.; Wang, Z.Y.; Turner, D.M.; McKenzie-Smith, G.; Kocher, S.D.; Falkner, A.L. SLEAP: Multi-animal pose tracking. *bioRxiv* **2020**. [CrossRef]
25. Mathis, A.; Biasi, T.; Schneider, S.; Yuksekgonul, M.; Rogers, B.; Bethge, M.; Mathis, M.W. Pretraining boosts out-of-domain robustness for pose estimation. In Proceedings of the IEEE/CVF Winter Conference on Applications of Computer Vision, Virtual Conference, 5–9 January 2021; pp. 1859–1868.
26. Johnston, B.; de Chazal, P. A review of image-based automatic facial landmark identification techniques. *EURASIP J. Image Video Process.* **2018**, *2018*, 1–23. [CrossRef]
27. Andriluka, M.; Pishchulin, L.; Gehler, P.; Schiele, B. 2D Human Pose Estimation: New Benchmark and State of the Art Analysis. In Proceedings of the IEEE Conference on Computer Vision and Pattern Recognition (CVPR), Columbus, OH, USA, 24–27 June 2014; pp. 3686–3693.
28. Sandler, M.; Howard, A.; Zhu, M.; Zhmoginov, A.; Chen, L. MobileNetV2: Inverted residuals and linear bottlenecks. In Proceedings of the IEEE Conference on Computer Vision and Pattern Recognition, Salt Lake City, UT, USA, 18–22 June 2018; pp. 4510–4520.
29. He, K.; Zhang, X.; Ren, S.; Sun, J. Deep residual learning for image recognition. In Proceedings of the IEEE Conference on Computer Vision and Pattern Recognition, Las Vegas, NV, USA, 26 June–1 July 2016; pp. 770–778.
30. Tan, M.; Le, Q. Efficientnet: Rethinking model scaling for convolutional neural networks. In Proceedings of the International Conference on Machine Learning PMLR, Long Beach, CA, USA, 9–15 June 2019; Volume 97, pp. 6105–6114.
31. Insafutdinov, E.; Pishchulin, L.; Andres, B.; Andriluka, M.; Schiele, B. DeeperCut: A Deeper, Stronger, and Faster Multi-person Pose Estimation Model. In *Computer Vision—ECCV 2016*; Springer International Publishing: Cham, Switzerland, 2016; pp. 34–50.
32. Ribeiro, M.S.; Sanger, J.; Zhang, J.; Eshky, A.; Wrench, A.; Richmond, K.; Renals, S. TaL: A synchronised multi-speaker corpus of ultrasound tongue imaging, audio, and lip videos. In Proceedings of the 2021 IEEE Spoken Language Technology Workshop (SLT), Shenzhen, China, 19–22 January 2021; pp. 1109–1116.
33. Eshky, A.; Ribeiro, M.S.; Cleland, J.; Richmond, K.; Roxburgh, Z.; Scobbie, J.; Wrench, A. UltraSuite: A repository of ultrasound and acoustic data from child speech therapy sessions. *arXiv*, 2019; arXiv:1907.00835.
34. Strycharczuk, P.; Ćavar, M.; Coretta, S. Distance vs time. Acoustic and articulatory consequences of reduced vowel duration in Polish. *J. Acoust. Soc. Am.* **2021**, *150*, 592–607. [CrossRef]
35. Fabre, D.; Hueber, T.; Girin, L.; Alameda-Pineda, X.; Badin, P. Automatic animation of an articulatory tongue model from ultrasound images of the vocal tract. *Speech Commun.* **2017**, *93*, 63–75. [CrossRef]
36. Nath, T.; Mathis, A.; Chen, A.C.; Patel, A.; Bethge, M.; Mathis, M.W. Using DeepLabCut for 3D markerless pose estimation across species and behaviors. *Nat. Protoc.* **2019**, *14*, 2152–2176. [CrossRef]
37. GetContours V3.5. Available online: https://github.com/mktiede/GetContours (accessed on 21 July 2021).
38. Scobbie, J.M.; Lawson, E.; Cowen, S.; Cleland, J.; Wrench, A.A. A Common Co-Ordinate System for Mid-Sagittal Articulatory Measurement. QMU CASL Working Papers WP-20. 2011. Available online: https://eresearch.qmu.ac.uk/handle/20.500.12289/3597 (accessed on 28 November 2021).
39. Eslami, M.; Neuschaefer-Rube, C.; Serrurier, A. Automatic vocal tract landmark localization from midsagittal MRI data. *Sci. Rep.* **2020**, *10*, 1–13. [CrossRef]
40. Lim, Y.; Toutios, A.; Bliesener, Y.; Tian, Y.; Lingala, S.G.; Vaz, C.; Sorensen, T.; Oh, M.; Harper, S.; Chen, W. A multispeaker dataset of raw and reconstructed speech production real-time MRI video and 3D volumetric images. *arXiv*, 2021; arXiv:2102.07896. [CrossRef] [PubMed]

sensors

MDPI

Article

FlexLip: A Controllable Text-to-Lip System

Dan Oneață [1,*], Beáta Lőrincz [2], Adriana Stan [3] and Horia Cucu [1,4]

1 Speech and Dialogue Research Lab, University "Politehnica" of Bucharest, 060042 Bucharest, Romania; horia.cucu@upb.ro
2 Faculty of Mathematics and Computer Science, "Babeș-Bolyai" University, 400347 Cluj-Napoca, Romania; beata.lorincz@ubbcluj.ro
3 Department of Communications, Technical University of Cluj-Napoca, 400114 Cluj-Napoca, Romania; adriana.stan@com.utcluj.ro
4 Zevo Technology, 077042 Roșu, Chiajna, Romania
* Correspondence: dan.oneata@speed.pub.ro

Abstract: The task of converting text input into video content is becoming an important topic for synthetic media generation. Several methods have been proposed with some of them reaching close-to-natural performances in constrained tasks. In this paper, we tackle a subissue of the text-to-video generation problem, by converting the text into lip landmarks. However, we do this using a modular, controllable system architecture and evaluate each of its individual components. Our system, entitled FlexLip, is split into two separate modules: text-to-speech and speech-to-lip, both having underlying controllable deep neural network architectures. This modularity enables the easy replacement of each of its components, while also ensuring the fast adaptation to new speaker identities by disentangling or projecting the input features. We show that by using as little as 20 min of data for the audio generation component, and as little as 5 min for the speech-to-lip component, the objective measures of the generated lip landmarks are comparable with those obtained when using a larger set of training samples. We also introduce a series of objective evaluation measures over the complete flow of our system by taking into consideration several aspects of the data and system configuration. These aspects pertain to the quality and amount of training data, the use of pretrained models, and the data contained therein, as well as the identity of the target speaker; with regard to the latter, we show that we can perform zero-shot lip adaptation to an unseen identity by simply updating the shape of the lips in our model.

Keywords: text-to-lip; speech synthesis; text-to-speech; speech-to-lip; zero-shot adaptation; generative models; deep learning; artificial intelligence; objective measures

Citation: Oneață, D.; Lőrincz, B.; Stan, A.; Cucu, H. FlexLip: A Controllable Text-to-Lip System. *Sensors* **2022**, *22*, 4104. https://doi.org/10.3390/s22114104

Academic Editors: Bruce Denby, Tamás Gábor Csapó and Michael Wand

Received: 16 May 2022
Accepted: 26 May 2022
Published: 28 May 2022

1. Introduction

Over the past few years, our society has constantly been increasing the amount of multimedia output. From online radio and television, to YouTube video bloggers and the popular Facebook Lives, professionals and nonexperts alike generate multimedia content at a tremendous pace. However, the costs and effort of generating high-quality multimedia content become increasingly significant, and many parties are already looking into deep learning for solutions to lessen the burden of pre- and postproduction, as well as end-to-end media generation. In the area of spoken content generation, text-to-speech (TTS) systems have already been, to a large extent, adopted by semiprofessional content creators, with the most-known platform for providing this service to its users being TikTok. However, when tackling the complete text-to-video synthesis, the solutions and quality of the available systems are not at the same level of integration into the media platforms. Even though there are numerous applications that it could address, such as anchor news delivery, video podcasts, gaming characters generation, and so on.

In this context, our work focuses on the task of rendering a video of a person delivering a spoken content starting from a given text, and optionally, a selected identity that can differ from the available training data. We decompose this complex task into a three module pipeline—(i) text-to-speech, (ii) speech-to-keypoints, and (iii) keypoints-to-video—and set out to derive a controllable and objectively measurable architecture for it. However, our work does not focus on the complete head movement and generation of facial characteristics, but rather it limits its scope to the generation of lip landmarks starting from an input text. As a result, we only tackle the first two modules of the complete pipeline described before and show an overview of our system in Figure 1. One reason for why we are not addressing the final module is the fact that up to this moment, there are no generally agreed upon objective measures for it, and most papers publish only perceptual, subjective evaluations.

Another important aspect of our work is the fact that most of the previous works address only the second module, i.e., speech-to-lips [1,2]. Having a controllable, easily adaptable TTS system integrated into the flow of the system can enable the end-user to control all aspects of the media generation system, including the spoken content and identity. As shown in Figure 1, the proposed architecture encompasses the ability to select various speech identities, alter the prosodic patterns, while also being able to control or disentangle the identity of the generated lips from the spoken one. For example, we could generate speech with the voice of former president Obam, using the lip shape of former president Trump and the face of former president Bush.

We can summarise the main contributions of our paper as follows:

- We propose a novel text-to-lips generation architecture, entitled FlexLip;
- We design its architecture as a flexible and highly controllable one;
- We analyse the effect of using synthesised speech as opposed to natural recordings;
- We propose a zero-shot adaptation of the speech-to-lips component;
- We show that by using as little as 20 min of spoken data, we can control the target speaker's identity;
- We also show that the controllability of the architecture enables us to perform more accurate objective measures of its performance.

The paper is organised as follows: Section 2 introduces the works related to our proposed method, with the method being described in Section 3. The experimental setup and results are presented in Sections 4 and 5, and their conclusions are drawn in Section 6.

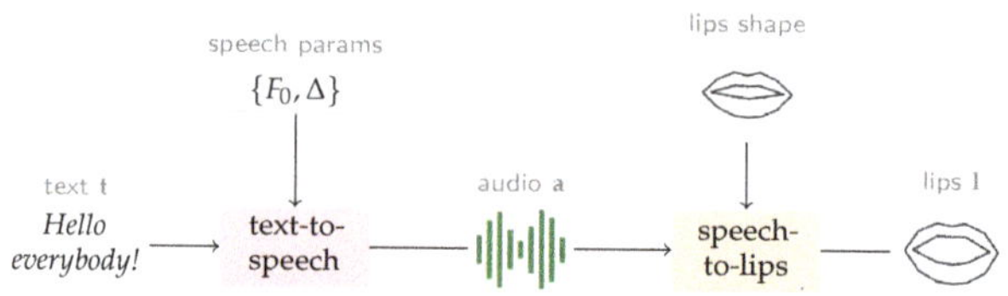

Figure 1. Schematic illustration of the proposed FlexLip pipeline. Our approach allows for a high degree of controllability by explicitly passing through the audio modality and permitting to specify speech parameters (fundamental frequency F_0, and phoneme durations Δ) and lips shape (as the mean shape upon which the learnt displacements are applied).

2. Related Work

The task of generating video (i.e., a talking-head video) starting from speech or text has recently gained interest in the research community next to other tasks converting one modality to another, such as image-to-text or video-to-text, also called image or video captioning. For addressing the transformation of text into video output, various pipelines and different types of latent spaces were proposed. All studies that approach text-to-video conversion use an initial text-to-speech system to generate speech or speech features [3,4]. When going from speech to video, some studies argue that having an intermediate representation of the face or mouth can heighten the performance of the system [3–5], while in other cases, the authors approach the issue in an end-to-end manner, going from speech directly to video [1,2]. There are also studies that focus solely on the speech-to-keypoints task [6,7], as this is regarded as being more difficult than the subsequent keypoints-to-video task, as there is no direct, one-to-one mapping from the text input to the individual video frames. One of the first attempts to create a complete text-to-video pipeline was introduced in [8]. A simple idea was explored, whereby mouth images extracted from a video sequence were reordered to match a new, unseen phoneme sequence derived from the input text. In following works, text-to-video synthesis was approached as an extension of text-to-speech synthesis. Concatenative speech synthesis was extended to map characters to visemes, defined as static mouth shapes [9] or as temporal movements of the visual speech articulators [10]. Appropriate visemes were chosen from a large dataset for a particular speaker and morphed from one to the other in order to generate smooth transitions between successive visemes. In a more recent work, Fried et al. [11] approached text-to-video synthesis from a slightly different perspective by proposing a method to edit an existing video in order to reflect a new text input. This method still relies on a viseme search, concatenation, and blending. To generate mouth movements matching the edited text, visemes present in other parts of the video were used.

Inspired by the application of face keypoints in video prediction, where the keypoints guide the generation of future frames [12], most recent methods in text-to-video synthesis use face or mouth keypoints as intermediate representations [3,4]. Zhang et al. [4] addressed the task in two steps: (i) transforming the text into a sequence of keypoints (which the authors denote as poses), using a dictionary of (phoneme, keypoints) pairs; (ii) using a GAN-based architecture to generate video from interpolated phoneme poses. Simultaneously, the text was transformed into speech by a text-to-speech synthesis system. Kumar et al. [3] were the first to propose a sequence of fully trainable neural modules to address text-to-video conversion in three main steps. First, a text-to-speech system was used for audio generation starting from characters, not phonemes. A second network was then employed to generate mouth region keypoints synchronised with the synthetic speech. Finally, a video generation network produced video frames conditioned on the mouth region keypoints.

A couple of recent studies focused on parts of the complete text-to-video pipeline performing video synthesis directly from speech features [1,2,5] or approached just the speech-to-keypoints task [6,7]. In the video synthesis systems presented in [1,5], the mapping between audio features and mouth shapes are learnt by recurrent or convolutional neural networks. The audio input is paired with either mouth region keypoints [5] or with images of the target face [1] and the lip-synced video is predicted by the network. The approach for audio-to-video generation described in [2] is based on a neural network that includes a latent 3D representation of the face. As the keypoint-based intermediate representation seems to be a common choice in previous studies and also paves the way for creating speaker-independent systems, Eskimez et al. [6] proposed an LSTM trained with 27 subjects to solve the task that is able to generalise to new subjects. Greenwood et al. [7] generated full-face keypoints, as opposed to only mouth region keypoints, for two subjects using a BiLSTM.

For the evaluation of video synthesis methods, both objective and subjective measures are employed. In the objective evaluation, various difference metrics are computed between the real (ground-truth) and generated videos. Chen et al. [13] used the mean squared error, the Frechet video distance, based on the distance between features of the real and generated videos reported in [12]. Human evaluations are frequently used as subjective measures to capture the visual quality of the generated videos [12,14]. In a very recent paper, Aldeneh et al. [15] proposed a perceptual evaluation model that can be used to automatically and accurately estimate the subjective perceptual score for a given lip motion sequence.

With respect to text-to-speech synthesis (TTS), there are numerous neural architectures that achieve close to natural synthetic speech quality. Further, if at the beginning of the deep learning era for TTS the research was oriented towards full end-to-end systems going from input characters to audio waveforms [16], in recent years, the focus shifted towards flexible and controllable architectures [17–19]. This type of architecture enables several factors of the synthetic speech to be easily tuned during inference. This is the case for the FastPitch architecture [18], in which the duration, energy, and pitch of the output speech can be specified or copied from a different audio input. In the context of evaluating text-to-lip models, being able to control the duration of the synthesised speech means that the original phoneme durations of a natural speech sample can be replicated. This can provide a better alignment between the keypoints predicted from natural and synthetic input speech, and no substantial additional error is introduced by the alignment, making the objective evaluation more accurate.

In this context, our work resembles relatively well the work of Kumar et al. [3], but bears a few important distinctions. First and most importantly, our aim was to assess in a thorough and objective manner the quality of each component of a text-to-video pipeline. As such, we did not address the keypoints-to-video task, which is inherently a subjective one. Different from Kumar et al. [3], who do not evaluate their system subjectively, nor objectively, we carefully designed the text-to-keypoints pipeline to allow for objective evaluation and performed the evaluation of each component independently and at the system-level as well.

In terms of evaluation of generated lip movements, our work is, to the best of our knowledge, the first one to evaluate objectively the output of the text-to-keypoints task. Many works evaluate the quality of the generated sequence when the system is fed natural speech [6,7,15], but none of these start with text as input. One of the problems is that most of the neural-based TTS systems do not enable the exact control of the duration of the output; therefore, there is no one-to-one correspondence between the ground-truth frames and the synthesised ones. With respect to this, we believe that managing to objectively assess the quality for the text-to-keypoints seen as a whole is one of the important contributions of our work.

3. Text-to-Lip System Description

Our text-to-lip system is composed of two independent modules: **a text-to-speech synthesis system** and a **speech-to-lip** one. This independence ensures a more controllable setup and each module can be easily replaced. The following sections describe the two modules and their training procedures, while also focusing on their controllability.

3.1. The Text-to-Speech Component

When generating lip movements from speech, the quality of the input speech is essential. If the input contains natural speech recordings, they should also be high-quality. Therefore, for our text-to-speech synthesis component (TTS), we selected one of the latest deep-neural-based architectures, able to generate speech that is very close to the natural one. The architecture is FastPitch [18], and aside from its high performance, it uses a fast-inference parallel architecture, and enables the control of the pitch and duration of the input phonemes. The latter feature facilitates the tweaking of the output such that the speech-to-lip module is better fitted to the target speaker. FastPitch is based on bidirectional Transformers, which make up the encoder and decoder sections of the network. Separate paths are allocated for the pitch and duration prediction, as well as (if this is the case) for a speaker embedding layer. The encoder predicts one Mel-spectrogram frame per phoneme, which is then augmented with the pitch information, and upsampled according to the duration predictor. The prediction is then passed through the decoder to obtain the smoothed, complete Mel-spectrogram.

The Mel-spectrogram is then transformed into a waveform with the help of the WaveGlow neural vocoder [20]. WaveGlow uses a normalising flow-based architecture inspired from Glow [21] and WaveNet [16], but eliminates the autoregressive nature of them. The architecture uses a single network trained to maximise the likelihood of the data and, based on its flow nature, enables the computation of the true distribution of the training data.

As high-quality TTS systems commonly require large amounts of training data, we also adopt a fine-tuning procedure for the FastPitch model. Two pretrained models were used:

a single speaker one, and a multispeaker one for which the network was not conditioned on the speaker identity. These models were then adapted to the target speaker using various amounts of speech data, as described in Section 5.

3.2. The Speech-to-Lips Component

This subsection describes the speech-to-lips component, which takes as input an audio of a person speaking and outputs the keypoints (the moving lips) that correspond to the spoken words. Since the task is a sequence-to-sequence one (we want to map a sequence of audio frames to a sequence of lip keypoints), we opt to implement the speech-to-lips module as a Transformer network, which has shown remarkable performance on many related tasks. The Transformer has two main components: an encoder module that uses self-attention layers to pool the input audio, and a decoder module that uses attention layers to aggregate information from both the encoded audio and previously generated lips. The decoder predicts the lips keypoints at each time step in an autoregressive manner, and is exposed to the entire input audio sequence.

Transferring representations. Given that the network processes audio streams, we decided to reuse the encoder architecture and its pretrained weights from a state-of-the-art speech recognition system. As such, we evaluate two variants of the network: one in which the encoder is frozen to the pretrained weights and we train only the decoder part; a second in which we train both components, the encoder and decoder. Note that training the decoder is mandatory because the speech recognition decoder is designed to output a sequence of characters, while in our task, the output is a sequence of keypoints.

Lips keypoints preprocessing. Video recordings of people talking involve variations in terms of their position, size, and head pose. Since this information affects the lips' coordinates but is irrelevant to the task at hand, we remove these variations from our training data by transforming the absolute coordinates of the lips into a normalised space. More precisely, we apply the following three transformations to the extracted face landmarks: translate such that they are centred on the lips; rotate such that the line connecting the eyes is horizontal; scale such that the distance between the eyes is constant (we set an arbitrary value of five).

A second preprocessing step consists in projecting the normalised lip coordinates (40 coordinates of the lips: 20 keypoints with the x and y coordinates each) to a lower-dimensional manifold using principal component analysis (PCA); we denote the principal components by $\mathbf{v}_1, \dots, \mathbf{v}_D$. We use an eight-dimensional space for projection ($D = 8$), which captures around 97% of the variation of the training data. Figure 2 illustrates the axes of variation captured by the selected principal components. The reconstruction for component i and magnitude s is given by $\mu + \sum s\mathbf{v}_i$, where μ is the mean lip shape and the scaling factor s ranges in $\{-1.5, 1.1, \dots, 1.5\}$. We observe that the principal components capture the following variations in the data: open versus closed mouth is modelled by the first and fifth components; 3D rotations (yaw, pitch, roll) are captured by the third (yaw), fourth (pitch), and sixth (roll) components; lip thickness varies across the second, fifth, seventh, and eight components.

To sum up, our method maps a stream of audio to a list of 8D points, which correspond to the PCA coefficients α. Note that both preprocessing steps are invertible; so, at test time, if we want to overlay the predicted lips on a given subject, we first reconstruct the lips $\mathbf{l}$ based on the predicted PCA coefficients

$$\mathbf{l} = \mu + \sum \alpha_i \mathbf{v}_i, \tag{1}$$

and then we reproject the normalised coordinates in the absolute coordinate space by inverting the scaling, rotation, and translation transforms.

Zero-shot speaker adaptation. When predicting the lip movements of an unseen speaker, we have observed that the lip dynamics are accurate, but unsurprisingly, their shape resembles the one of the trained subject. We propose a method to adapt the lip shape to the one of the new person by replacing the lips mean in the PCA reconstruction with the lips of the target speaker:

$$\mathbf{l}' = \mu' + \sum \alpha_i \mathbf{v}_i, \tag{2}$$

where the coefficients α are obtained in the same way as before, but the mean μ' is updated, and is computed from the target, unseen speaker. This operation is inexpensive (as the mean can be estimated on a few frames) and does not require retraining. Figure 3 shows the lip shapes of three speakers and how they vary along the first principal component. Even if the principal components were estimated on data from the first speaker, these qualitative results suggest that the variations obtained for new speakers are plausible.

Figure 2. Axes of lips variation. On each row, we show the variation of the lips captured by one of the top eight principal components. The reconstructions are obtained by adding the scaled principal component to the mean lip shape. The scaling factor ranges from −1.5 to 1.5, as indicated on the top of the columns.

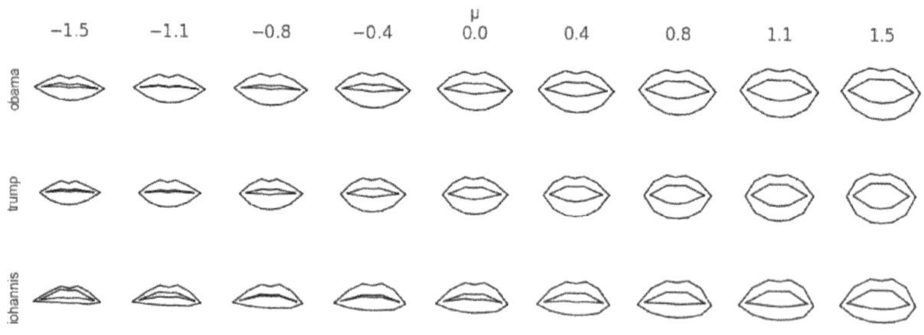

Figure 3. Lip shapes. The middle column, denoted by μ, represents the shapes of the lips for three speakers (one along each row). The columns show the variation of the shapes along the first principal component $\mathbf{v}$; more precisely, the lips in row r and column c are computed as $\mu_r + \sum s_c \mathbf{v}$, with μ_r being the (mean) lip shape and s_c a scaling factor that ranges in $\{-1.5, 1.1, \ldots, 1.5\}$. The PCA was fitted on data from the first speaker.

4. Experimental Setup

This section presents the datasets used for training and evaluating the proposed methods (Section 4.1) and implementation details regarding the two components of our system (Section 4.2).

4.1. Datasets

In order to evaluate a text-to-lip system, we need a three-modality dataset: video, audio, and text. However, there are very few large, freely available datasets that accomplish this requirement. In this context, we decided to build our own dataset comprising high-quality data available on YouTube. For the audio part, we also relied on well-established speech datasets. All the datasets used in our experiments are described in the following subsections.

4.1.1. Obama Dataset

The former president of the United States, Barack Obama, has many videos on YouTube in which he addresses the nation in a systematic fashion: front-facing the camera and speaking clearly, with most of the videos being recorded in his office. From this series of weekly addresses, we downloaded a set of 301 videos, which were originally introduced by Suwajanakorn et al. [5] and subsequently used by Kumar et al. [3]. The YouTube videos come along with both manual and automatically generated closed captions. While the automatic captions are better aligned to the speech than the manual captions, the latter are more accurate and also include punctuation. The punctuation is essential for splitting the audio into sentence-length chunks, required by the TTS system. We split the audio into sentences based on the end-of-sentence punctuation marks encountered in the manual captions. This procedure leads to a set of around 10 k audio–video chunks and their approximate transcripts. The duration of the chunks is between 1 and 20 s long with transcripts between 15 to 500 characters. Very short and very long utterances were discarded, as the TTS architecture has problems attending to short sequences, as well audio sequences longer than 30 s. Some text examples are listed below:

Under my Administration, we're producing more oil than at any other time in the last eight years.

It had the support of 52 Democrats and 22 Republicans.

It's our job as citizens to make sure we keep pushing this country we love toward our most cherished ideals—that all of us are created equal, and all of us deserve an equal shot.

The total duration of the data is around 17 h, with most of the videos having a resolution of 720p and a frame rate of 30 frames per second. The audio is sampled at 44.1 kHz and has a bit depth of 16. To speed up the preprocessing (in particular the face landmark extraction), we downscaled the videos to 360p; for the speech-to-lip and TTS systems training, we downsampled the audio data to 22 kHz and maintain the 16 bps. Although the videos are recorded in quiet conditions, some reverberation and background noise are present; therefore, we preprocessed the audio through the Postfish tool using the default parameters (https://github.com/ePirat/Postfish (accessed on 15 May 2022)). Volume normalisation and silence trimming were performed using the Librosa tool (https://librosa.org/doc/latest/index.html (accessed on 15 May 2022)).

The dataset was then split into train–validation–test subsets at video-level, meaning there are no samples from a video that are independently assigned to the different splits. The test set was composed of 500 utterance-length samples, randomly selected from 30 videos, and was manually checked for alignment and transcription errors.

Text processing and data selection. Although the manual transcripts seemed to be of very high-quality, combined with the splitting algorithm, we noticed that the correspondence between the audio and the text was not completely accurate. Therefore, we performed a series of postprocessing steps. The first step included the normalisation of all nonalphabetic symbols present in the captions, such as numbers, currency symbols, and so on. A first pass was performed using the Num2Words Python package (https://pypi.org/project/num2words/ (accessed on 15 May 2022)); then, the entire dataset was manually checked for nonalphabetic symbols. The second step involved the use of an automatic speech recognition (ASR) system [22] to transcribe the audio chunks. A word error rate (WER) measure was computed between the ASR results and the closed captions. High WER utterances were discarded. Even with a highly accurate ASR system, errors can still propagate to the transcripts. As a result, we ran an additional quality check based on the TTS system's loss. We used an intermediate checkpoint trained on the entire Obama data, set the batch size to 1, froze the layers, and ran the low WER utterances through the architecture. Audio chunks exhibiting a low loss measure were considered as having the correct transcripts, as the TTS system's loss function includes the alignment of audio-to-text. From this step, a set of 3281 samples were retained. In order to extend the dataset as much as possible, we also manually checked 1480 samples and added them to the training set. The total duration of the final audio dataset is 8 h.

Due to the rather complex grapheme-to-phoneme rules present in English, which can make it hard for a TTS system to learn the appropriate alignment and pronunciation, we performed the phonetic transcription of the text prompts associated with the audio data. The front-end tool of the Festival TTS system [23] was used to generate the phonetic representations of the transcripts.

4.1.2. Trump Dataset

To test the capabilities of our speech-to-lip system to adapt to unseen speakers, we collected a small dataset of a different speaker: the former president, Donald Trump. We manually inspected multiple videos of him found on YouTube and finally selected two that satisfy the following criteria: are reasonably long; he is the single speaker appearing; his head is mostly forward-facing, without extreme pose variations. The first video (https://www.youtube.com/watch?v=KJTlo4bQL5c (accessed on 15 May 2022)) is around one hour long and represents his speech at the Conservative Political Action Conference; this video was used for training. The second video (https://www.youtube.com/watch?v=xrPZBTNjX_o (accessed on 15 May 2022)) is almost ten minutes long and represents his presidential address on the COVID-19 pandemic; this video was used for testing. As for the Obama dataset, we downscaled the video to a resolution of 360p. As we did not use this speaker for the text-to-speech component, we did not extract the textual component of the data.

4.1.3. Datasets for TTS and ASR Models' Pretraining

Both text-to-speech and speech-to-lip components use pretrained models and fine-tuning towards the target speaker. The following datasets were used for the model pretraining.

LibriSpeech is a corpus of approximately 1000 h of 16-kHz, multispeaker, read English speech derived from read audiobooks of the LibriVox project, and has been carefully segmented and aligned [24]. LibriSpeech was used in our experiments to pretrain the ASR-based feature extractor in the speech-to-lip module.

LibriTTS [25] is derived from the LibriSpeech corpus and includes multispeaker English data of approximately 585 h sampled at a 24-kHz sampling rate. The LibriTTS corpus is designed for TTS research and has normalised text transcripts, as well as an automatic selection of samples that do not contain severe background noise and reverberation. LJSpeech dataset [26] is a high-quality single female speaker dataset containing around 24 h of read speech audio clips. The data are transcribed and aligned at utterance level. LibriTTS and LJSpeech were used to pretrain the TTS models.

4.2. Implementation Details

Text-to-speech. For the TTS models' training process, we focused on the Obama dataset. In our initial experiments, we found that training a system only on the Obama data yields subpar results in terms of naturalness of the synthetic output (see Table 1, system id O-8). Therefore, we decided to pretrain two large models on the LJSpeech and LibriTTS datasets, and to fine-tune them towards our target data. The models were trained over 1000 epochs and use the phonetic transcription provided by the Festival tool. In the LibriTTS model, although it contains multiple speakers, we did not condition the output on the speaker id, but rather aimed at obtaining an eigen voicelike model. For this model, as the data are recorded in semiprofessional environments, we also performed a dereverberation preprocessing step. We can summarise the two pretrained models as follows:

- **LJ-24h**—trained on the 24 h of LJSpeech dataset recordings, which translate into 13,077 utterances;
- **LT-47h**—trained on a subset of 31,526 utterances (around 47 h belonging to 558 speakers) from the LibriTTS dataset.

Starting from the phonetic transcription, dereverberation process, and pretrained models, we carried on the training procedure for several TTS systems using the Obama dataset. The systems pertain to the different pretrained models, the use of dereverberation for the Obama samples, and the amount of data selected to perform fine-tuning. The list of Obama TTS systems is presented in Table 1, along with their objective evaluation in terms of word error rate (WER) and cosine similarity to the natural recordings. The fine-tuning was run over the data for an additional 500 epochs.

Table 1. TTS models trained for the target speaker and the objective evaluation over the test set in terms of WER and cosine similarity to the natural recordings. The dereverberation column refers to the target speaker's data. The cosine similarity for the `Natural` and `Natural-dvb` rows is computed as an intradata measure by comparing random sample pairs from within the speech set. The arrows in the table header indicate the direction of best performance for the respective measure. Boldface numbers highlight the best results for the respective column.

	Training Data					Cosine
ID	Duration (h)	No. of Utts	Dereverb	Init.	WER ↓	Similarity ↑
Natural	8	4761	no	–	9.32	0.684
Natural-dvb	8	4761	yes	–	9.22	0.683
O-8	8	4761	no	–	14.72	0.673
LJ-8	8	4761	no	LJ-24h	8.48	0.697
LT-8	8	4761	no	LT-47h	**7.31**	0.709
LJ-8-dvb	8	4761	yes	LJ-24h	11.45	**0.723**
LT-8-dvb	8	4761	yes	LT-47h	11.42	0.683
LJ-1	1	545	no	LJ-24h	10.50	0.690
LT-1	1	545	no	LT-47h	8.76	0.690
LJ-1-dvb	1	545	yes	LJ-24h	9.48	0.722
LT-1-dvb	1	545	yes	LT-47h	8.64	0.713
LJ-0.3	0.3	175	no	LJ-24h	12.98	0.679
LT-0.3	0.3	175	no	LT-47h	9.68	0.677
LJ-0.3-dvb	0.3	175	yes	LJ-24h	11.43	0.704
LT-0.3-dvb	0.3	175	yes	LT-47h	9.08	0.681

Speech-to-lip. The input to the speech-to-lip network are Mel filterbank features extracted from the audio files. The output consists of lips landmarks extracted automatically from the corresponding video. Concretely, we used the `dlib` library [27] to extract facial landmarks, which produced 68 landmarks for each frame; we keep those twenty landmarks corresponding to the lips (indices 48–68). Occasionally, `dlib` detects zero or more than one face (most of the time erroneously since there is a single face appearing in the video shots); we deal with these exceptions as follows: if there is no face detected, we interpolate in time based on the neighbouring faces; if there is more than one face detected, we keep the one with the largest confidence. Next, we project the absolute coordinates into a normalised space by translating, rotating, and scaling (translate to centre the lips, rotate to have the eye-line horizontal, scale to ensure that the distance between eyes is five pixels). Finally, the lips are projected to a low-dimensional space using a principal component analysis (PCA) model fitted on a subset of landmarks from the train split of the Obama dataset. We use the top eight principal components, which capture about 97% of the total data variation.

When training the speech-to-lip network, in order to have balanced batches and improve the speed, each video shot is split into eleven-second chunks with an overlap of one second. At test time, if we are given an audio file longer than eleven seconds, we carry out a similar procedure as for training: we split the audio file into eleven-second chunks with a second overlap; predict the lips for each of the chunks separately; then, average out the overlapping regions (the averaging operation is carried in the low-dimensional space of the PCA).

The training is set to 100 epochs and the final model averages the weights of the top ten best models on the validation set [28]. We use a warm-up learning scheduler, which increases the learning rate linearly for 5000 steps up to 0.02 and then decreases it as a function of $1/\sqrt{s}$, where s is the step value. The batch size is set to 6 samples.

5. Experimental Results

This section presents quantitative results for the two components described in the paper (the text-to-speech module, in Section 5.1, and the speech-to-lips module, in Section 5.2) as well as their integration (in Section 5.3). We also provide qualitative samples corresponding to these results at the following link: https://zevo-tech.com/humans/flexlip/ (accessed on 15 May 2022).

5.1. Objective Evaluation of the TTS Models

The quality of a TTS system in general pertains to its naturalness and intelligibility, as well as speaker similarity. To evaluate the TTS systems, we analysed the objective measures of word error rate (WER) and a speaker–encoder-based cosine similarity over the entire Obama test set. These two measures have recently been found to have a high correlation to the perceptual measures obtained in listening tests [29,30]. The WERs are extracted based on the automatic transcripts provided by the SpeechBrain ASR system [22], while the cosine similarity uses SpeechBrain's speaker embedding network. Both the ASR and the speaker embedding networks are based on an ECAPA-TDNN architecture [31]. For the speaker similarity measure, we averaged the cosine similarity measure values between the synthesised samples and their natural counterparts. To estimate the cosine similarity over the natural samples, we employed an intradata evaluation and compared random pairs of samples. This means that the linguistic content is different, and it can affect the estimation of the speaker similarity, as these types of speaker embedding networks do not truly factor out the spoken content and background conditions. The results are shown in Table 1. We explore several different dimensions of the TTS system's training procedure: (i) using only the target speaker's data (id: `O-8`) vs. using a pretrained model (ids: `LT-*` and `LJ-*`); (ii) using dereverberation over the target data; (iii) number of training samples used from the target speaker. We should point out the fact that the 1- and 0.3-hour subsets are selected from the manually checked training set, and that the dereverberation refers to the target speaker's data, not the pretraining data. By inspecting the results, the first interesting thing to notice is that the WER over the natural samples (id: `Natural`) is worse than the best-performing system (id: `LJ-8`). However, during the manual check of the test set, we noticed that some of the samples contained high background noise and reverberation. This can definitely affect the performance of the ASR system. Although, even after applying the dereverberation method over the natural samples (id: `Natural-dvb`), the results were similar. We could explain this by the fact that the background noise is one of the major causes of degradation, while for the TTS system, the different background conditions are, in principle, averaged out within the model.

Dereverberation. Being also within the area of background conditions, the dereverberation algorithm was employed as a measure to improve the quality of the output synthetic speech; the cosine similarity measures support this preprocessing step. All systems trained with the dereverberated data exhibit higher cosine similarity measures with the natural samples. On the other side, in terms of WER, it seems that only when using the manually checked speech data does the dereverberation improves the overall results. When using all the data available (ids: `LJ-8` and `LT-8`), the results over the dereverberated data are not as good as for the original dataset. The interpretation of these results can be based on the fact that, although the full target speaker dataset may still contain some transcription errors and various background noises, the high amount of available data is able to leverage the errors. It is also true that, although the dereverberation algorithm ensures a better perceptual quality of the audio samples, it may introduce signal-level artefacts that will interfere with the model training step. When using a smaller amount of the target speaker's data (i.e., 20 min or 1 h), any improvement of the training data is directly transposed into the output of the synthesis system, and the dereverberated versions of these systems perform better than their non-dereverberated counterparts.

Pretrained models. With respect to the use of pretrained models, when using only the target speaker's data (id: `O-8`), the WER and cosine similarity results are less-performing than any of the fine-tuned models. It also appears that, even though the complete 8-h dataset from the target speaker may still contain alignment errors between the audio and the transcript, these are not reflected in the overall results of the `LJ-8` and `LT-8` systems. This concludes the fact that having large amounts of data can average out some of the errors in the transcript. However, using only an eighth of the speech data (i.e., 1 h) can nearly match the top-line results of our TTS systems (see system id `LT-1`). Another result of our analysis is the fact that having multiple speakers in the pretrained model can provide a better starting point for our target speaker adaptation—comparing `LJ-*` with `LT-*` systems in terms of WER—and improve its intelligibility. However, it does not influence the speaker similarity, where the results are rather similar.

Amount of training data. As a general conclusion, using as little as 20 min of transcribed data can achieve similar results as the top-line systems. This means that in scenarios where only limited data are available, given a pretrained multispeaker model, good-quality TTS systems for the target speaker can still be obtained.

5.2. Evaluating the Speech-to-Lips System

This section presents an empirical evaluation of the speech-to-lips networks and their variants. We measure the performance of the systems by mean squared error (MSE) between the ground-truth lips and the predicted lips averaged across the number of keypoints, frames, and video segments. Unless specified otherwise, the MSE is computed in the normalised and low-dimensional space (eight-dimensional).

Transferring representations. Instead of training the speech-to-lip network from scratch (from random initial weights), we incorporate learned audio representations into the speech-to-lip module by transfer learning. We initialise the audio encoder from a state-of-the-art ASR system and evaluate two variants: either keep the encoder frozen or allow to update its weights together with the decoder's. The results are presented in Table 2. We observe that the best results are obtained when the audio encoder is initialised from the ASR and is fine-tuned with the rest of the system. This approach has conceptual advantages over the other two variants: compared with a fully random initialisation, it has the benefit of reusing learnt information; compared with the frozen pretrained encoder, it has the advantage of being more flexible.

Speaker adaptation and zero-shot adaptation. In the next set of experiments, we investigate the best ways of reusing a pretrained speech-to-lip model for an unseen speaker. We explore three adaptation strategies: (i) Applying the pretrained model "as is" on audio data from the new speaker; (ii) Fine-tuning the pretrained model on a small amount of data from the new speaker (we attempt with datasets of 5, 10, 20, and 40 min); (iii) performing zero-shot adaptation by updating the PCA mean with the lip shape of the unseen speaker (as described in Section 3.2).

Table 2. Transferring representations. Evaluation of the speech-to-lip system in terms of the mean squared error (MSE) on the Obama test set. We consider three combinations of encoding the audio input in terms of initialisation (either random or from a pretrained automatic speech recognition system) and whether we train this component or keep it frozen. The arrows in the table header indicate the direction of best performance. Boldface numbers highlight the best results for the respective column.

Encoder		
Init.	**Train**	**MSE ↓**
random	yes	0.130
from ASR	frozen	0.132
from ASR	yes	**0.064**

The impact of data. To understand how the amount of training data affects the speech-to-lip model, we performed a systematic study in which we trained the model on varying quantities of data: from the full training dataset (of 16 h) to smaller fractions (eight hours, four hours, and so on until only 15 min). The subsampling of the training data was carried at the video level, and not at the chunk level. We believe that this setup is more realistic, as it emulates the scenario in which we have access only to a few videos. Note, however, that this subsampling strategy might yield less-diverse training samples than subsampling chunks from the entire training dataset. For these experiments, we adjusted the warm-up cycle of the learning rate for the smaller datasets. More precisely, we linearly scaled the number of warm-up steps with the fraction of the data used.

The quantitative results are presented in Figure 4. We notice two regimes: (i) using at least four hours of video seems to yield reasonable performance, close to that obtained using the entire dataset of sixteen hours; (ii) using less than one hour of data, the results are degraded.

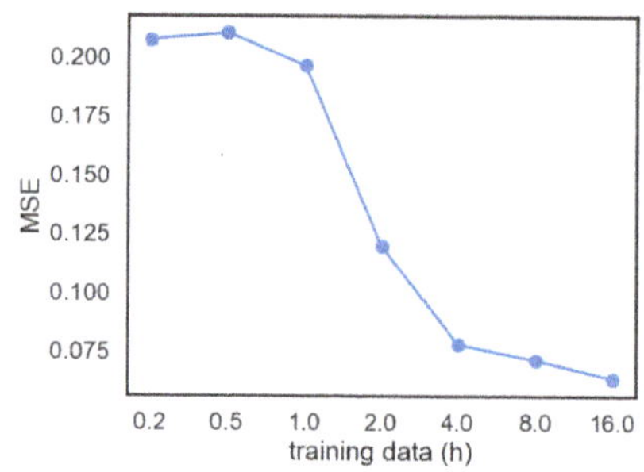

Figure 4. The impact of training data. We report the mean squared error (MSE) on the Obama test set for multiple speech-to-lip networks trained on varying fractions of data—1/64, 1/32, 1/16, and so on—up to the entire dataset, which consists of around 16 h of data. All models have their encoder initialised from a pretrained ASR and fully trained (not frozen).

We start from the best speech-to-lip model trained on the full Obama data, with the encoder initialised from an ASR and fully fine-tuned (the model corresponding to the last row in Table 2). The evaluation setup follows the previous one, but differs in a couple of aspects:

- Data: for evaluation and fine-tuning, we used the Trump dataset, which was trawled from the internet, as described in Section 4.1.2. For simplicity, we evaluate on chunks of data (eleven-second chunks with overlap of one second), preprocessed in the same way as for training; based on the experiments on the Obama dataset, we have found very similar results for the per-chunk and per-sentence evaluations.

- Evaluation metrics: in addition to the mean squared error (MSE) computed in the normalised PCA space (8-dimensional; 8D), we also report the MSE computed in the original, 40-dimensional (40D) space. This second evaluation is especially relevant for the zero-shot adaptation method, which only affects the reconstructed lips.

The results are presented in Table 3. We observe that the first adaptation approach, which directly uses the Obama-trained model, yields a performance of 0.345 MSE 8D or 0.071 MSE 40D. Unsurprisingly, the results are much worse than what we observed when applying the model to the in-domain Obama dataset, 0.064 MSE 8D (as shown in Table 2), likely due to the mismatch of the datasets. The performance can be improved by the second adaptation method, fine-tuning on the speaker-specific data, which leads to better results with more data, reaching the best value of 0.096 MSE 8D and 0.021 MSE 40D (shown in bold in Table 3). These results are close to those obtained on the Obama dataset with around two–four hours of data (see Figure 4), showing that starting from a pretrained model can alleviate the need of large quantities of training data. Finally, we see that the proposed zero-shot adaptation method yields a significant improvement over the baseline: the error halves from 0.071 to 0.034 MSE 40D (the figure in italics from Table 3), which is close to the best result we achieve, of 0.021 MSE 40D. Note that this performance improvement is obtained without performing any training on the new speaker: we just update the PCA mean at test time.

As an additional experiment, we attempt to combine the second and third approaches for adaptation: we update the mean of the PCA to Trump's lips shape also for the fine-tuned models. While the results are better than the zero-shot variant, they do not improve over the fine-tune-only variant that uses Obama's lips shape. We believe that this happens because the fine-tuning process helps in adjusting for the new speaker's lip shape, which then—when changed at test time—negatively affects the system.

Table 3. Speaker adaptation and zero-shot adaptation. We report two variants of the mean squared error (MSE) on the Trump test set: one computed in the 8-dimensional PCA space (MSE 8D), the other computed in the reconstructed 40-dimensional original space (MSE 40D). The table presents results for three adaptation methods, as follows: 1. The first two entries from the first row correspond to the system pretrained on the Obama dataset. 2. The entries in rows 2–5 correspond to the pretrained model fine-tuned on various amounts of Trump data. 3. The third column corresponds to updating the PCA mean at test time to Trump's shape; the first entry in the third column (shown in italics) represents the proposed zero-shot speaker adaptation. The arrows in the table header represent the direction of the best performance for the respective measure. Boldface numbers highlight the best results for the respective column.

		MSE ↓ 8D	**MSE ↓ 40D**	
	Training		**PCA Mean**	
	Data		**Obama**	**Trump**
1	obama	0.345	0.071	*0.034*
2	trump 5 m	0.109	0.024	**0.024**
3	trump 10 m	0.101	0.022	0.032
4	trump 20 m	0.099	0.022	0.031
5	trump 40 m	**0.096**	**0.021**	0.030

5.3. Evaluating the Complete Text-to-Keypoints System

In this section, we evaluate our text-to-lip method in an end-to-end manner—that is, given a text input, we want to asses the quality of the generated lips. A major challenge of this joint evaluation is that the generated lips are not guaranteed to be synchronised with the ground-truth lips, since the synthesised audio is not necessarily synchronised with the natural audio. To address this issue, we propose two approaches, both aimed at aligning the intermediary audio representation. The first approach uses dynamic time warping (DTW) to align the Mel-frequency cepstral coefficient (MFCC) representation of the two audio sources. We use forty-dimensional MFCCs, and to facilitate the transfer of the alignment at the lip level, we extract the MFCCs using a hop length that yields the same number of coefficient vectors as the number of video frames. In the second approach, we make crucial use of our model's ability to control the phoneme-level durations within the TTS system. More precisely, we set the durations to those obtained by running a forced-aligner over the phonetically transcribed evaluation text and its corresponding natural audio.

For the current evaluation, we consider the best TTS models as determined from the objective evaluation, i.e., LJ-8 and LT-8 (see Table 1) in terms of WER. We consider that the intelligibility of the speech is more likely to affect the correct lip movement, as opposed to having a speech input, which has a smaller speaker similarity measure. To estimate an upper bound of the performance, we also measure the performance obtained by starting with natural audio. As in the previous experiments, we report the mean squared error (MSE) computed in the 8D PCA space between the generated lips and the automatically extracted landmarks from the original video sequence.

The results are shown in Table 4. We can observe that using a DTW-based alignment, the estimated MSE measure is worse than the one obtained when using the original phone durations. This means that being able to control this particular aspect of the generated audio enables us to perform a more accurate evaluation of the speech-to-lip component. With respect to the natural vs. synthesised speech, the differences are not substantial, 0.094 vs. 0.064 in favour of natural speech. These differences partially pertain to the fact that the forced aligner is not perfect, and slight alignment errors are still present between the natural and synthesised audios. However, it is impossible to evaluate how much of the total error is determined by the misalignments versus the lip landmark generation network. Perceptual differences over the lip generation performance between the two types of speech inputs can also be analysed from our samples' page (https://zevo-tech.com/humans/flexlip/ (accessed on 15 May 2022)).

Table 4. From text to keypoints. We report the mean squared error on the Obama test set with respect to the use of natural vs. synthetic audio data. We also report the MSE values over the Dynamic Time Warping (DTW) alignments of the synthetic audio to the natural audio vs. controlling the phoneme durations (phone Δ) within the TTS system. The arrow in the table header represent the direction of best performance for the respective measure.

Audio	Alignment	MSE ↓
Natural	–	0.064
TTS · LJ-8	DTW	0.179
TTS · LT-8	DTW	0.181
TTS · LJ-8	phone Δ	0.095
TTS · LT-8	phone Δ	0.094

6. Conclusions

In this paper, we proposed a flexible text-to-speech-to-lip pipeline that allows the user to control various facets of its outputs: the phone durations, the pitch contour of the voice, the audio altogether, the shape of the lips. Our model is based on two components: a text-to-speech network and a speech-to-lip network. For the text-to-speech component, we proposed using the FastPitch [18] architecture, which we carefully evaluated in multiple settings (involving the quantity and quality of the training data) and showed that the synthetic speech of the best models is intelligible and resembles the original speaker's voice. For the speech-to-lip part, we used a Transformer network to map the Mel-spectrogram representation of the audio to PCA-encoded lips, which capture the dynamics of the lip movements. We have made the observation that the shape is encoded by the PCA mean and this can be easily replaced at test time (in a zero-shot adaptation setting), yielding results close to those obtained when retraining with 40 min of data. Finally, we have made crucial use of the controllability of our pipeline to carry out an objective end-to-end evaluation—by setting the phone durations to match the natural audio, we managed to obtain synchronised generated and natural lips, ensuring a relevant score. Importantly, these results have shown that the speech-to-lip module is robust to using synthetic data at input, as the performance of the full pipeline is close to that of the speech-to-lip with natural audio. As future work, we plan to extend our pipeline to the video domain (with a keypoints-to-video component) and use the proposed objective evaluation approach to evaluate objectively the output of the end-to-end system.

Author Contributions: Methodology, D.O., B.L., A.S. and H.C.; software, D.O., B.L., A.S. and H.C.; writing—review and editing, D.O., B.L., A.S. and H.C. All authors have read and agreed to the published version of the manuscript.

Funding: This research was funded by Humans DNA and Zevo Technology. This work was supported in part by a grant of the Romanian Ministry of Education and Research, CNCS-UEFISCDI, project number PN-III-P1-1.1-PD-2019-0918, within PNCDI III.

Institutional Review Board Statement: Not applicable.

Informed Consent Statement: Not applicable.

Data Availability Statement: The processed datasets used in this paper can be obtained from the authors.

Acknowledgments: All the authors contributed equally to this work.

Conflicts of Interest: The authors declare no conflict of interest.

Sample Availability: Qualitative samples are available on FlexLip's webpage: https://zevo-tech.com/humans/flexlip/ (accessed on 15 May 2022).

References

1. Chung, J.S.; Jamaludin, A.; Zisserman, A. You said that? *arXiv* **2017**, arXiv:1705.02966.
2. Thies, J.; Elgharib, M.; Tewari, A.; Theobalt, C.; Nießner, M. Neural voice puppetry: Audio-driven facial reenactment. In Proceedings of the European Conference on Computer Vision, Virtual, 23–28 August 2020; pp. 716–731.
3. Kumar, R.; Sotelo, J.; Kumar, K.; de Brébisson, A.; Bengio, Y. ObamaNet: Photo-realistic lip-sync from text. *arXiv* **2017**, arXiv:1801.01442.
4. Zhang, S.; Yuan, J.; Liao, M.; Zhang, L. Text2Video: Text-driven Talking-head Video Synthesis with Phonetic Dictionary. *arXiv* **2021**, arXiv:2104.14631
5. Suwajanakorn, S.; Seitz, S.M.; Kemelmacher-Shlizerman, I. Synthesizing Obama: Learning lip sync from audio. *Acm Trans. Graph.* **2017**, *36*, 1–13. [CrossRef]
6. Eskimez, S.E.; Maddox, R.K.; Xu, C.; Duan, Z. Generating talking face landmarks from speech. In Proceedings of the International Conference on Latent Variable Analysis and Signal Separation, Guildford, UK, 2–6 July 2018; pp. 372–381.
7. Greenwood, D.; Matthews, I.; Laycock, S. Joint learning of facial expression and head pose from speech. In Proceedings of the Interspeech, Hyderabad, India, 2–6 September 2018.
8. Bregler, C.; Covell, M.; Slaney, M. Video Rewrite: Driving Visual Speech with Audio. In Proceedings of the 24th Annual Conference on Computer Graphics and Interactive Techniques, Los Angeles, CA, USA, 3–8 August 1997; pp. 353–360.
9. Ezzat, T.; Poggio, T. Visual Speech Synthesis by Morphing Visemes. *Int. J. Comput. Vision* **2000**, *38*, 45–57. [CrossRef]
10. Taylor, S.L.; Mahler, M.; Theobald, B.J.; Matthews, I. Dynamic Units of Visual Speech. In Proceedings of the ACM SIGGRAPH/Eurographics Symposium on Computer Animation, Lausanne, Switzerland, 29–31 July 2012; Eurographics Association: Goslar, Germany, 2012; pp. 275–284.
11. Fried, O.; Tewari, A.; Zollhöfer, M.; Finkelstein, A.; Shechtman, E.; Goldman, D.B.; Genova, K.; Jin, Z.; Theobalt, C.; Agrawala, M. Text-Based Editing of Talking-Head Video. *ACM Trans. Graph.* **2019**, *38*, 1–14. [CrossRef]
12. Kim, Y.; Nam, S.; Cho, I.; Kim, S.J. Unsupervised keypoint learning for guiding class-conditional video prediction. *arXiv* **2019**, arXiv:1910.02027.
13. Chen, L.; Wu, Z.; Ling, J.; Li, R.; Tan, X.; Zhao, S. Transformer-S2A: Robust and Efficient Speech-to-Animation. *arXiv* **2021**, arXiv:2111.09771.
14. Villegas, R.; Yang, J.; Zou, Y.; Sohn, S.; Lin, X.; Lee, H. Learning to generate long-term future via hierarchical prediction. In Proceedings of the International Conference on Machine Learning, Sydney, Australia, 6–11 August 2017; pp. 3560–3569.
15. Aldeneh, Z.; Fedzechkina, M.; Seto, S.; Metcalf, K.; Sarabia, M.; Apostoloff, N.; Theobald, B.J. Towards a Perceptual Model for Estimating the Quality of Visual Speech. *arXiv* **2022**, arXiv:2203.10117.
16. van den Oord, A.; Dieleman, S.; Zen, H.; Simonyan, K.; Vinyals, O.; Graves, A.; Kalchbrenner, N.; Senior, A.; Kavukcuoglu, K. WaveNet: A Generative Model for Raw Audio. *arXiv* **2016**, arXiv:1609.03499.

17. Ren, Y.; Hu, C.; Tan, X.; Qin, T.; Zhao, S.; Zhao, Z.; Liu, T.Y. FastSpeech 2: Fast and High-Quality End-to-End Text to Speech. In Proceedings of the International Conference on Learning Representations, Virtual, 3–7 May 2021.
18. Łańcucki, A. Fastpitch: Parallel text-to-speech with pitch prediction. In Proceedings of the ICASSP 2021–2021 IEEE International Conference on Acoustics, Speech and Signal Processing (ICASSP), Toronto, ON, Canada, 6–11 June 2021; pp. 6588–6592.
19. Beliaev, S.; Ginsburg, B. TalkNet 2: Non-Autoregressive Depth-Wise Separable Convolutional Model for Speech Synthesis with Explicit Pitch and Duration Prediction. *arXiv* **2021**, arXiv:2104.08189.
20. Prenger, R.; Valle, R.; Catanzaro, B. Waveglow: A flow-based generative network for speech synthesis. In Proceedings of the ICASSP 2019-2019 IEEE International Conference on Acoustics, Speech and Signal Processing (ICASSP), Brighton, UK, 12–17 May 2019; pp. 3617–3621.
21. Kingma, D.P.; Dhariwal, P. Glow: Generative Flow with Invertible 1 × 1 Convolutions. In Proceedings of the Advances in Neural Information Processing Systems, Montréal, QC, Canada, 3–8 December 2018; Volume 31.
22. Ravanelli, M.; Parcollet, T.; Plantinga, P.; Rouhe, A.; Cornell, S.; Lugosch, L.; Subakan, C.; Dawalatabad, N.; Heba, A.; Zhong, J.; et al. SpeechBrain: A General-Purpose Speech Toolkit. *arXiv* **2021**, arXiv:2106.04624.
23. Taylor, P.; Black, A.W.; Caley, R. The architecture of the Festival speech synthesis system. In Proceedings of the Third ESCA/COCOSDA Workshop (ETRW) on Speech Synthesis, Blue Mountains, Australia, 26–29 November 1998.
24. Panayotov, V.; Chen, G.; Povey, D.; Khudanpur, S. Librispeech: An ASR corpus based on public domain audio books. In Proceedings of the 2015 IEEE International Conference on Acoustics, Speech and Signal Processing (ICASSP), South Brisbane, Australia, 19–24 April 2015; pp. 5206–5210. [CrossRef]
25. Zen, H.; Clark, R.; Weiss, R.J.; Dang, V.; Jia, Y.; Wu, Y.; Zhang, Y.; Chen, Z. LibriTTS: A Corpus Derived from LibriSpeech for Text-to-Speech. *arXiv* **2019**, arXiv:1904.02882.
26. Ito, K.; Johnson, L. The LJ Speech Dataset. 2017. Available online: https://keithito.com/LJ-Speech-Dataset/ (accessed on 15 May 2022).
27. King, D.E. Dlib-ml: A Machine Learning Toolkit. *J. Mach. Learn. Res.* **2009**, *10*, 1755–1758.
28. Izmailov, P.; Podoprikhin, D.; Garipov, T.; Vetrov, D.; Wilson, A.G. Averaging weights leads to wider optima and better generalization. In Proceedings of the Uncertainty in Artificial Intelligence, Monterey, CA, USA, 6–10 August 2018.
29. Taylor, J.; Richmond, K. Confidence Intervals for ASR-based TTS Evaluation. In Proceedings of the Interspeech, Brno, Czech Republic, 30 August–3 September 2021.
30. Dehak, N.; Kenny, P.J.; Dehak, R.; Dumouchel, P.; Ouellet, P. Front-end factor analysis for speaker verification. *IEEE Trans. Audio Speech Lang. Process.* **2010**, *19*, 788–798. [CrossRef]
31. Desplanques, B.; Thienpondt, J.; Demuynck, K. ECAPA-TDNN: Emphasized Channel Attention, Propagation and Aggregation in TDNN Based Speaker Verification. *arXiv* **2020**, arXiv:2005.07143.

sensors

MDPI

Article

A Transformer-Based Neural Machine Translation Model for Arabic Dialects That Utilizes Subword Units

Laith H. Baniata [1], Isaac. K. E. Ampomah [2] and Seyoung Park [1,*]

[1] School of Computer Science and Engineering, Kyungpook National University, 80 Daehak-ro, Buk-gu, Daegu 41566, Korea; laith@knu.ac.kr
[2] Department of Computer Science, Durham University, Stockton Road, Durham DH1 3LE, UK; Isaac.k.ampomah@durham.ac.uk
* Correspondence: seyoung@knu.ac.kr

Abstract: Languages that allow free word order, such as Arabic dialects, are of significant difficulty for neural machine translation (NMT) because of many scarce words and the inefficiency of NMT systems to translate these words. Unknown Word (UNK) tokens represent the out-of-vocabulary words for the reason that NMT systems run with vocabulary that has fixed size. Scarce words are encoded completely as sequences of subword pieces employing the Word-Piece Model. This research paper introduces the first Transformer-based neural machine translation model for Arabic vernaculars that employs subword units. The proposed solution is based on the Transformer model that has been presented lately. The use of subword units and shared vocabulary within the Arabic dialect (the source language) and modern standard Arabic (the target language) enhances the behavior of the multi-head attention sublayers for the encoder by obtaining the overall dependencies between words of input sentence for Arabic vernacular. Experiments are carried out from Levantine Arabic vernacular (LEV) to modern standard Arabic (MSA) and Maghrebi Arabic vernacular (MAG) to MSA, Gulf–MSA, Nile–MSA, Iraqi Arabic (IRQ) to MSA translation tasks. Extensive experiments confirm that the suggested model adequately addresses the unknown word issue and boosts the quality of translation from Arabic vernaculars to Modern standard Arabic (MSA).

Keywords: neural machine translation (NMT); transformer; Arabic dialects; modern standard Arabic; subword units; multi-head attention; shared vocabulary; self-attention

Citation: Baniata, L.H.; Ampomah, I.K.E.; Park, S. A Transformer-Based Neural Machine Translation Model for Arabic Dialects That Utilizes Subword Units. *Sensors* **2021**, *21*, 6509. https://doi.org/10.3390/s21196509

Academic Editors: Bruce Denby, Tamás Gábor Csapó and Michael Wand

Received: 27 August 2021
Accepted: 26 September 2021
Published: 29 September 2021

1. Introduction

The area of Machine Translation (MT) is undergoing unbelievable development thanks to deep learning and artificial neural network models. Although a few years ago, machine translation research tried to produce a high-quality translation for the most popular and resourceful languages, today's level of translation quality has increased the need and significance of low-resource languages and the solution of further and more interesting translation tasks [1]. In particular, even national language varieties such as Arabic dialects, which are practiced by large populations (450 million) in the Arab world, lands as a spoken verity of modern standard Arabic (MSA) and has been largely ignored by industry and research. At the moment, commercial machine translation services do not provide translation services for any Arabic vernaculars. Conventional translation systems that perform translation from Arabic dialects to MSA generate inconsistent outputs such as mixing lexical parts. These systems translate parts of the source sentence twice and do not produce high translation quality. Moreover, in Arabic, a linguistic phenomenon known as diglossia occurs, in which language speakers practice local vernaculars for informal environments and they practice modern standard Arabic language for formal contexts. For example, communities in Morocco use both "standard" Arabic and Maghrebi vernacular, depending on the context and situation. This Maghrebi vernacular reflects their own identity, history, lived experiences, and culture. Dialects by region are immense, such as Levantine,

Maghrebi, Yemeni, Iraqi, Nile basin (Egypt and Sudan), and Gulf. Still, Arabic vernaculars also change even within individual Arabic-speaking countries. A further difficulty is mixing Arabic dialects and modern standard Arabic language together. To illustrate the importance of dealing with Arabic vernaculars, Ethnologue reported that Arabic has the 5th largest number of L1 speakers scattered all over 21 regional vernaculars. There are four types of machine translation: statistical machine translation (SMT), rule-based machine translation (RBMT), hybrid machine translation and neural machine translation (NMT).

Traditional methods such as statistical machine translation (SMT) require powerful computing devices. SMT is not suitable for managing the problem of word order, one of the Arabic vernacular's main syntactic problems. To study the word order, we need to know where the verb, object and subject in the phrases are. According to the research studies, languages can be categorized as verb-object-subject VOS (Arabic), subject-object-verb SOV (Hindi language), subject-verb-object SVO (English), and other languages such as Arabic vernaculars that allow the free word order feature. The word order does not only convey any information about the subject and the object but possible different information (old and new). These profound differences pose a challenge to the statistical translation systems due to the fact that as sentences become lengthier, they do not just contain an object, verb, and a subject, but instead, the sentence will have a complex structure made up of several parts. In the case of Neural Machin Translation (NMT) systems, the encoder compresses the input sequence into a single vector representation as noted in the encoder-decoder structure, where the decoder uses this vector representation to produce the output sequence. However, this structure has the disadvantage that input sequence information is lost and the quality of translation declines when the input sentence is longer. Furthermore, the lack of standardized spelling for Arabic dialects presents a challenge in developing an NMT models for these vernaculars. The lack involves morphological dissimilarities which are apparent by using affixes and suffixes that are not used in MSA. Basically, for NMT systems training, we need large amounts of annotated data, which is not possible in languages with low resources such as Arabic vernaculars. Moreover, the quality of translation is decreasing alongside a decrease in the amount of the training data for low resource languages.

In Arabic dialects, the translation of rare words is a clear problem. Typically, there are 30,000–50,000 words confined to the neural model's vocabulary. Nonetheless, translation is an open-vocabulary problem, mainly in languages that use productive word-formation processes such as compounding and agglutination; models of translation require methods below word level. For instance, in word-level NMT systems, the translation for the out-of-vocabulary words was discussed via back-off to a dictionary lookup [2,3]. We note that these techniques usually make incorrect assumptions in reality. For instance, due to the differences in the morphological synthesis between Arabic vernacular and modern standard Arabic language, one-to-one connection between source words and target words is not constantly occurring. Furthermore, word-level NMT systems are ineffective in translating and generating unseen words. One of the approaches is to copy unknown words into the target text as done by [2] and [3]. It is a suitable strategy for names, but it requires transliteration and morphological changes, particularly when the characters are different. In the case of transformer model that was proposed newly [4], it has outperformed recurrent neural network (RNN)-based models [5–7] and convolutional neural network (CNN)-based models [8] on various translation tasks, drawing the attention of MT researchers. The Transformer model, which applies a self-attention approach to measure the strength of a relationship within two words in a sentence, has contributed to raising performance in MT and various natural languages processing tasks, for instance, semantic role labeling and the language modeling. The techniques to tackle the difficulties of Arabic vernaculars translation are under research and investigation. There has been no earlier research project that concentrated exclusively on developing a Transformer-based NMT model running from Arabic vernaculars to modern standard Arabic language at the level of subword units.

A Transformer-based NMT model is presented in the current research, using subword units to perform translation tasks from various Arabic vernaculars to modern standard Arabic language. Moreover, this research study introduced and developed a Word-Piece model to create subword units for the Arabic dialects. Experiments showed that machine translation tasks, computed using Bilingual Evaluation Understudy (BLEU) metric and human evaluation metric, have been enhanced on the performance of Arabic vernaculars to modern standard Arabic language. Furthermore, we found that the proposed NMT subword model based on transformers achieves higher efficiency for the translation of scarce words in comparison with models that have a large vocabulary and back-off dictionaries. The model can produce new words that are not seen during training time. Moreover, the proposed Transformer-based NMT subword model achieved high translation accuracy per sequence for Arabic dialects. Additionally, the research investigated the impact of training the model with subword embeddings and with different dimensions. Moreover, this research study investigated the influence of utilizing subword units on the Arabic dialect's translation quality. This research project investigated the impact of training the model with a different number of encoders and decoders and with a different number of attention heads in the self-attention (MHA) sub-layer in the decoder and encoder.

2. Related Work

Despite that machine translation research area has been investigated for several years and decades, the majority of research effort has focused on high-resource translation pairs, for instance, French–English and German–English which have many free parallel datasets. Nevertheless, most language pairs in the world do not have large parallel data. Research attention in these low-resource translation settings has been growing during the last five years. Translations from and to written language varieties are mainly based on phrase-based SMT systems, such as those for Croatian, Serbian and Slovenian [9], Hindi and Urdu [10], and Arabic vernaculars [11]. Pourdamghani et al. [12] developed an unsupervised deciphering design to translate similarly associated languages with no need for parallel training data. Costa-jussà [13] showed the comparison of the Catalan–Spanish language pairs amongst rule-based systems, phrase-based systems, and NMT systems. The performance of NMT is better and more reliable than other systems when an in-domain test set is applied. Experiments in the out-of-domain test dataset have shown that better performance was provided by the rule-based method from Spanish language to Catalan language and phrase-based method from Catalan language to Spanish language. In order to translate texts from Kurman to Sorani, Hassani [14] introduced and developed an Intralingual MT model. The model performed a word-to-word translation either direct or literal translation between Kurman and Sorani dialects. The outcomes have been estimated by native speaker evaluators. The experiments confirmed, according to human raters, that this strategy can produce significantly clear results. Experiments also revealed that this strategy could be regarded as a fundamental solution to the lack of corpus problem.

The first NMT system that was trained to translate among language varieties was presented by Costa-jussà et al. [15]. The authors utilized language variety pairs, European Portuguese and Brazilian Portuguese for experiments, as well as a corpus of subtitles for neural machine translation training. The authors gained an additional 0.9 BLEU points for translating from European Portuguese to Brazilian Portuguese compared to the SMT system trained on similar data and an additional 0.2 BLEU points when translating in reverse direction. The results show that the neural machine translation model offers more reliable translation in terms of BLEU scores and seven native speakers' evaluation than the SMT model. Lakew et al. [16] investigated NMT training difficulties from English into special pairs of language varieties, analyzing parallel texts and low-resource situations, both labeled as well as unlabeled. The authors conducted experiments from English to two languages, European Brazilian Portuguese and European Canadian French and two standardized pairs, from Croatian language to Serbian and from Bahasa Indonesia to Malay. The researchers demonstrate that a significant BLEU score increases over basic models when translation into related languages is learned as a multilingual task with shared representations.

The main focus of research for Arabic vernaculars has been on SMT and rule-based methods. PADIC is a multi-dialect-Arabic corpus that was introduced by Meftouh et al. [17]. The PADIC corpus includes MSA, Levantine vernaculars (Syrian and Palestinian) and Maghrebi vernaculars (Tunisian and Algerian). In comparison to many other approaches, diverse experiments were applied on different SMT models with all language pairs (vernaculars and standard Arabic). In changing the smoothing methods, the researchers examined the influence of the language model on MT by interpolating them with a larger one. The most reliable translation outcomes were obtained in Algerian vernacular, which is not remarkable because there is no closeness between the Algerian vernacular and the MSA; therefore, the SMT model during training could not capture the whole semantic and syntactic features of Algerian vernacular. It also was noticed that because of the closeness of the vernaculars, MT performances within Palestinian and Syrian were relatively high. As far as MSA is concerned, Palestinian vernacular has achieved the most reliable results of MT. Sadat et al. [18] presented an approach to do translation of the Tunisian social media vernacular into modern standard Arabic. This system depends on a bilingual lexicon that was designed for this translation task. A collection of syntactic mapping rules alongside a disambiguation phase is used to choose the most appropriate translation phrases, depending on language model for MSA. The translation system should be noted as word-based. By using a test dataset of 50 sentences of Tunisian vernacular, it achieves a BLEU score [19] of 14.32 (the reference was done by hand). Bakr et al. [20] proposed a comprehensive system for translating Egyptian vernacular phrases to enunciated versions of modern standard Arabic phrases. The authors applied the statistical method to tokenize and tag Arabic phrases. The technique for producing diacritics for the target phrases in MSA was explicitly selected based on important rules. The research was assessed using a dataset that contains 1000 Egyptian vernacular sentences where the training set is 800 and the test set is 200. The method obtained a performance of 88% when translating vernacular words to modern standard Arabic words and an accuracy of 78% when generating the words in their correct order.

The majority of the methods discussed earlier concentrated on SMT system and rule-based system. The rule-based translation method has a notable shortcoming: the development of the before-mentioned methods requires a significant quantity of time. It is essential to adjust the rules to raise the rule-based MT quality, which needs an exceptional degree of lingual understanding. The statistical methods require high computing devices and these methods are unable to manage one of the Arabic vernacular syntactic problems: the problem of word order. There have been relatively few publications in NMT discussing the translation of closely related languages. Multitasking is widely regarded as a highly effective technique for boosting the effectiveness of translation for Arabic vernaculars. A new study that investigates NMT for Arabic vernaculars was first introduced by Baniata et al. [21]. For translation from Arabic vernaculars to modern standard Arabic, the researchers presented a multi-task neural machine translation system. The suggested system is based upon multitask learning, where the language pairs share a single decoder and every source language has a separate encoder. The practical experiments demonstrate that by employing small amount of training dataset, the multitask NMT model can generate a correct MSA phrase and produce a translation with very good quality and learning the predictive information of various targets at the same time. Among many methods to translate Arabic dialects, one of the most significant is the incorporation of outer knowledge into the neural network models for Arabic dialects. Baniata et al. [22] proposed a Multitask NMT model that shares an encoder between two types of tasks; Arabic vernacular to modern standard Arabic translation task and POS task on segment level. Between translation tasks, the system shares two layers; shared layer and invariant layer. By alternatively training translation and POS tagging tasks, the proposed model may exploit distinctive knowledge and enhance the translation effectiveness from Arabic vernaculars to modern standard

Arabic. Practical experiments involve translation tasks from Levantine Arabic to modern standard Arabic and from Maghrebi Arabic to modern standard Arabic.

Nguyen et al. [23] created a lexical semantic framework for unique features of Korean text as an information database to develop a morphological analysis and word sense disambiguation system called Utagger. Moreover, the authors created a corpus for Korean–Vietnamese where they utilized the word segmentation algorithm RDRsegmenter for Vietnamese text and Utagger for Korean text. This research team was able to build a bidirectional Korean–Vietnamese NMT system, using the encoder-decoder approach with attention. These experimental findings showed that the usage of UTagger and RDRsegmenter in the Korean–Vietnamese NMT system might increase its performance, obtaining exceptional outcomes from Korean to Vietnamese with BLEU score of 27.79 and TER score of 58.77 and in reverse way a BLEU score of 25.44 and TER score of 58.72. Park et al. [24] proposed the first ancient Korean NMT system based on the use of a Transformer. The method improves translator performance by instantly generating a draft translation for different ancient documents that remain untranslated. Moreover, shared vocabulary and the entity restriction byte pair encoding is a new subword tokenization approach that was proposed by the authors recently. This approach depends on the textual characteristics of ancient Korean sentences. By using this proposed approach, the effectiveness of the traditional subword tokenizing approaches such as the byte pair encoding will rise by 5.25 BLEU points. Additionally, several decoding algorithms such as the ensemble models and n-grams blocking contribute an additional 2.89 BLEU points to the performance. Luo et al. [25] suggested an NMT model in which the network is trained sequentially on not closely related high resource language pairs, intermediate language pairs which is related and low resource language pairs. These parameters are transferred and tuned from one layer to another for initialization step in the same way. Thus, the hierarchical transfer learning design unites data amounts benefits of languages with large resources with grammatical propinquity benefits the related language. For data preprocessing, the researchers applied byte pair encoding and character level embedding, which completely address the issue of shortage of vocabulary (OOV). Experiments analyzing Uygur–Chinese and Turkish–English translations illustrate the suggested method's superiority over the neutral machine translation model with parent–child framework. Few publications have been published on the subject of MT for Arabic vernaculars that employ subword units.

Aqlan et al. [26] suggested employing a romanization method that turns Arabic texts into subword units. The authors analyzed the impact of this strategy on Neural MT performance in various segmentation settings and measure the findings to methods trained on modern standard Arabic. Additionally, the authors combine Romanized Arabic text as an input component for Arabic-sourced neural machine translation compared to well-known components, including lemma, POS tags, and morph characteristics. The experiments that were performed on Arabic–Chinese translation show that recommended methodologies address the unknown word issue and improve the translation quality for the Arabic source language. This work carries out further experiments with the NMT system and develops it on Chinese–Arabic translation. Prior to conducting the experiments, the researchers created a criteria for filtering the text in the parallel corpus to remove the noise. Included sentence patterns have been shown to improve the performance of MT, particularly SMT and RNN-based NMT [27–29]. Further, Strubell et al. [30] have enhanced a Transformer-based SRL design by adding dependency formations of phrases into self-attention, which is named linguistically-informed self-attention (LISA). In LISA, one of the attention heads of a multi-head self-attention system is trained with constraints based on dependency relations to attend to syntactic parents for each token.

3. Background

Recently, the NMT has been offered as an exciting framework that has the possibility to overcome the shortcomings of the standard SMT methods. The strength of the NMT approaches is their capability in learning the mapping from the input text to the corresponding output text directly, in an end-to-end pattern. Neural models in the domain are not new, as Neco et al. [31] proposed an approach years ago. Other models [32,33] were introduced later, but Chao et al. [6] and Sutskever et al. [7] were the first to design a robust machine translation system. Peyman et al. [34] presented an encoder-decoder structure in which two RNNs are trained to maximize a target sequence's conditional probability (possible translation). y = $y_1, \ldots, y_m$, given a source sentence x = $x_1, \ldots, x_n$. Sequentially, the input words are processed until the end of the input string is reached. The encoder reads the input sequence and turns it into fixed length representation. Every time in step t an input word is received; the hidden state is updated. Equation (1) illustrates this process:

$$h_t = f(E_x\ [x_t], h_{t-1}) \tag{1}$$

where $h_t \in R^d$ is the hidden state (vector) at the time step t and $f(.)$ is a recurrent function such as the long short-term memory (LSTM) [35] or the gated recurrent unit (GRU). $f(.)$ is reasonable for updating the hidden state of the layer and other associated unit (if there are any, such as memory unit, etc). $E_x \in R^{|V_x| \times d}$ is an embedding matrix for the source symbols (d is the embedding size). The embedding matrix is a lookup table whose cells are treated as a network parameters and updated during training. The embedding (numerical vector) for the vth word in v_x (vocabulary) resides in the vth row of the table. In the next step, the model undertakes processing for all words in the source sentence; h_n is a summary of input sequence, referred to as context vector (c). Another RNN is initialized by c and seeks to produce a target translation. There is one word sampled from a target vocabulary v_y at each step of the process. The decoder conditions the probability of picking a target word y_t on the context vector, the last predicted target symbol, and the decoder's sate. This can be expressed in Equation (2):

$$y_t = g(E_y[y_{t-1}], S_t, c) \tag{2}$$

$S_t = f(E_y[y_{t-1}], S_{t-1}, c)$ where S_t is the hidden state of the decoder. Since we compute the probability of choosing y_t as the target word, $g(.)$ should give a value in the range [0, 1]. The most common function for $g(.)$ is Softmax. The encoder and decoder RNNs are trained together to maximize the log probability of generating a target translation and are given an input sequence x, so the training standards can be defined as in Equation (3):

$$\max_{\theta} \frac{1}{K} \sum_{k=1}^{k} \log(y_k | x_k) \tag{3}$$

where θ signifies a set of network parameters and K denotes the training set's size. As previously noted, the recurrent functions used in encoder-decoder models are not conventional mathematical functions.

4. The Proposed Transformer Based-NMT Model for Arabic Dialects That Utilizes Subword Units

Even while the translation is considered to be an open vocabulary issue, systems of NMT always work with word vocabularies that are fixed (names, numbers, dates, etc.). To address out-of-vocabulary (OOV) words, there are two broad approaches. One strategy is to try to copy and obtain scarce words from source language and place them in the target language (since most of the scarce words are numbers or names so the right translation is a copy), either through the use of an attention mechanism [2], external alignment approach [3], or through the use of a complex special purpose pointing framework [36]. An extra group of methods is the subword units such as the combined word/characters or more knowledgeable subwords [37]. Subword segmentation is fundamentally an algorithm used under the assumption that a word consists of a combination of several subwords. Even Arabic dialects and MSA are languages based on Arabic characters, and many words are made of subwords. Therefore, breaking into subword units through suitable subword detection can reduce the number of vocabularies and efficiently reduce sparsity. In addition to reducing sparsity, the most representative outcome of subword segmentation is a useful coping with unknown (UNK) tokens. The majority of the deep learning NLP algorithms, including Natural Language Generation (NLG), take sentences as inputs simply as word sequences. So, when UNK appears, the probability of the language model in the future is very ruined. Therefore, it is difficult to encode or generate suitable sentences. Especially for sentence generation such asautomatic machine translation, it is more difficult because it predicts the next word based on the previous word. However, through subword units' utilization, it is possible to create a combination of known tokens by dividing UNKs such as new words or typo into units of subwords or characters. In this way, by eliminating UNK itself, you can completely cope with UNK and boosts the translation quality of Arabic vernaculars. Many word patterns were generated from the same origin and root in Arabic vernaculars and modern standard Arabic, fragmenting the data and generating scattered data.

Various research papers introduced statistical segmentation for the Arabic language in order to divide words down into their morphemes, which are the smallest meaningful unit in the language. This subtask is a fundamental part of a variety of natural language processing applications. For example, machine translation (MT) is powerful for the representation of the input and needs consistency across test and train data. Thus, by segmenting Arabic vernaculars and MSA words into subword units, the affixes and suffixes that are attached to the words are separated and the proposed model will capture more semantic and syntactic features of the input source sentence and produce a high-quality MSA sentence. This research paper developed an Arabic dialects Transformer-based NMT model that utilizes AD and MSA Subwords units to translate from different Arabic vernaculars to MSA. We created the model depend on the Transformer model introduced recently by Vaswani et al. [4]. For the proposed Transformer-Based NMT Subword model, as illustrated in Figure 1, both the decoder and encoder consist of a stack of 12 layers. Every layer has two different sub-layers: multi-head attention sub-layer and position wise feed forward sublayer (FFN). The encoder and the decoder in the suggested Transformer NMT model architecture for Arabic dialects make use of an attention model and feed-forward net to create sequences of changeable lengths without the need to use the RNN unit or CNN unit. The operation of attention across the various layers is based on multi-head attention (see Section 4.1). An input sequence of symbol representations (source sentence) $X = (x_1, x_2, \ldots, x_{nenc})T$ is mapped to an intermediate vector. Next, the decoder creates an output sequence (target sentence) $Y = (y_1, y_2, \ldots, y_{ndec})T$, given the intermediate vector. Because the transformer architecture does not contain convolutional or recurrent structure, it encodes positional word information as sinusoidal positional encodings:

$$P_{(pos\,,2i)} = \sin\left(pos/10000^{2i/d}\right) \tag{4}$$

$$P_{(pos\,,2i+1)} = \cos\left(pos/10000^{2i/d}\right) \tag{5}$$

where pos is position, i is considered to be the dimension, and d is the dimension of the intermediate representation. At the first layer of both encoder and the decoder, the positional encodings computed by Equations (4) and (5) are summed to the input embeddings. The encoder subnetwork comprises a stack of L similar layers so that L is set to different numbers 12, 8, and 4.

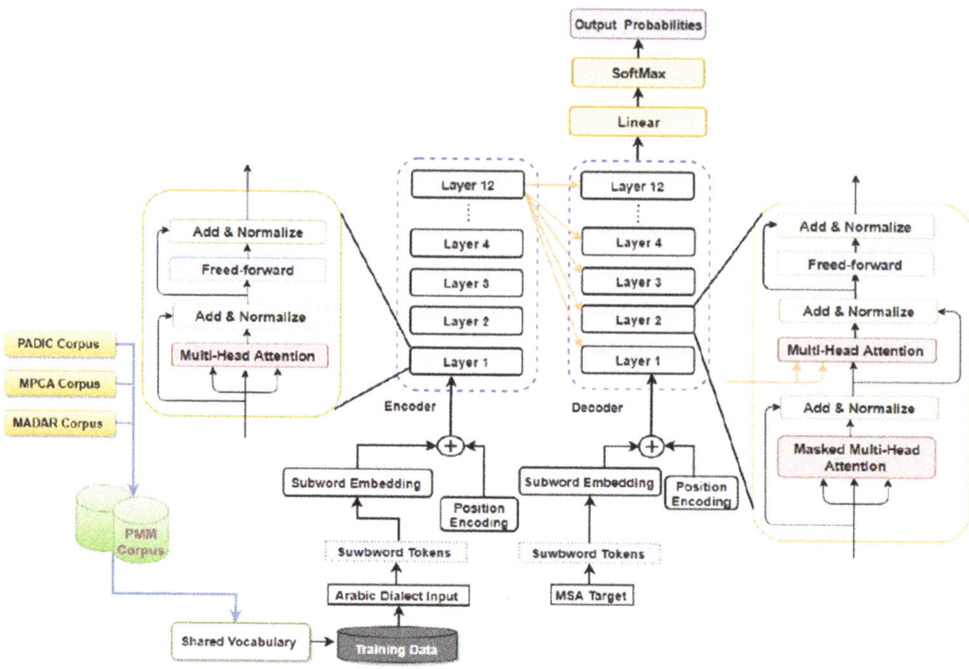

Figure 1. The Architecture of Transformer Based-Neural Machine Translation Subword Model for Arabic dialects.

Every encoding layer has two layers: a multi-head attention sub-layer and position wise feed forward sub-layer. To ease training and improve performance, residual connection mechanism [38] and a layer normalization unit (LayerNorm) [39] are employed around each sublayer. Formally, the outcome of every layer l (H_e^l) is calculated as below:

$$S_e^l = LayerNorm\left(MHA\left(H_e^{l-1}, H_e^{l-1}, H_e^{l-1}\right) + H_e^{l-1}\right), \tag{6}$$

$$H_e^l = LayerNorm\left(FFN\left(S_e^l\right) + S_e^l\right), \tag{7}$$

where S_e^l is considered to be the output from multi-head attention sublayer calculated based upon source sentence representation of previous encoding layer $(l-1)$. Moreover, the decoder consists of a stack of L similar layers in which L is set to different numbers 12, 8, and 4. Unlike the encoder, every layer in the decoder consists of three sublayers, a multi-head attention sublayer and a position wise feed forward sublayer. However, the encoder decoder multi-head attention sublayer is placed between them. The (encoder-decoder) multi-head attention sublayer is utilized to perform attention calculations for the output of encoder H_e^L particularly, output of every decoding layer l (H_d^l) is computed as:

$$S_d^l = LayerNorm\left(MHA\left(H_d^{l-1}, H_d^{l-1}, H_d^{l-1}\right) + H_d^{l-1}\right), \tag{8}$$

$$E_d^l = LayerNorm\left(MHA\left(S_d^l, H_e^L, H_e^L\right) + S_d^l\right), \tag{9}$$

$$H_d^l = LayerNorm\left(FFN\left(E_d^l\right) + E_d^l\right) \tag{10}$$

where S_d^l is considered to be the output of multi-head attention sub-layer computed from target representation from previous decoder layer $(l-1)$. E_d^l is considered to be the output of the encoder decoder MHA sub-layer generated based upon S_d^l and H_e^L. The top-level layer output (H_d^L) of the decoder is used by a linear transformation layer to generate the target sequence. Specifically, the linear transformation layer via Softmax activation computes probability distribution of the output for target vocabulary.

4.1. Multi-Head Attention (MHA)

A neural attention mechanism is an essential feature of the seq2seq structure, which is used to solve a variety of sequence generating challenges, including document summarization [40] and NMT. The Transformer Based-NMT subword model perform the scale dot product attention function as shown in Figure 2. This takes three vectors as inputs, the queries Q, values V and keys K. It outlines the provided query and key–value pairs to an output weighted sum of the values. The weights show the association among every query and key. An attention is illustrated below:

$$Attention(Q,K,V) = softmax(\alpha)V \tag{11}$$

$$\alpha = score(Q,K) \tag{12}$$

$$score(Q,K) = \frac{Q \times K^T}{\sqrt{d_k}} \tag{13}$$

where $k \in R^{J \times d_k}$ is the key, $V \in R^{J \times d_v}$ is the value $Q \in R^{Z \times d_k}$ is a query. Z and J are considered to be the lengths of sequences expressed by Q and K, respectively. d_k and d_v are considered to be the dimension of value and key vectors, respectively. The query dimension is expressed by d_k to perform the dot product calculation.

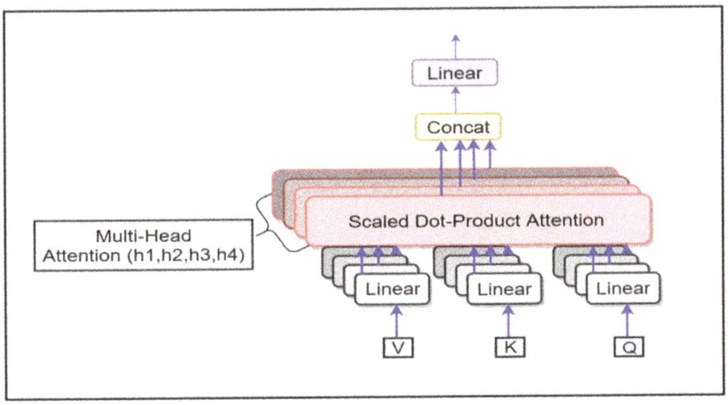

Figure 2. The Multi-Head Attention consist of several attention layers running in parallel.

The division of $Q \times K^T$ by $\sqrt{d_k}$ is performed to measure the output of the product operation so maintaining the calculation Vaswani et al. [4]. The overall attention weight distribution is obtained by applying the $softmax(.)$ operation to the attention score $\alpha \in R^{Z \times J}$. For better performance, the transformer architecture uses MHA, which comprises N_h (number of head attentions) measured dot product attention operations. Provided the Q, K, and V, multi-head attention computation is shown below:

$$MHA(Q,K,V) = O, \tag{14}$$

$$O = HW_o, \tag{15}$$

$$H = concat(head_1, head_2, \ldots\ldots, head_{N_h}) \tag{16}$$

$$head_h = Attention\left(QW_h^Q, KW_h^k, VW_h^v\right) \tag{17}$$

where QW_h^Q, KW_h^k and VW_h^v are projections of query, key and value vectors for hth head, respectively. These projections are made with metrics $W_h^Q \in R^{d_{model} \times d_k}$, $W_h^k \in R^{d_{model} \times d_k}$, $W_h^v \in R^{d_{model} \times d_v}$. The inputs to the MHA(.) are $K \in R^{J \times d_{model}}$, $V \in R^{J \times d_{model}}$ and $Q \in R^{Z \times d_{model}}$. $head_h \in R^{J \times d_v}$ is the output of measured dot product calculation for the hth head. The N_h measured dot product operation are united by using the concatenation function $concat(.)$ to generate $H \in R^{Z \times (N_h . d_v)}$. Eventually, the output $O \in R^{Z \times d_{model}}$ is produced from the projections of H utilizing the weight matrix $W_o \in R^{(N_h . d_v) \times d_{model}}$. The MHA contains the same number of parameters as vanilla attention if

$$d_k = d_v = \frac{d_{model}}{N_h} \tag{18}$$

4.2. Segmentation Approach: Wordpiece Model

The subword units are the best approach for handling the problems and challenges of Arabic dialects. This study uses the WordPiece Model (WPM) implementation, which was originally used by Google to tackle a Japanese–Korean segmentation challenge [41]. This method is entirely data-driven and guarantees that any possible Arabic dialect sequences are segmented in a deterministic way. It is similar to the approach used in Neural Machine Translation [37] to address rare words. To begin the process of the random words, we divide these words to word pieces using a trained wordpiece method. Prior to training the model, accurate word boundary symbols are added to ensure that original word sequence is extracted without ambiguity of word piece sequence. At the time of decoding, this model produces the wordpiece sequence, where this wordpiece sequence is reshaped to an identical word sequence. The example below illustrates the sequence of the word and equal word piece sequence for a sentence in Levantine Arabic vernacular:

Word: "وين فيي اركب عباص مخرج المدينة"
Word Translation: "Where can I take a bus to the city exit"
Wordpieces: "_وين _في ي _اركب _ع باص _مخرج _المدينة"

As illustrated in the above example: The Arabic word in LEV dialect "فيي" " I can " is decomposed into two-word pieces " في" "particle that derives a preposition from فيي" and "_ ي," " particle that derives a suffix from فيي" while the word " عباص" " a bus " is decomposed into word pieces " ع" " particle that derives an affix from عباص" and "_ باص" "particle that derives a noun from عباص". The remaining of the words are maintained as single word pieces. Wordpiece design is constructed by employing the data driven method which maximizes the language model probability for the training data given a word description. In the availability of parallel corpus for training and set of tokens R, the optimization challenge is by choosing R word pieces in a way where the final corpus contains the fewest word pieces when segmented by the word piece method. A unique token is utilized at the beginning of words rather than two ends. Additionally, the number of primary characters is decreased to a changeable number based upon the data. Moreover, the remaining characters are mapped to a particular unknown alphabet to avoid connecting word piece vocabulary with divided characters.

We noticed that when exploiting a vocabulary within 100,000-to-24,000 word pieces, it leads to a high BLEU score and fast decoding quality for all language pairs that were evaluated. It is advantageous in translation to copy scarce names or numbers from source language to target language in a direct way. We utilized a shared word piece approach for the source language (Arabic dialect) and the target language (modern standard Arabic) to facilitate this type of direct copying. When this strategy is applied, the same string is segmented precisely in the same way in the source and target sentences, which makes it more straightforward for the model to copy these tokens. Word pieces accomplish stability between words' efficiency and alphabets' flexibility. The motivation behind can be summarized in two points. The first point illustrates that the processing for the source language (Arabic vernaculars) and the target language (MSA) is done by exploiting the shared vocabulary approach. The encoder of the proposed model shares the same vocabulary with the decoder. By applying the shared vocabulary approach between decoder and encoder, the proposed model substitutes the words in input sentence with translation words in target language. The second point, as illustrated by Lample et al. [42], a shared vocabulary enhances the alignment of embedding vectors. We also notice that when we use wordpieces, our suggested model achieves superior overall BLEU scores, which is most likely due to the proposed model's ability to deal with an effectively unlimited vocabulary without simply relying on characters. The latter would require more computation and greatly increase average lengths for output and input sequences.

5. Experimental Results

Multiple experiments were conducted to evaluate the proposed Transformer-based NMT subword system on a variety of translation tasks. The introduced model is evaluated on the basis of its ability to translate from Arabic vernaculars to modern standard Arabic. Practical experiments were carried out with five different dialects of Arabic: Levantine, Nile Basin, Maghrebi, Gulf and Iraqi. Levantine Arabic is an Arabic dialect spoken widely in Jordan, Syria, Lebanon and Palestine. The Maghrebi variety is commonly practiced in Algeria, Morocco, Tunisia, Libya. Arabic in Nile Basin is a popularly spoken dialect used in Egypt, Sudan and South Sudan. Gulf Arabic is a spoken dialect commonly spoken in KSA, UAE, Qatar, Oman, Kuwait and Bahrain. Iraqi Arabic is a dialect spoken in Iraq. For the language of low resources, the proposed Transformer-Based NMT subword model will be applied.

5.1. Data

For the translation tasks, we grouped the Maghrebi vernaculars (Moroccan vernacular, Algerian vernacular, Tunisian vernacular and Libyan vernacular) unitedly from PADIC corpus [17], MPCA corpus [43] and MADAR corpus [44] into a single corpus, we will name it PMM-MAG. The Levantine vernaculars (Jordanian vernacular, Syrian vernacular, Lebanese vernacular and Palestinian vernacular), which are grouped collectively from PADIC corpus, MPCA corpus and MADAR corpus are grouped into a single corpus, and we will name it PMM-LEV. Furthermore, we concatenated Nile Basin Dialects (Egyptian vernacular, Sudanese vernacular) from MADAR corpus and the Gulf Dialects (Saudi Dialect, Omani Dialects and Qatari Dialect) are concatenated together from the same corpus. Moreover, we used the MADAR Corpus for the translation task of the Iraqi dialect. Figure 3 presents an example translation graph with nodes and dotted edges. We will use this graph as our running example. The Transformer NMT subword system was trained on 36,850 sentence pairs for Levantine vernacular, 54,736 sentence pairs for Maghrebi vernacular, 18,000 sentence pairs for Nile Basin Dialect, 18,000 sentence pairs for Gulf vernacular and 5000 sentence pairs for Iraqi vernacular. Textual information was gathered from many resources such as television episodes, films and social media. Regarding the test dataset, the proposed system was tested on 3000 sentence pairs for Levantine vernaculars, 3000 sentence pairs for Maghrebi vernacular, 2000 sentence pairs for Nile Basin vernacular, 2000 sentence pairs for Gulf vernacular, and 1000 sentence pairs for Iraqi vernacular. Moreover, the suggested system was trained with 13,805 sentence pairs for Levantine vernacular and it was trained on 17,736 sentence pairs for Maghrebi vernacular from the same corpus that was utilized by Baniata et al. [22] and tested on 2000 sentence pairs for Levantine vernacular and 2000 sentence pairs for Maghrebi vernacular.

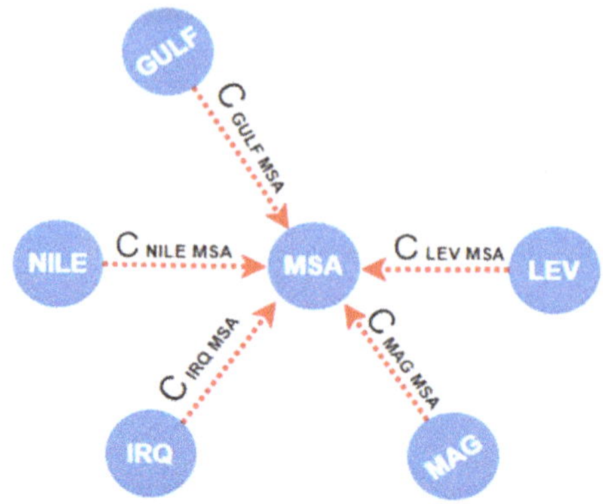

Figure 3. Translation graph: Arabic dialects (nodes), Parallel Corpora (dotted edges).

The parallel corpus for each Arabic dialect was divided into two parts: 80% for training and 20% for testing. Additionally, each Arabic dialect's test set was drawn from the same domain. The corpora employed in this study contain unprocessed information that may affect the proposed model's performance. Therefore, all categories of Arabic vernaculars and modern standard Arabic sentences have undergone pre-processing. Hashtags, punctuation, non-Arabic characters and diacritics were excluded in Arabic vernaculars and modern standard Arabic. Additionally, the orthographic normalization process was applied. As an example, the characters آإأ were converted to the ا alphabet. Stop word removal or stemming have not been applied. The modern standard Arabic (MSA) includes various tokens found in Arabic dialects (AD), and the AD Sentences are shorter than those found in MSA. Three cross-validation strategies are commonly applied to determine a predictor's predicted success rate: jackknife test approach, K-fold cross validation and independent dataset test. Among these procedures, the jackknife test is considered to be the least arbitrary and the most thematic, and as a result, it is popularly known and regularly chosen by researchers in order to evaluate the quality of different predictors. However, this strategy consumes the time and consumes the source because its estimated standard error tends to be slightly larger than other methods. Moreover, the jackknife test performs poorly when the estimator is not sufficiently smooth. Accordingly, K-fold cross-validation is applied in this paper, which sets K to 2 to create a train/ test split to assess the proposed Transformer-Based NMT Subword model. To avoid model overfitting, we used the Early stopping option where the patience parameter is set to 3 epochs and the model checkpoint is used to save the best weights for the evaluation of the proposed model.

5.2. Model Setup

The proposed model was developed using Python, Keras and TensorFlow. The experiments on the LEV–MSA, MAG–MSA, GULF–MSA, Nile–MSA and IRQ–MSA translation tasks are conducted based on these basic and advanced configurations where the subword embedding dimension has the three values which are 1024, 512 and 256, hidden state has two values which are 1024 and 512 and the attention heads are set to two values which are 8 and 4. The position-wise FFN has a filter of dimensions 512 and 1024. The proposed Arabic dialect Transformer-based NMT Subword model trained on the LEV–MSA, MAG–MSA, GULF–MSA, Nile–MSA and IRQ–MSA translation tasks consists of a 12, 8 and 4-layer encoder subnetwork. Moreover, it consists of a 12, 8 and 4-decoder subnetwork.

5.3. Traning and Inference

For the translation tasks LEV–MSA, MAG–MSA, GULF–MSA, Nile–MSA, and IRQ–MSA, the proposed model is trained for 13k iterations with batch size 2048 tokens and the maximum sequence length is set to 100 subword tokens. Moreover, the maximum subword tokens length is set to 150 subword tokens. The optimizer employed to train the model in this research study is Adam optimizer [45] with ($\beta1 = 0.9$, $\beta2 = 0.98$, $\epsilon = 1 \times 10^{-9}$). Further, the number of epochs is set to 6 for all translation tasks. By following So et al. [46], a single cosine cycle with a warm up is applied for the learning rate schedule algorithm. The target sentences are produced by using beam search during the inference stage. For the LEV–MSA, MAG–MSA, GULF–MSA, Nile–MSA, IRQ–MSA, LEV–MSA (Baniata et al. [22] Corpus) and MAG–MSA (Baniata et al. [22] Corpus) translation tasks, beam size of 6 and length penalty of 1.1 are applied. The study used a shred vocabulary for the source language and target language. This research study employed 21,000-subword vocabularies for Levantine vernacular (LEV)—MSA translation task, 21,000 subword vocabularies for Maghrebi vernacular (MAG)–MSA translation task, 21,000 subword vocabularies for the Nile Basin Arabic (NB)—MSA translation task, 21,000 subword vocabularies for the Gulf Arabic (Gulf)—MSA translation task and 9235 subword vocabularies for the Iraqi Arabic (IRQ)—MSA translation task. Twenty-nine thousand five hundred subword vocabularies are employed for the corpus applied by Baniata et al. [24] on the Levantine Arabic (LEV)–MSA translation task and on the Maghrebi vernacular (MAG)–MSA translation task. Relu dropout value and attention dropout value are 0.1. The suggested model proved to be very fast and required 268 s per epoch for MAG–MSA task (Baniata et al. [22] Corpus), 419 s per epoch for LEV–MSA task (Baniata et al. [22] Corpus), 251 s per epoch for the MAG–MSA task, 216 s per epoch for Nile–MSA task, 254 s per epoch for LEV–MSA task, 231 s per epoch for Gulf–MSA task, 152 s per epoch for IRQ–MSA task. For each translation task, the proposed model is trained to minimize cross-entropy loss.

5.4. Results

5.4.1. Automatic Metric

Many practical experiments were carried out with the proposed Transformer-based NMT model through exploiting subword units and shared vocabulary for Arabic vernaculars and MSA. The proposed Transformer-based subword model was experimented with different subword embeddings to find the most efficient subword embedding dimension of the proposed model. Moreover, the proposed model was trained with different number of encoders and decoders and with different number of heads in multi-head attention sublayer to find the most efficient number of attention heads for the proposed model. The translation quality is reported based on the sacreBLEU. SacreBLEU is a standard BLEU [19] implementation that manages WMT datasets, creates scores on detokenized outputs and reports a string encapsulating BLEU parameter, helping the generation of sharable, comparable BLEU scores. This section shows the performance evaluation of the suggested transformer-based NMT subword model for Arabic dialects on five Arabic dialects translation tasks. The findings of the LEV–MSA translation task are summarized in Table 1. For the MAG–MSA, Nile–MSA, Gulf–MSA and Iraqi–MSA translation tasks, Tables 2–5 illustrate the results. Table 1 shows Transformer-Based NMT subword model results with different settings on the test dataset for the LEV–MSA translation task. The results in Table 1 show the effectiveness of the proposed transformer machine translation subword model. As illustrated from Table 1, the suggested model achieved an outstanding 63.71 BLUE score where the number of encoders' layer and decoders' layer value is 12, number of attention heads is 4 and the subword embedding value is 512. The findings are obvious as a result of the close connection between Levantine vernacular and modern standard Arabic, and the fact that both languages share a large number of vocabularies. It can be noticed from Table 1 that the experiments' settings with low dimensions of subword embeddings achieved better BLUE score results in comparison to experiments with a high dimension of subword embeddings.

Table 1. Results of the Transformer-Based NMT Subword Model on PMM-LEV corpus for LEV–MSA translation task, where SW-E-D is the Subword embedding dimension, FS is the filter size, EL is the encoder layer, DL is the decoder layer and AH is the attention heads number.

SW-E-D	FS	EL	DL	AH	BLEU
512	1024	4	4	4	61.65
512	1024	8	8	4	63.56
512	1024	12	12	4	63.71
1024	1024	4	4	4	59.68
1024	1024	4	4	8	59.53
512	512	4	4	4	60.04

Table 2. Results of the Transformer-Based NMT Subword Model on PMM–MAG corpus for MAG–MSA translation task, where SW-E-D is the Subword embedding dimension, FS is the filter size, EL is the encoder's layers, DL is the decoder's layers and AH is the attention heads number.

SW-E-D	FS	EL	DL	AH	BLEU
512	1024	4	4	4	59.46
512	1024	8	8	4	63.02
512	1024	12	12	4	65.66
1024	1024	4	4	4	59.54
1024	1024	4	4	8	62.17
512	512	4	4	4	56.68

Table 3. Results of the Transformer-Based NMT Subword Model on MADAR–Nile Basin corpus for NILE–MSA translation task, where SW-E-D is the Subword embedding dimension, FS is the filter size, EL is the encoder layer, DL is the decoder layer and AH is the attention heads number.

SW-E-D	FS	EL	DL	AH	BLEU
512	1024	4	4	4	47.51
512	1024	8	8	4	48.19
512	1024	12	12	4	47.58
1024	1024	4	4	4	42.02
1024	1024	4	4	8	44.08
512	512	4	4	4	47.52

Table 4. Results of the Transformer-Based NMT Subword Model on MADAR–Gulf corpus for GULF–MSA translation task, where SW-E-D is the Subword embedding dimension, FS is the filter size, EL is the encoder layer, DL is the decoder layer and AH is the attention heads number.

SW-E-D	FS	EL	DL	AH	BLEU
512	1024	4	4	4	47.26
512	1024	8	8	4	46.66
512	1024	12	12	4	47.18
1024	1024	4	4	4	43.48
1024	1024	4	4	8	43.68
512	512	4	4	4	46.35

Table 5. Results of the Transformer-Based NMT Subword Model on MADAR–Iraqi corpus for IRQ–MSA translation task, where SW-E-D is the Subword embedding dimension, FS is the filter size, EL is the encoder layer, DL is the decoder layer and AH is the attention heads number.

SW-E-D	FS	EL	DL	AH	BLEU
512	1024	4	4	4	56.50
512	1024	8	8	4	49.03
512	1024	12	12	4	47.14
1024	1024	4	4	4	25.51
1024	1024	4	4	8	40.17
512	512	4	4	4	55.23

Table 2 illustrates the findings of the suggested model on test dataset for the MAG–MSA translation task. We observed that the proposed Transformer-based NMT subword model, as shown and highlighted with bold text in Table 2 is able to translate the Maghrebi Arabic sentences to MSA with a 65.66 BLEU score. It is clear that Maghrebi vernacular is a combination of many diverse languages such as the Berber language, African Romance, old Arabic expressions, Turkish language, Spanish, Italian and Niger Congo languages, as well as some new vocabularies borrowed from French and English. Therefore, the proposed Transformer-based NMT subword system was able to capture the semantic and syntactic features of Maghrebi dialect and improved the translation performance on the (MAG)–MSA translation task despite that the Maghrebi dialect is not close to the MSA in terms of expressions and vocabularies. Conventional NMT models [21,22] were not able to produce a high translation quality for the Maghrebi vernacular because Maghrebi dialect has expressions from many other languages. By utilizing subword units as an input to the encoder, there will be a sharing of information between the subwords forms and words forms and the model will generate MSA sentences with high quality.

Table 3 reveals the proposed model results with diverse settings on the test dataset for the Nile–MSA translation task. As seen from Table 3, the proposed system achieved a 48.19 BLUE score and the model was able to translate the Egyptian and Sudanese sentences to MSA correctly. Tables 4 and 5 present satisfied results on Gulf–MSA, Iraqi–MSA translation tasks, respectively. The proposed model proved to produce high translation quality of Arabic Gulf sentences with a 47.26 BLEU score and 56.50 BLEU score on the Iraqi–MSA translation task. Furthermore, the model was applied on Maghrebi–MSA, Levantine–MSA parallel Corpora used by Baniata et al. [22]. It can be shown from Tables 6 and 7 that the proposed Transformer-based NMT subword system has achieved 57.85 and 57.92 BLUE scores on Maghrebi–MSA, Levantine–MSA translation tasks, respectively. Therefore, it can be summarized as illustrated in Tables 6–8 that the proposed Transformer-Based NMT model that utilizes Arabic dialects subword units outperforms the multitask NMT system with part of speech tags that was proposed by Baniata et al. [22] on Maghrebi–MSA, Levantine–MSA translation tasks. These results were an indication of the effectiveness of utilizing subwords units and the usage of the shared vocabulary between Arabic dialect and MSA in the proposed model. Exploiting subwords' units of Arabic dialects (Levantine, Maghrebi, Nile Basin, Gulf and Iraqi) as an extra feature in the Transformer-based machine translation model is advantageous for word order problem and it generated a high quality MSA sentences. By using subword units and Self-attention (multi-Head attention), the proposed model could better represent and obtain more semantic features from AD source language and solve the grammatical problem for AD: the word ordering issue.

Table 6. Results of the Transformer-Based NMT Subword Model on corpus used by Baniata [22] for MAGHREBI–MSA translation task, where SW-E-D is the Subword embedding dimension, FS is the filter size, EL is the encoder layer, DL is the decoder layer and AH is the attention heads number.

SW-E-D	FS	EL	DL	AH	BLEU
512	1024	4	4	4	57.06
512	1024	8	8	4	57.41
512	1024	12	12	4	57.85
1024	1024	4	4	4	37.15
1024	1024	4	4	8	49.47
512	512	4	4	4	55.14

Table 7. Results of the Transformer-Based NMT Subword Model on corpus used by Baniata [22] for LEVANTINE–MSA translation task, where SW-E-D is the Subword embedding dimension, FS is the filter size, EL is the encoder layer, DL is the decoder layer and AH is the attention heads number.

SW-E-D	FS	EL	DL	AH	BLEU
512	1024	4	4	4	56.38
512	1024	8	8	4	53.98
512	1024	12	12	4	57.92
1024	1024	4	4	4	44.13
1024	1024	4	4	8	56.49
512	512	4	4	4	55.46

Table 8. Results of Multi-Task NMT Model with POS tagging using FAST Text Embedding that was proposed by Baniata [22].

Model	Pairs	Epochs	Accuracy	BLEU
NMT+POS_LEV	LEV–MSA	90	-	43.00
NMT+POS_LEV	MSA-ENG	50	-	30.00
POS_LEV	POS_LEV	40	98%	-
NMT+POS_MAG	MAG–MSA	50	-	34.00
NMT+POS_MAG	MSA-ENG	30	-	29.00
POS_MAG	POS_MAG	20	99%	-

5.4.2. Human Evaluation

The human evaluation experiments confirm the results that were obtained by the automatic evaluation. The pilot rating experiments were selected [15]. Participants were requested to evaluate the translations on a 1 to 7 Likert metric. We evaluated the translation quality for LEV–MSA, MAG–MSA, Nile–MSA, Gulf–MSA and Iraqi–MSA tasks asking seven speakers who know modern standard Arabic and understand every Arabic vernacular to evaluate sentences generated from the proposed transformer NMT subword model. We offered to the speakers a segment in LEV, MAG, Gulf, Nile and Iraqi and one translation in MSA for each Arabic dialect. We selected at random 100 segments and divided them into five subsets of twenty segments each. We give every annotator a subset and request them to evaluate the translations considering adequacy and fluency applying Likert metric from 1 to 7. The average results that were produced through every model using pilot rating experiments are illustrated in Tables 9 and 10. The average results pointed out that native speakers have positive and real judgment regarding the translations generated through the proposed Transformer-Based NMT subword model. The average score on the LEV–MSA translation task captured by multitask NMT part of speech tags system that was proposed by Baniata et al. [22] was 5.9. Furthermore, the average score on the MAG–MSA translation task captured through multitask NMT with part of speech tags system was 4.4. The average score on LEV–MSA translation task captured through the Transformer-Based NMT subword model is 6.0 and the average score obtained for the MAG–MSA is 6.2. Moreover, the average score for Gulf–MSA, Nile–MSA and Iraqi–MSA translation tasks obtained by the proposed model is 5.85, 5.8, 6.10, respectively. The findings of pilot rating experiments give confirmation that the Transformer-Based NMT subword model (T-NMT-Subword) generates better translation quality than the multitask NMT with part of speech tags system for all translation tasks.

Table 9. Human Evaluation Scores -Pilot Rating Experiments (PRE).

Model	Pairs	Average Score
Transformer-NMT-Subword	LEV–MSA	6.35
Transformer-NMT-Subword	MAG–MSA	6.3
Transformer-NMT-Subword	Gulf–MSA	5.85
Transformer-NMT-Subword	Nile–MSA	5.8
Transformer-NMT-Subword	IRQ–MSA	6.1

Table 10. Human Evaluation Scores -PRE for Levantine Arabic (LEV) and Maghrebi Arabic (MAG).

Model	Pairs	Average Score
Transformer-NMT-Subword	LEV–MSA	6.0
Transformer-NMT-Subword	MAG–MSA	6.2
Multi-Task Learning-NMT [22]	LEV–MSA	1.4
Multi-Task Learning-NMT [22]	MAG–MSA	1.3
MTL-NMT+POS [22]	LEV-MAG	5.9
MTL-NMT+POS [22]	MAG–MSA	4.4

6. Analysis

This analysis clarifies the positive effect of utilizing subword units on Arabic vernaculars to modern standard Arabic translation task efficiency. Table 11 shows sample translations from the Transformer-based NMT subword model on the LEV–MSA, MAG–MSA, Gulf–MSA, Nile–MSA, and Iraqi–MSA translation tasks. Due to the lack of standardization for the Arabic vernaculars, Conventional NMT methods for Arabic dialects are incapable of translating parts of input source sentences. Affixes and clitics are not obtained and captured effectively without the need to utilize the Subword units and multi-head attention. In Table 11, the proposed Transformer-Based NMT subword model translated 100% of Maghrebi vernacular sentences correctly. For Levantine Arabic, the proposed model translated 99% of whole Levantine Arabic phrases properly with the exact meaning regardless the word بعض الاصدقاء"some friends" in the generated sentence is not the same word اصدقائي"my friends" in the reference sentence, but it gives the same meaning. For the Gulf Arabic, the proposed model was able to translate 95% of the Gulf Arabic sentence fluently except the words مثل ما تشوف"as you see ", the proposed model could not translate them well and translated it to لن يكون"it will not be "rather than انت ترى"you see". Furthermore, the Transformer-based NMT subword model achieves the overall best translation performance for Nile–Arabic and Iraqi–Arabic sentences. The translation quality of the suggested Transformer-Based NMT subword system for Arabic vernaculars has enhanced in comparison to Multitask NMT system (with POS tags) that was recently suggested by Baniata et al. [22] that applied experiments on the same corpus as seen in Table 12. The proposed Transformer-Based NMT subword model translated the source sentences of Maghrebi dialect and Levantine dialect to MSA fluently and with high translation quality without any translation mistakes.

Table 11. Translation Examples for MAG, LEV, NILE, GULF and IRQ.

Source Language: MAG (Maghrebi)	واخا بلحاق هاد الشي بزاف وانا اصلا واكل بزاف
English Translation (MAG)	Yeah, but that's so much and I ate lot
Target Language: MSA	اجل ولكن هذا كثير جدا ولقد شبعت بالفعل
Transformer-NMT Subword Model	اجل ولكن هذا كثير جدا ولقد شبعت بالفعل
English translation for output of the Transformer-NMT Subword model	Yes, but that's too much and I'm already full
Source Language: LEV (Levantine)	انا مع صحابي
English Translation (LEV)	I am with my dudes
Target Language: MSA	انني مع اصدقائي
Transformer-NMT Subword Model	انا مع بعض الاصدقاء
English translation for output of the Transformer-NMT Subword model	I am with some Friends
Source Language: GULF	عندك جوازك وتذكرتك؟ مثل ما تشوف هذي سوق حرة
English Translation (GULF)	You got your passport and your ticket? as you see, this is a duty-free market
Target Language: MSA	هل معك جواز السفر والتذكرة ؟ انت ترى فهذا متجر معفى من الرسوم
Transformer-NMT Subword Model	هل لديك جواز سفرك وتذكرتك ؟ لن يكون عندك متجر معفى من الرسوم
English translation for output of the Transformer-NMT Subword model	Do you have your passport and ticket? You see, this is a duty-free shop
Source Language: Nile (Egypt, Sudan)	عاوز اعمل مكالمة لليابان
English Translation (Nile)	I wanna do a call to Japan
Target Language: MSA	اريد الاتصال هاتفيا باليابان
Transformer-NMT Subword Model	اريد الاتصال هاتفيا باليابان
English translation for output of the Transformer-NMT Subword model	I want to do a phone call to Japan
Source Language: IRQ (Iraqi)	احس ببرودة ومعدتي تاذيني كلش
English Translation (IRQ)	I feel cold and my stomach is hurting me a lot
Target Language: MSA	اشعر ببرودة وتؤلمني معدتي جدا
Transformer-NMT Subword Model	اشعر ببرودة وتؤلمني معدتي جدا
English translation for output of the Transformer-NMT Subword model	I feel cold and my stomach hurts so much

Table 12. Translation Examples for Maghrebi Arabic and Levantine Arabic, Baniata [22] Corpus.

Source Language: MAG (Maghrebi)	الى نتي قبلتي تضحي ب هاد الطريقة ف حتى هو خاصو يقدر هاد الامر و ميتخلاش عليك
English Translation (MAG)	If you accept to sacrifice in this way, he is also obliged to appreciate this matter and not abandon you
Target Language: MSA	اذا انت قبلت بان تضحي بهذه الطريقة فهو كذلك مجبر على ان يقدر هذا الامر و الا يتخلى عنك
Transformer-NMT Subword Model	اذا انت قبلت بان تضحي بهذه الطريقة فهو كذلك مجبر على ان يقدر هذا الامر و الا يتخلى عنك
English translation for output of the Transformer-NMT Subword model	If you accept to sacrifice in this way, he is also obliged to appreciate this matter and not abandon you
Source Language: LEV (Levantine)	اه ريحتها زي ريحة العطر منيحة في اليوم الاول بس بعد هيك
English Translation (LEV)	yeah, it smells like perfume, good on the first day, but later on
Target Language: MSA	نعم رائحتها كرائحة العطر جيدة في اليوم الاول لكن فيما بعد
Transformer-NMT Subword Model	نعم رائحتها كرائحة العطر جيدة في اليوم الاول لكن فيما بعد
English translation for output of the Transformer-NMT Subword model	Yes, it smells as good as perfume on the first day, but later on

The influence of utilizing subword units and shared vocabulary between Arabic dialects and MSA and applying the self-attention mechanism on translation quality for Arabic vernaculars is significantly evident. The proposed Transformer-Based NMT Subword model is well-suited to handle the issue of free word ordering and create a right context and a correct order for the target language sentences, as illustrated in Tables 11 and 12. Furthermore, the proposed system can obtain excellent translation effectiveness with various language pairs, as illustrated in Tables 11 and 12 for MAG–MSA, LEV–MSA, Nile–MSA, Gulf–MSA and IRQ–MSA tasks. The proposed model was evaluated on the same parallel corpus that was applied by Baniata et al. [22] for Maghrebi Arabic and Levantine Arabic. Compared to the multitask NMT with part of speech tags system that was trained on the same parallel corpus, the suggested Transformer NMT subword system scored an effectiveness of 56.49 BLEU score for translating from Levantine Arabic vernacular to MSA and 57.06 BLEU score for translating from Maghrebi Arabic vernacular to MSA. The findings show that the proposed Transformer NMT subword model performs outstanding translation quality than the multitask NMT with part of speech (POS) tagging system by evaluating the systems using BLEU score and experiments of human assessment.

The Transformer-based NMT subword model obtained remarkable results BLEU score for all Arabic dialects in comparison with various NMT systems, as shown in

Figures 4 and 5. It should be noted that the representation of source language learned via the proposed model significantly improves the translation effectiveness for language pairs. Generally, the suggested model is capable to produce fluent sentences in MSA language and transfer the information regarding the subject, object and verb for a free word order language such as Arabic vernacular. This section presents more analysis to know the effect of utilizing subword units on the performance of the proposed Transformer NMT subword system. This analysis includes (a) contribution of different numbers of encoder layers to the translation effectiveness of the proposed system (b) impact of translation quality regarding source sentence length, (c) impact of varying beam size concerning the effectiveness of the suggested model, (d) the impact of the encoder's self-attention and, (e) quantitative analysis of the proposed model. These analyses are applied to the MAG–MSA because of the dataset's size and the number of layers used to train the model.

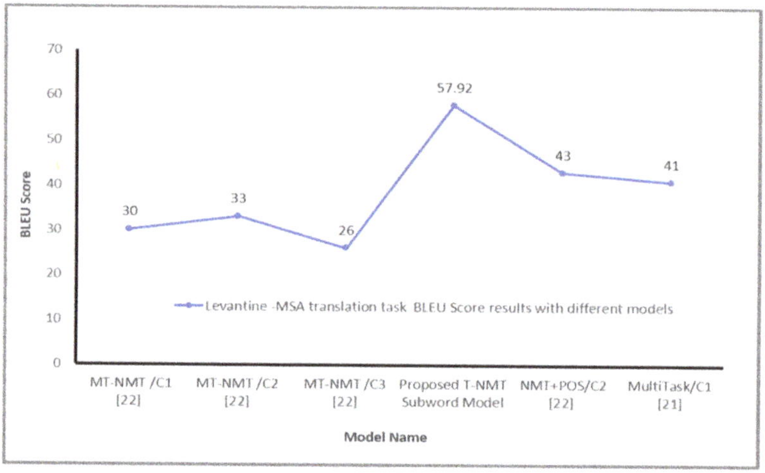

Figure 4. Levantine Arabic-MSA BLEU Score with different models, where C 1 is a random embedding, C 2 a pre-trained/Fast-text, C 3 a pre-trained/Polyglot.

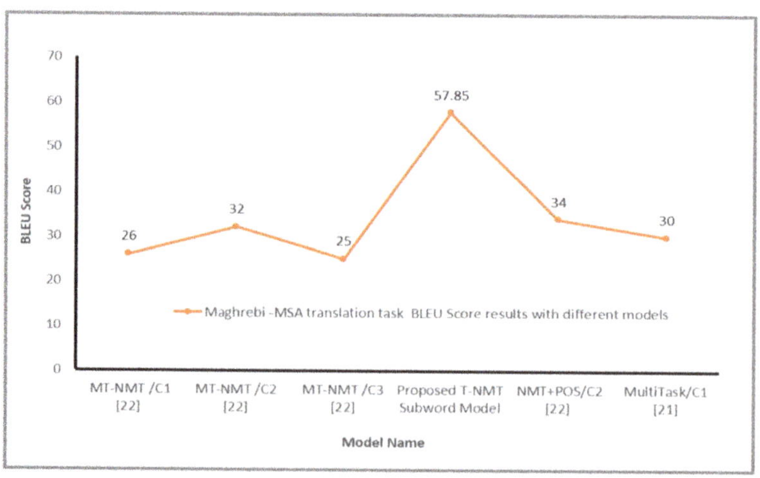

Figure 5. Maghrebi Arabic-MSA BLEU Score with different models, where C 1 is a random embedding, C 2 a pre-trained/Fast-text, C 3 a pre-trained/Polyglot.

6.1. Impact of Hayperparameter n

Tables 1–7 show that performing the Transformer-Based NMT subword model across various source representations obtained from several encoding layers significantly enhances the performance of the proposed model for all Arabic vernaculars. n represent the number of encoding layers in the proposed Transformer-Based NMT subword model. This part studies the impact of varying the value n (using only the representations from top n encoding layers). The proposed Transformer based NMT subword model is trained with various values of **n** where **n** is set to 4, 8 and 12. Where n = 4 indicates a setup for a simple model, **n** = 8 indicates a configuration for a medium-sized model and n = 8 indicates a configuration for a large model. As seen from Table 2, for example, in the Maghrebi Arabic-MSA task (and in all translation tasks), there is (in most cases) a significant change in BLEU score as the value of n changes.

6.2. Length of Source Sentence

Obtaining contextual information and long-distance dependencies between the source sentence's tokens can considerably increase the performance of longer sentences' translation. As mentioned by (Luong et al. [47]) sentences that have the same lengths (number of source tokens) are collected together. The grouping is arranged by the lengths of source sentences (the number of subword tokens in every source sentence) over the MAG–MSA test set. We selected the MAG–MSA translation task to investigate the translation quality of long sentences because of the large size of the MAG–MSA corpus. The comparison in this research project is based upon these lengths: >50, 40–50, 30–40, 20–30, 10–20 and <10. Regarding every length interval, BLEU score metric is computed for the output of the suggested Transformer-Based NMT subword model. As can be noted in Figure 6, the proposed model performance improves while the input sentence lengths increase, particularly for the lengths between 40 and 50 subword tokens and for the lengths larger than 50 subword tokens with 61.76 and 62.17 BLEU scores, respectively. The proposed model, through the use of self-attention sublayers, is capable of modeling or obtaining contextual knowledge and dependencies within the tokens regardless of their distance or location within Arabic dialect sentence input. However, the performance of the suggested model decreased for very short sentences that have lengths smaller than 10 (in terms of the number of subword tokens). The performance increased for sentences with lengths larger than 50 (in terms of the number of subword tokens). Moreover, the proposed model performed inadequately on few numbers of short sentences with length less than 10 subword tokens with the lowest BLEU score (23.37). This occurs because these very short AD sequences contain only subword tokens (suffixes, affixes and morphemes) and they are cannot be aligned to the corresponding words in the target language. Overall, the performance of the proposed Transformer-Based NMT subword model obtained across the different groups motivates the hypothesis that employing subword units and using shared vocabulary between source language (Arabic dialect) and target language (MSA) enhances the encoder's self-attention sublayers' effectiveness in effectively capturing the global dependencies between words in the input sentences.

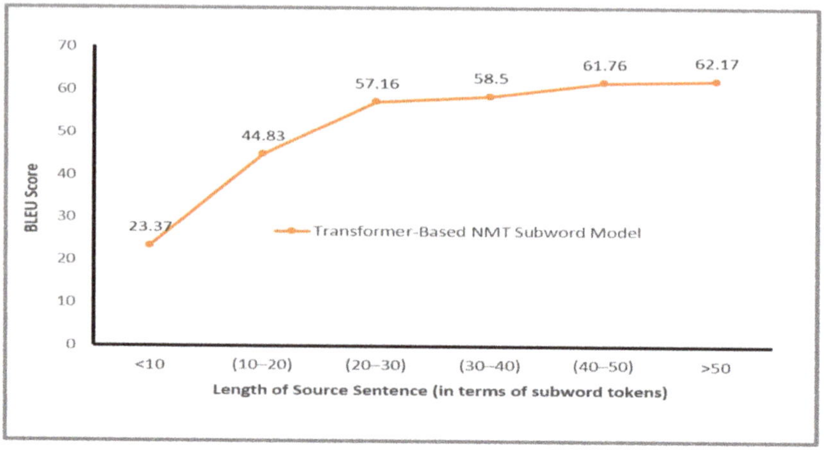

Figure 6. BLEU Score on MAG–MSA test dataset for the Transformer-NMT Subword model with respect to different source sentence length.

6.3. Beam Size Evaluation

A comparison is conducted for the performance of the proposed model by modifying the beam size. The experiments are carried out based on the Transformer NMT subword model for Arabic vernaculars. We employed Wordpiece model for subword tokenization. Beam size has a large impact on the decoding speed (as words per second) and translation quality (in BLEU). Table 13 shows the outcomes of the experiments on the MAG–MSA translation task and results show that the optimum performance (in BLEU) of the proposed model is achieved when beam size has the value 6 and the fastest decoding speed of the model is obtained with beam size 6.

Table 13. Change in BLEU Score According to Beam Size.

Beam Size	BLEU
1	52.82
2	58.10
3	59.69
4	60.31
5	61.70
6	65.66
7	61.22
8	61.35
9	60.72
10	61.36

6.4. The Effect of the Encoder Self Attention

The encoding layers' effectiveness is determined by the capability of several heads of the multi-head attention sub-layer placed inside each layer to capture important structural information. These attention heads, to varying degrees, capture structural information. As remarked by Raganato et al. [48] and Vig et al. [49], some of the heads in multi-head attention sublayer hold the long-distance relations among input token. Other heads in the multi-head attention sub-layer hold the short distance relations among input tokens. This makes the suggested Transformer-Based NMT Subword model obtain the structural and fundamental characteristics efficiently for the input source sentence of Arabic vernacular to increase the effectiveness [48]. As previously stated, the use of Subword units influences how the source language information is handled through the layers of the encoder. As mentioned by Vig et al. [49], this approach is examined through computing two things;

the attention entropy and the attention distance spanned through several attention heads within every encoding layer multi-head attention sublayer. The mean distance $\overline{D}_h^l$ that is spanned by attention head h for encoding layer l is calculated a weighted average distance within tokens pairs every sentence of a given corpus X. So:

$$\overline{D}_h^l = \frac{\sum_{x \in X} \sum_{i=1}^{|x|} \sum_{y=1}^{i} w_{i,j}^h \cdot (i-j)}{\sum_{x \in X} \sum_{i=1}^{|x|} \sum_{y=1}^{i} w_{i,j}^h} \tag{19}$$

where $w_{i,j}^h$ is attention weight from the input token x_i to x_j for attention head $h.i$ and j signifies the tokens' places x_i and x_j in source sentences. By performing aggregation for the attention distance for every head, the mean attention distance spanned $\overline{D}^l$ with reference to the encoding layer l is computed as:

$$\overline{D}^l = \frac{1}{N_h} \cdot \sum_{h=1}^{N_h} \overline{D}_h^l \tag{20}$$

where N_h indicates the number of attention heads that are used within the layer. Mean attention distance provides no information about how the attention weight is distributed through the input tokens for a particular attention head. Attention head that has greater mean attention distance may concentrate on sequences of same tokens that are separated [49,50]. To estimate the dispersion or concentration pattern for attention head h inside layer l for input token x_i, entropy of attention distribution [50], $E_h^l\ (x_i)$ for attention head h is calculated as:

$$E_h^l\ (x_i) = -\sum_{j=1}^{i} w_{i,j}^h \log w_{i,j}^h \tag{21}$$

The mean entropy of the attention distribution for encoding layer l is computed similarly to the attention distance spanned as:

$$E^l\ (x_i) = \frac{1}{N_h} \sum_{h=1}^{N_h} E_h^l\ (x_i) \tag{22}$$

Attention heads with larger entropy have a more distributed attention pattern, whereas attention heads with a lower entropy have a more focused attention weight distribution. Attention distance and the entropy of attention analysis are conducted based on the attention weights produced for a random 2000 sentences from the MAG–MSA task's test split (PMM-MAG Corpus). Figure 7 presents mean attention distance span and the mean entropy of attention distribution for every attention head for every encoding layer of the proposed Transformer NMT subword model for Arabic vernaculars. As remarked, some heads focus on short-distance relations among input tokens, other self-attention heads obtain the long-distance relations between input tokens. Furthermore, the entropy of the attention distribution changes through layers. Moreover, the entropy of the attention distribution change for the attention heads within the same layer. The mean average attention distance and entropy for all heads in the multi-head attention through the encoder layers are shown in Figure 8. As shown in Figure 7, for the suggested model, most of the attention heads that have a high mean attention span and much more stable values of attention distribution are located in fourth layer. Despite, a large mean attention distance does not indicate stable attention distribution. The preceding layers have multiple attention heads that have high value of distance span but significantly less consistent attention weights distribution. For instance, in the second layer, attention heads 1 and 4 have the largest mean attention spans (3.09 and 3.24, respectively), but the lowest mean entropy scores (0.22 and 0.23). As Vig et al. [49] highlight, attention heads that have large value of mean attention distance span concentrate their attention to word in repeated sentences that occurs in various places within the source sentence. This can justify their reduced entropy of weight distribution over the sequence of input tokens. Attention heads with a stable or less stable weight distribution and low attention distance span focus significantly more on nearby tokens. Those attention heads that have changeable mean attention distance and changeable entropy enable suggested transformer NMT subword model for Arabic vernaculars to learn efficiently changeable structural information across its layers. This demonstrates the Transformer-Based architecture's superiority over seq2seq architectures such as RNN and CNN.

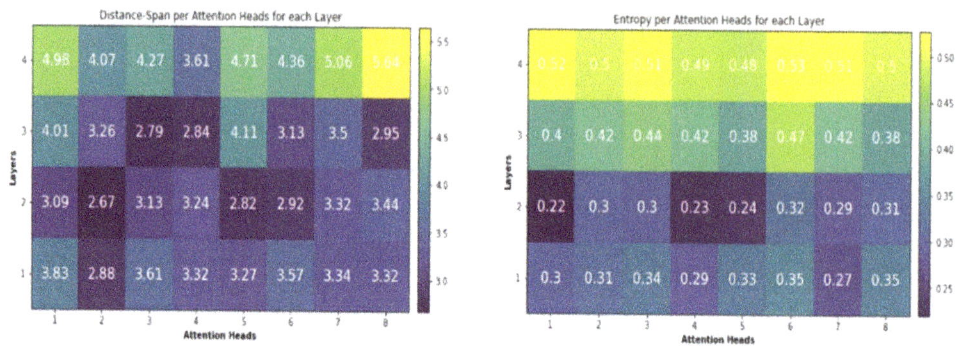

Figure 7. Variation of mean attention distance span and attention distribution entropy with respect to the encoding layers and the attention heads for the suggested model.

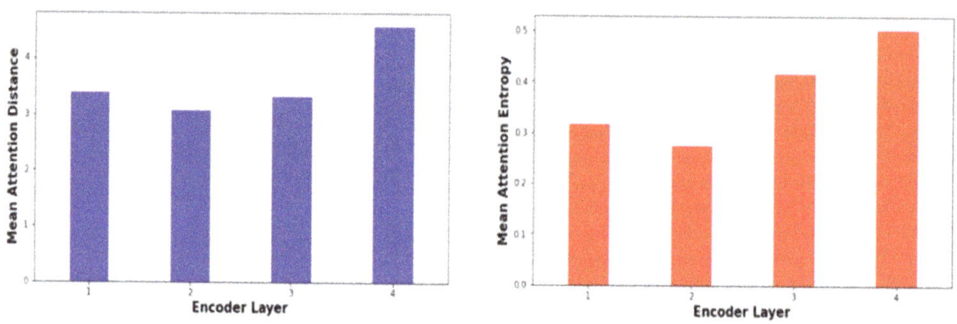

Figure 8. Variation of the average mean attention distance and variation of the average entropy of head attention distribution with respect to each encoder layer for the suggested model.

In the proposed model, the Subword units have a strong influence on multi-head attention sub-layer inside the encoding layer. As illustrated in Figures 7 and 8, exposing the encoder layers to the decoder network, enables the encoder subnetwork to learn the source information in a more customized manner. Figure 8 illustrates the change in average mean attention distance span and entropy of the attention weight distribution for different attention heads over various layers of the encoder. As illustrated in Figure 8, the proposed Transformer based subword model concentrated the attention heads that has a shorter attention span over the layers $l \leq 3$. These layers are employed to learn the short rang contextual and local knowledge within the neighborhood of input source tokens. The upper layers learn the long-distance interaction within the input source tokens. Generally, utilizing subword units explained how the source information (Arabic dialects) is captured over several attention heads and layers in the encoder as revealed by the entropy of attention weight distribution and attention distance. This improves the proposed Transformer-based NMT subword model's performance at learning the source semantic information that is required to enhance the quality of translation.

6.5. Quantitative Analysis

The proposed model provides a method of examining the alignment of words in generated translation with words in source sentence. This method is performed through the visualization of annotation weights as illustrated in Figure 9. Every row of the matrix in every plot indicates the weights linked with the annotations where the x-axis represents the input sentence (Maghrebi Arabic) and the y-axis represents the generated sentence in MSA. This reveals which locations in the source sentence were rated more significant when the target word was generated. As illustrated in Figure 9, the alignment of words between Maghrebi Arabic (MAG) and MSA is primarily monotonic. Along the diagonals of each matrix, we see strong weights. Although, we see several non-trivial, non-monotonic alignments. Typically, nouns and adjectives are ordered differently in MAG and MSA, as illustrated in the upper left part of Figure 9. From Figure 9, we see that the model correctly translates a MAG dialect sentence (اسكر عليا برا تقدر تفتح بابي لو سمحت) "it has been closed from outside, can you open my door please" into MSA sentence (لقد اغلق علي الباب هل يمكن ان تفتح بابي من فضلك ؟) "the door has been closed while I am inside, can you open my door please". The proposed transformer-based NMT subword model was able to correctly align (اسكر عليا برا) "it has been closed from outside" with (لقد اغلق علي الباب) which means "the door has been closed", the proposed model was able to understand the contextual clues (الباب) "the door" of Maghrebi Arabic and translated the MAG sentence to MSA correctly. Additionally, the model often handles source and target sentences of variable length. Furthermore, sub-word units approach can be applied on various Arabic NLP tasks such as sentiment analysis [51] and text summarization.

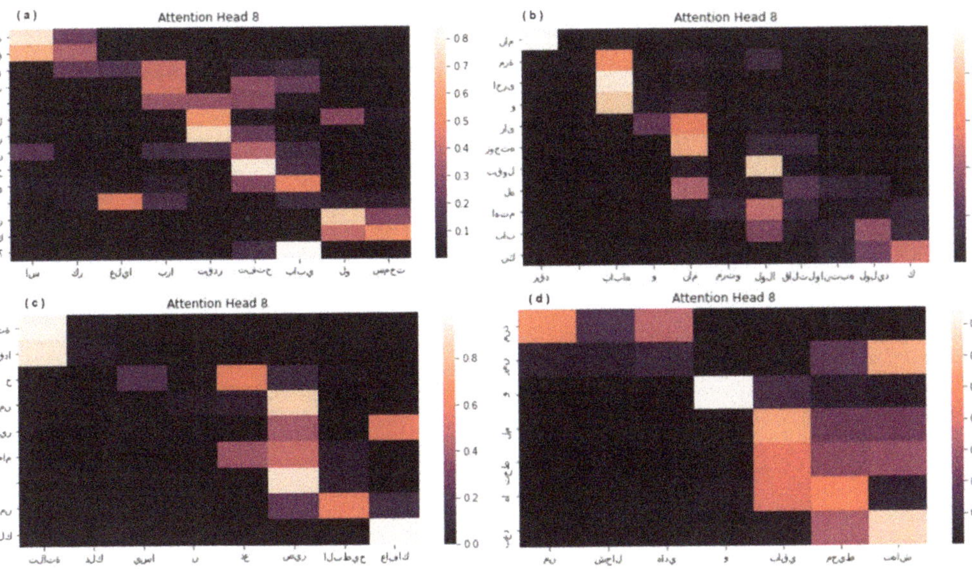

Figure 9. (**a**–**d**) Four Sample Alignments.

7. Conclusions

This research project introduced a Transformer-Based NMT model for Arabic dialects that utilize Subword units. Through training the suggested model on translation tasks from diverse Arabic dialects to MSA, the model's translation performance was significantly improved. Utilizing various source representations captured by stacked encoding layers enhance the efficiency of the transformer NMT subword system. The findings of this research project confirm that the Transformer NMT subword model that exploits subword units enhanced the translation BLEU score for MAG–MSA, LEV–MSA, Nile–MSA, Gulf–MSA and IRQ–MSA tasks. The utilization of subword units by using the Wordpiece Model showed that this method is promising and significant for low-resource languages such as Arabic vernaculars. Additionally, using a changeable number of heads in self attention sublayer and training the model with a different number of encoders and decoders improved the quality of translation from Arabic vernaculars to modern standard Arabic. Experimental results on MAG–MSA, LEV–MSA, Nile–MSA, Gulf–MSA and IRQ–MSA translation tasks showed that the proposed model improved the BLEU score's effectiveness in comparison to other NMT systems. However, the experimental analysis performed reveals that performance gain is reliant on the value of the number of encoding layers considered. Additional analysis reports that increasing the layers of encoder and decoder subnetworks adjust how the local and general contextual knowledge is obtained and captured by employing multi-head attention sublayer used within each encoding layer. The current proposed Transformer-Based NMT subword model can deal with the issue of low availability of Arabic dialects training data. Additionally, the suggested system addressed the Arabic dialect's grammatical problem; free word ordering. The proposed model with subword units' utilization is effective and suitable to perform machine translation for low resource languages such as Arabic vernaculars.

Author Contributions: L.H.B. and I.K.E.A. conceived and designed the methodology and experiments; L.H.B. performed the experiments; I.K.E.A. performed the Visualization; S.P., I.K.E.A. and L.H.B. analyzed the data; L.H.B. wrote the paper; S.P. reviewed and edited the paper. All authors have read and agreed to the published version of the manuscript.

Funding: This research received no external funding.

Institutional Review Board Statement: Not Applicable.

Informed Consent Statement: The study did not involve humans.

Data Availability Statement: The datasets generated during the current study are available in [Transformer_NMT_AD] repository (https://github.com/laith85/ (accessed on 26 September 2021). Transformer_NMT_AD).

Acknowledgments: This study was supported by the BK21 FOUR project (AI-driven Convergence Software Education Research Program) funded by the Ministry of Education, School of Computer Science and Engineering, Kyungpook National University, Korea (4199990214394).

Conflicts of Interest: The authors declare no conflict of interest.

References

1. Bentivogli, L.; Bisazza, A.; Cettolo, M.; Federico, M. Neural versus phrase-based MT quality: An in-depth analysis on English-German and English-French. *Comput. Speech Lang.* **2019**, *49*, 52–70. [CrossRef]
2. Jean, S.; Cho, K.; Memisevic, R.; Bengio, Y. On Using Very Large Target Vocabulary for Neural Machine Translation. In Proceedings of the 53rd Annual Meeting of the Association for Computational Linguistics and the 7th International Joint Conference on Natural Language Processing, Beijing, China, 26–31 July 2015; Volume 1, pp. 1–10.
3. Luong, M.T.; Sutskever, I.; Le, Q.V.; Vinyals, O.; Zaremba, W. Addressing the rare word problem in neural machine translation. In Proceedings of the 53rd Annual Meeting of the Association for Computational Linguistics and the 7th International Joint Conference on Natural Language Processing, Beijing, China, 26–31 July 2015; pp. 11–19.
4. Vaswani, A.; Shazeer, N.; Parmar, N.; Uszkoreit, J.; Jones, L.; Gomez, A.N.; Kaiser, L.; Polosukhin, I. Attention is all you need. In Proceedings of the 31st Conference on Neural Information Processing Systems (NIPS), Long Beach, CA, USA, 4–9 December 2017; pp. 5998–9008.
5. Bahdanau, D.; Cho, K.; Bengio, Y. Neural machine translation by jointly learning to align and translate. In Proceedings of the 3rd International Conference on Learning Representations (ICLR), San Diego, CA, USA, 7–9 May 2015.
6. Cho, K.; van Merrienboer, B.; Gulcehre, C.; Bahdanau, D.; Bougares, F.; Schwenk, H.; Bengio, Y. Learning phrase representations using RNN encoder-decoder for statistical machine translation. In Proceedings of the 2014 Conference on Empirical Methods in Natural Language Processing (EMNLP), Doha, Qatar, 25–29 October 2014; pp. 1724–1734.
7. Sutskever, I.; Vinyals, O.; Le, Q.V. Sequence to sequence learning with neural networks. In Proceedings of the 27th International Conference on Neural Information Processing Systems, Montreal, QC, Canada, 8–13 December 2014; Volume 2, pp. 3104–3112.

8. Gehring, J.; Auli, M.; Grangier, D.; Yarats, D.; Dauphin, Y.N. Convolutional sequence to sequence learning. In Proceedings of the 34th International Conference on Machine Learning, Sydney, NSW, Australia, 6–11 August 2017; Volume 70, pp. 1243–1252.
9. Popović, M.; Arcan, M.; Klubička, F. Language Related Issues for Machine Translation between Closely Related South Slavic Languages. In Proceedings of the Third Workshop on NLP for Similar Languages, Varieties and Dialects (VarDial3), Osaka, Japan, 12 December 2016; pp. 43–52.
10. Durrani, N.; Sajjad, H.; Fraser, A.; Schmid, H. Hindi-to-Urdu Machine Translation through Transliteration. In Proceedings of the 48th Annual Meeting of the Association for Computational Linguistics, Uppsala, Sweden, 11–16 July 2010; pp. 465–474.
11. Harrat, S.; Meftouh, K.; Smaili, K. Machine translation for Arabic dialects. *Inf. Process. Manag.* **2019**, *56*, 262–273. [CrossRef]
12. Pourdamghani, N.; Knight, K. Deciphering Related Languages. In Proceedings of the Empirical Methods in Natural Language Processing, Copenhagen, Denmark, 7–11 September 2017; pp. 2513–2518.
13. Costa-Jussà, M.R. Why Catalan-Spanish neural machine translation? Analysis, comparison and combination with standard rule and phrase-based technologies. In Proceedings of the Fourth Workshop on NLP for Similar Languages, Varieties and Dialects, Valencia, Spain, 3 April 2017; pp. 55–62.
14. Kurdish, H.H. Inter dialect machine translation. In Proceedings of the Fourth Workshop on NLP for Similar Languages, Varieties and Dialects, Valencia, Spain, 3 April 2017; pp. 63–72.
15. Costa-Jussà, M.R.; Zampieri, M.; Pal, S. A Neural Approach to Language Variety Translation. In Proceedings of the Fifth Workshop on NLP for Similar Languages, Varieties and Dialects, Santa Fe, NM, USA, 20 August 2018; pp. 275–282.
16. Lakew, S.M.; Erofeeva, A.; Federico, M. Neural machine translation into language varieties. In Proceedings of the Third Conference on Machine Translation, Brussels, Belgium, 31 October–1 November 2018; pp. 156–164.
17. Meftouh, K.; Harrat, S.; Jamoussi, S.; Abbas, M.; Smaili, K. Machine translation experiments on padic: A parallel Arabic dialect corpus. In Proceedings of the 29th Pacific Asia conference on language, information and computation, Shanghai, China, 30 October–1 November 2015.
18. Sadat, F.; Mallek, F.; Boudabous, M.; Sellami, R.; Farzindar, A. Collaboratively Constructed Linguistic Resources for Language Variants and their Exploitation in NLP Application—The case of Tunisian Arabic and the social media. In Proceedings of the Workshop on Lexical and Grammatical Resources for Language Processing, Dublin, Ireland, 24 August 2014; pp. 102–110.
19. Papineni, K.; Roukos, S.; Ward, T.; Zhu, W.-J. BLEU: A method for automatic evaluation of machine translation. In Proceedings of the 40th annual meeting of the Association for Computational Linguistics, Philadelphia, PA, USA, 7–12 July 2002; pp. 311–318.
20. Abo Bakr, H.; Shaalan, K.; Ziedan, I. A hybrid approach for converting written Egyptian colloquial dialect into diacritized Arabic. In Proceedings of the 6th International Conference on Informatics and Systems, Cairo, Egypt, 27–29 March 2008.
21. Baniata, L.H.; Park, S.; Park, S.-B. A Neural Machine Translation Model for Arabic Dialects That Utilizes Multitask Learning (MTL). *Comput. Intell. Neurosci.* **2018**, *2018*, 10. [CrossRef] [PubMed]
22. Baniata, L.H.; Park, S.; Park, S.-B. A Multitask-Based Neural Machine Translation Model with Part-of-Speech Tags Integration for Arabic Dialects. *Appl. Sci.* **2018**, *8*, 2502. [CrossRef]
23. Nguyen, Q.; Vo, A.; Shin, J.; Tran, P.; Ock, C. Korean-Vietnamese Neural Machine Translation System with Korean Morphological Analysis and Word Sense Disambiguation. *IEEE Access* **2019**, *7*, 32602–32616. [CrossRef]
24. Park, C.; Lee, C.; Yang, Y.; Lim, H. Ancient Korean Neural Machine Translation. *IEEE Access* **2020**, *8*, 116617–116625. [CrossRef]
25. Luo, G.; Yang, Y.; Yuan, Y.; Chen, Z.; Ainiwaer, A. Hierarchical Transfer Learning Architecture for Low-Resource Neural Machine Translation. *IEEE Access* **2019**, *7*, 154157–154166. [CrossRef]
26. Aqlan, F.; Fan, X.; Alqwbani, A.; Al-Mansoub, A. Arabic Chinese Neural Machine Translation: Romanized Arabic as Subword Unit for Arabic-sourced Translation. *IEEE Access* **2019**, *7*, 133122–133135. [CrossRef]
27. Chen, K.; Wang, R.; Utiyama, M.; Liu, L.; Tamura, A.; Sumita, E.; Zhao, T. Neural machine translation with source dependency representation. In Proceedings of the EMNLP, Copenhagen, Denmark, 7–11 September 2017; pp. 2513–2518.
28. Eriguchi, A.; Tsuruoka, Y.; Cho, K. Learning to parse and translate improves neural machine translation. In Proceedings of the 55th Annual Meeting of the Association for Computational Linguistics, Vancouver, BC, Canada, 30 July–4 August 2017; pp. 72–78.
29. Wu, S.; Zhang, D.; Zhang, Z.; Yang, N.; Li, M.; Zhou, M. Dependency-to-dependency neural machine translation. *IEEE/ACM Trans. Audio Speech Lang. Process.* **2018**, *26*, 2132–2141. [CrossRef]
30. Strubell, E.; Verga, P.; Andor, D.; Weiss, D.; McCallum, A. Linguistically-informed self-attention for semantic role labeling. In Proceedings of the EMNLP, Brussels, Belgium, 31 October–4 November 2018; pp. 5027–5038.
31. Neco, R.P.; Forcada, M.L. Asynchronous translations with recurrent neural nets. In Proceedings of the International Conference on Neural Networks, Houston, TX, USA, 9–12 June 1997; pp. 2535–2540.
32. Schwenk, H.; Dchelotte, D.; Gauvain, J.L. Continuous space language models for statistical machine translation. In Proceedings of the 21st COLING/ACL, Sydney, NSW, Australia, 17–21 July 2006; pp. 723–730.
33. Kalchbrenner, N.; Blunsom, P. Recurrent continuous translation models. In Proceedings of the EMNLP, Seattle, WA, USA, 18–21 October 2013; pp. 1700–1709.
34. Passban, P.; Liu, Q.; Way, A. Translating low-resource languages by vocabulary adaptation from close counterparts. *ACM Trans. Asian Low Resour. Lang. Inf. Process.* **2017**, *16*, 1–14. [CrossRef]
35. Hochreiter, S.; Schmidhuber, L. Long short-term memory. *Neural Comput.* **1997**, *9*, 1735–1780. [CrossRef] [PubMed]
36. Gülçehre, C.; Ahn, S.; Nallapati, R.; Zhou, B.; Bengio, Y. Pointing the unknown words. In Proceedings of the 54th Annual Meeting of the Association for Computational Linguistics, Berlin, Germany, 7–12 August 2016; pp. 140–149.

37. Sennrich, R.; Haddow, B.; Birch, A. Neural machine translation of rare words with subword units. In Proceedings of the 54th Annual Meeting of the Association for Computational Linguistics, Berlin, Germany, 7–12 August 2016; pp. 1715–1725.
38. He, K.; Zhang, X.; Ren, S.; Sun, J. Deep residual learning for image recognition. In Proceedings of the IEEE CVRP, Las Vegas, NV, USA, 27–30 June 2016; pp. 770–778.
39. Ba, J.L.; Kiros, J.R.; Hinton, G.E. Layer normalization. In Proceedings of the Advances in NIPS 2016 Deep Learning Symposium, Barcelona, Spain, 5–10 December 2016.
40. Al-Sabahi, K.; Zuping, Z.; Nadher, M. A hierarchical structured self attentive model for extractive document summarization (HSSAS). *IEEE Access* **2018**, *6*, 24205–24212. [CrossRef]
41. Schuster, M.; Nakajima, K. Japanese and Korean voice search. In Proceedings of the ICASSP, Kyoto, Japan, 25–30 March 2012; pp. 5149–5152.
42. Lample, G.; Conneau, A. Cross-lingual language model pretraining. In Proceedings of the 33rd Conference on Neural Information Processing Systems, Vancouver, BC, Canada, 8–14 December 2019.
43. Bouamor, H.; Habash, N.; Oflazer, K. A Multidialectal Parallel Corpus of Arabic. In Proceedings of the LREC, Reykjavik, Iceland, 26–31 May 2014; pp. 1240–1245.
44. Bouamor, H.; Habash, N.; Salameh, M.; Zaghouani, W.; Rambow, O.; Abdulrahim, D.; Obeid, O.; Khalifa, S.; Eryani, F.; Erdmann, A.; et al. The madar arabic dialect corpus and lexicon. In Proceedings of the LREC, Miyazaki, Japan, 7–12 May 2018; pp. 3387–3396.
45. Kingma, D.P.; Ba, J. Adam: A method for stochastic optimization. In Proceedings of the International Conference on Learning Representations (ICLR), San Diego, CA, USA, 7–9 May 2015.
46. So, D.; Le, Q.; Liang, C. The Evolved Transformer. In Proceedings of the 36th International Conference on Machine Learning, Long Beach, CA, USA, 9–15 June 2018; pp. 5877–5886.
47. Luong, M.-T.; Pham, H.; Manning, C.D. Effective approaches to attention-based neural machine translation. In Proceedings of the Empirical Methods in Natural Language Processing, Lisbon, Portugal, 17–21 September 2015; pp. 1412–1421.
48. Raganato, A.; Tiedemann, J. An analysis of encoder representations in transformer-based machine translation. In Proceedings of the 2018 Empirical Methods in Natural Language Processing Workshop BlackboxNLP: Analyzing and Interpreting Neural Networks for NLP, Brussels, Belgium, 1 November 2018; pp. 287–297.
49. Vig, J.; Belinkov, Y. Analyzing the Structure of Attention in a Transformer Language Model. In Proceedings of the Second BlackboxNLP Workshop on Analyzing and Interpreting Neural Networks for NLP, Florence, Italy, 1 August 2019; pp. 63–76.
50. Ghader, H.; Monz, C. What does Attention in Neural Machine Translation Pay Attention to? In Proceedings of the 8th IJCNLP, Taipei, Taiwan, 27 November–1 December 2017; pp. 30–39.
51. Alali, M.; Mohd Sharef, N.; Azmi Murad, M.A.; Hamdan, H.; Husin, N.A. Narrow Convolutional Neural Network for Arabic Dialects Polarity Classification. *IEEE Access* **2019**, *7*, 96272–96283. [CrossRef]

MDPI
St. Alban-Anlage 66
4052 Basel
Switzerland
Tel. +41 61 683 77 34
Fax +41 61 302 89 18
www.mdpi.com

Sensors Editorial Office
E-mail: sensors@mdpi.com
www.mdpi.com/journal/sensors

www.ingramcontent.com/pod-product-compliance
Lightning Source LLC
Chambersburg PA
CBHW040807250926
51850CB00007BA/1099
* 9 7 8 3 0 3 6 5 6 9 3 8 3 *